Science: Philosophy, History and Education

Efforts to leverage the history and philosophy of science (HPS) to improve science education have a long and productive history. This book series serves as a venue for science education and HPS scholars to continue this tradition. Thoughtful consideration of the synergistic relationships among HPS and science education can improve science teaching and learning, science education policy and outreach, and the teaching and learning of HPS. Science education efforts to improve teaching and learning about the nature of science (NOS) should obviously be informed by HPS scholarship, but HPS also offers much for improving science teaching more generally. For instance, HPS plays an important role in teaching and learning for authentic conceptual understanding; making clear how such understanding requires *all* students to at times abandon everyday reasoning when learning particular science ideas. Science education scholarship, in turn, can assist efforts among HPS scholars to promote public engagement with science and HPS. Recent emphasis on engineering and technology in science education calls for historians and philosophers in those disciplines to contribute their expertise in promoting a robust STEM education and avoiding undesired and unanticipated problems. This series is directed toward publishing authoritative books that overtly address how history and philosophy can and should inform science education. This series complements the journal *Science & Education* (http://www.springer.com/journal/11191). Questions regarding author ideas for book proposals should be directed to the Series Editor, Michael Clough (mclough@tamu.edu). Book proposals are to submitted to the Publishing Editor: Claudia Acuna (Claudia.Acuna@springer.com).

Glenn Dolphin

Teaching Geology Using the History and Philosophy of Science

Enhancing Conceptual Understanding

With Contrib. by Dr. Robert Kelly and Dr. Nicole Ladue

Springer

Glenn Dolphin
Department of Earth, Energy, and the Environment
University of Calgary
Calgary, AB, Canada

With Contrib. by
Robert Kelly
University of Calgary
Calgary, AB, Canada

Nicole Ladue
Northern Illinois University
DeKalb, IL, USA

ISSN 2520-8594 ISSN 2520-8608 (electronic)
Science: Philosophy, History and Education
ISBN 978-3-031-68666-5 ISBN 978-3-031-68667-2 (eBook)
https://doi.org/10.1007/978-3-031-68667-2

This Springer imprint is published by the registered company Springer Nature Switzerland AG
The registered company address is: Gewerbestrasse 11, 6330 Cham, Switzerland

For my three muses: May Lan Rose, Jia Huan Pearl, and Jia Le Ruby

Preface

The focus of this book originated in the Harlequin Room, a dimly lit cocktail lounge with a dozen or so life-size harlequin-style statues lining the walls. I was giving my first presentation about teaching earth science at a state science conference held at a resort in the Catskill Mountains. The resort had been a prime vacation destination 30 or 40 years prior. At this time, in the late 1990s, it was all but hanging on. I was presenting to colleagues from all over New York State about some teaching strategies I had incorporated into my ninth-grade earth science classes. A major component of that teaching was using the history of geology to teach concepts like geologic history, geologic time, and plate tectonics. I was a new teacher and not expecting much of an audience: What could I possibly tell teachers with 10 or 20 years, or even 30 years of experience?

To my great surprise, the room filled passing its stated capacity with standing room only. I talked about how I engaged students by recounting the developmental trajectory of our understanding of the history and dynamics of the earth. The feedback from the attendees was extremely positive. They really appreciated this novel approach to teaching and provided the positive reinforcement I needed. During the 15 years that followed, I became very interested in how students learn; I eventually resigned as a public school teacher, entered a PhD program in Science Education, and then started teaching at a university where I was assigned an introductory geology course to teach. The course is for non-science majors and usually has 200–300 students enrolled. Here I was able to fully develop the historical contextualization of the content, since I was no longer beholden to a mandated curriculum, as I had been in the public school system. Moving from classes of 20 students to classes of 200 along with other differences from my high school teaching experience required many iterations to create what appears in this book. This book documents that journey—of my trying out classroom activities, evaluating them, and subsequently revising them to improve or hone student learning outcomes.

During the five iterations of teaching the course, I collected data about student learning to measure the efficacy of the instruction aimed at geoscience conceptual development and nature of science (NOS) understandings. I used the data to make decisions about subsequent iterations of the course—in essence, to improve it. After five iterations of teaching the course, I determined that students' conceptual understanding of both geology content and NOS was far more robust. Thus, I thought sharing features of the course that promoted that robust understanding was important.

Importantly, I want to show interested readers that simultaneously teaching both core geology content and NOS is possible and promotes a deeper understanding of fundamental geology ideas. My review of the literature reveals that those using the history and philosophy of science (HPS) in science courses are usually only aiming to teach about NOS. This is fine, except that HPS then seems like extra instruction, an add-on, just one more thing to teach in an already full curriculum. Indeed, it makes teaching HPS—like the historical vignettes placed in the sidebars of textbooks to give context to the main text—appear to be superfluous. Readers then often skip over this content or see it as of little importance.

I also wanted to document the iterative process of design thinking I employed that helped shape the course and informed the decisions to include different evidence-based teaching strategies. To that end, the book unfolds in two overarching parts:

- **Part I:** This part can be used as a "how-to" manual: It serves as a blueprint of a successful course design that others could reproduce in their own unique context. In particular, it shows the reader how to engage in the process of design engineering to develop and implement any course.
- **Part II:** This latter part turns toward evaluation as a process of reflective practice. Based on analyses of the course and student learning outcomes, I then considered how to revise my teaching in light of that data. The analysis is quite detailed at times and reflects my thought processes and how it led to my understanding of student learning that informs iterations of the course. This approach demonstrates how teachers can assess the effectiveness of their course design, a method that could be applied to a geology course or some other domain of study.

In sum, my hope is that practicing teachers will use this book to reflect upon and perhaps reconsider their teaching practice. Certainly, adopting all that appears in this book is not necessary. Some may like the idea of trying a strategy here and another strategy there depending on a given context or picking parts of a historical narrative to illuminate one specific concept. This book will also be useful to science teacher educators, preservice teachers, and other beginning science teachers to illustrate a way to think about teaching that is quite different from the traditional lecture format, while also focusing on reflective practice.

The book is also intended to be a meaningful resource for geoscience education researchers and NOS researchers who may find insights from a long-term investigation into the efficacy of novel teaching strategies. The research has straightforward questions and an outline of the different approaches and methodologies for answering those questions. Additionally, the research highlights the use of different validated instruments, such as the following: Geological Concept Inventory; click-on diagrams that use a classroom response system; the Views on Science–Technology–Society (VOSTS) survey; and open-ended questions and qualitative analysis for understanding students' developing understanding of NOS.

Notably, the qualitative research aspect showcases distinct theoretical frameworks—phenomenography and conceptual metaphor theory—and includes how these frameworks influence data analysis. Part II is thus also important for the science education researcher because it provides the context through which I analyzed and interpreted the data. This fuller picture will help practitioners and researchers to make sense of the analysis, especially where the results were unexpected.

This book came about because I believe documenting the processes of instructional design, development, and evaluation is important. Science educators, science teacher educators, and science education researchers could benefit from these documented experiences. Fundamentally, the book offers a map for teaching introductory geoscience concepts; more than that, however, it describes the lengthy and thoughtful process involved in creating such instruction, evaluating it, and then amending the instruction in light of that evaluation. In other words, reaching for superlative of teaching is an iterative endeavour. I am also struck by the fact that this approach appears to produce more geoscience-literate students (and by extension, members of the public) and could work to encourage more diverse people to enter STEM fields.

Calgary, AB, Canada

Glenn Dolphin

Acknowledgments

This work was really a labour of love. As with any large project, it did not happen in a vacuum. Many people were involved, contributing at various times to the final product. Many thanks to Dr. Robert Kelly for "giving me permission" and direction to transform my teaching. I also thank Haboun and Lin at the Taylor Institute for Teaching and Learning at the University of Calgary for helping me think about and develop the rubrics for the activities I incorporated in class. I thank Dr. Nicole LaDue for giving me the opportunity to work with the click-on diagrams and for helping me with statistics. I owe Donna-Lee Wybert, my editor, a great deal of gratitude for making the manuscript readable. Finally, my wife Lu offered unwavering support that allowed me the time and space to complete the planning, implementation, evaluation, and writeup of this project.

Contents

Part I Designing the Course with Teaching the Nature of Science in Mind

1 Earth Science Education: Rationales and Current State 3
1.1 Introduction: State of the Earth 3
1.1.1 Our Home in Space 4
1.1.2 Natural Hazards 5
1.1.3 Mineral Resources: Consumption and Extraction 7
1.1.4 Energy Resources: Consumption and Extraction 9
1.1.5 Endangered Soils 10
1.1.6 Diminishing Water Supply 11
1.2 Geoscience Education: Essential, But Neglected in Science Literacy 12
1.2.1 Geoscience Literacy: Fundamental Ideas 13
1.2.2 Geoscience Teaching and Learning Standards 15
1.2.3 Geoscience Education Research Also Weighs In 16
1.3 Geoscience Education: The Role of History and Nature of Science 19
1.3.1 Characterizing NOS for Science Education 21
1.3.2 Ideas of Science 22
1.3.3 Knowledge of the Nature of Whole Science 22
1.3.4 Features of Science 23
1.3.5 Family Resemblance Approach 23
1.3.6 Questions Rather Than Tenets 24
1.4 Incorporating the History and Nature of Geoscience: Helping Students Think Like a Geologist and Recognize Geological Expertise 27
1.4.1 Using History to Teach NOS 29
1.4.2 Case Studies as a Vehicle for History 30
1.5 The Structure of This Book 33
References 34

2 **Transformative Teaching: From Educator to Design Engineer** 43
Glenn Dolphin and Robert Kelly
2.1 Introduction 44
2.2 Surveying the Educational Landscape: Design Constraints 46
2.3 Cultivating High Levels of Interactivity: Interpretation 49
2.4 Planting Seeds for Learning: Ideation to Early Prototyping 50
2.5 Tending to the Seedlings: Prototypes and Experiments with Strategies 54
2.6 Summary 57
References 59

3 **Biographies of Three Different Concepts** 65
3.1 Introduction: The Structure of the Historical Narratives 66
3.2 The Earth's Biography: Historical and Contingent 68
3.2.1 Signs of Earth's History: Learning to Read the Rocks 68
3.2.2 Eternalism: The Present Becomes the Key to the Past 70
3.2.3 Constructing Earth's History: Giving Fossils Meaning 72
3.2.4 Using Fossils to Make Maps 74
3.2.5 Discerning Ice Ages Fortifies the Notion of a Historic Earth 75
3.3 The Age of the Earth: The Discovery of Deep Time 77
3.3.1 Breaking with Tradition: Empiricism and a "Bottom-Up" Approach 78
3.3.2 Enter Thermodynamics: A Top-Down Approach 79
3.3.3 How the Discovery of Radioactivity Changed the Dating Game 81
3.4 Developing a Theory of the Earth: Why Are There Still Mountains? 83
3.4.1 Initial Thoughts Concerning the Origin of Continents and Oceans 83
3.4.2 Drifters versus Fixists 87
3.4.3 The Role of Geophysics to Understand Earth's Dynamics 89
3.5 Summary 99
References 100

4 **Transformative Teaching Strategies** 103
4.1 Introduction: How Learning Theory Guides the Teaching Strategies 103
4.2 The Foundation of the Course: Team-Based Learning (TBL) 104
4.3 Case-Based (Problem-Based) Learning 105
4.4 Inquiry-Based Learning 108
4.4.1 Visible Geology Activity 110
4.4.2 Extended Seismic Activity 112
4.4.3 Seafloor Data Activity 113
4.4.4 Age of the Earth Activity 114
4.4.5 Models of Earth Dynamics Activity 114

4.5 Model-Based Learning (MBL) . . . 114
4.6 Using Wikipedia to Highlight Women of Geology . . . 119
4.7 Textbook Aligned with Curriculum . . . 121
4.7.1 Limitations of Textbooks . . . 121
4.7.2 Addressing the Limitations . . . 125
4.8 Explicit, Reflective NOS Instruction . . . 126
4.8.1 Informing Students How They Learn . . . 127
4.8.2 Geology as a Historical and Hermeneutic Science . . . 129
4.8.3 Resolving Controversy in Geology . . . 131
4.8.4 Cognitive Resources . . . 133
4.8.5 The Role of Serendipity . . . 134
4.8.6 Prematurity in Geologic Discovery . . . 135
4.8.7 Models and Modeling . . . 135
References . . . 138

Part II Measuring Shifts in Students' Content Knowledge and Nature of Science Understandings

5 Measuring Student Learning . . . 149
Glenn Dolphin and Nicole Ladue
5.1 Introduction: Student Understandings of Content and NOS . . . 149
5.1.1 Student Participants and Learning Context . . . 150
5.2 Geology Content Assessment—Click-On Diagrams . . . 153
5.2.1 Methods . . . 155
5.2.2 COD Results and Discussion . . . 156
5.3 Geology Content Assessment—Geoscience Concept Inventory (GCI) . . . 164
5.3.1 Methods . . . 165
5.3.2 GCI Results and Discussion . . . 167
5.4 NOS Assessment—Views on Science–Technology–Society (VOSTS) . . . 173
5.4.1 Methods . . . 174
5.4.2 VOSTS Results and Discussion . . . 175
5.5 Conclusions . . . 191
References . . . 193

6 Students' Understandings of Science and Scientists . . . 197
6.1 Introduction . . . 197
6.2 Methodology . . . 199
6.2.1 Study Participants . . . 199
6.2.2 Data Collection . . . 199
6.2.3 Theoretical Framework . . . 200
6.2.4 Data Analysis . . . 201
6.2.5 Verification Strategies . . . 204
6.2.6 Researcher's Perspective . . . 205

6.3 Findings 206
6.3.1 Similarities Among Responses 206
6.3.2 Differences from Pre- to Post-Instruction Responses 211
6.4 Interpretation: How Students' Thoughts About Science Changed 225
6.4.1 From Science Is Situational to Doing Science. 225
6.4.2 From Routine to Creative. 231
6.4.3 Scientific Knowledge: From an Always Changing Truth to Grades of Reliability 233
6.4.4 Scientists: From an Idealized to Humanized 235
6.5 Discussion: Outcome Space and the Shift in Understanding 237
References. 241

7 Students' Metaphors for Science 245
7.1 Introduction: Metaphor and Thinking 245
7.1.1 Diving Deep into Conceptual Metaphor Theory 246
7.1.2 Metaphor in Science and Science Education. 247
7.2 Methodology 250
7.2.1 Direct Metaphors 250
7.2.2 Indirect Metaphors. 251
7.3 Findings 252
7.3.1 Science Knowledge Development Is Moving Along a Path 253
7.3.2 Science Knowledge Is an Edifice. 255
7.3.3 Science Knowledge Is a Landscape. 256
7.3.4 Ideas Are Objects. 258
7.3.5 Science Knowledge Is a Liquid/Ocean 259
7.3.6 Scientific Theory Is a Puzzle 261
7.3.7 Science Knowledge Is a Plant 262
7.4 Discussion: Metaphors of Science Processes and Products 263
7.4.1 Framing Science Knowledge. 263
7.4.2 Metaphors for Teaching and Learning Science 265
7.4.3 Summary: The "What" and "How" of Science 267
References. 271

8 Concluding Remarks 275
8.1 Introduction 275
8.2 The Importance of Geoscience Education. 276
8.3 Design Engineering and Reflective Practice 278
8.4 The Three Narratives. 280
8.5 Transformative Teaching Strategies. 282
8.6 Measuring Learning. 283
8.7 Final Words on the Results 286
8.8 Final Words on the Process 289

Index. 293

Part I
Designing the Course with Teaching the Nature of Science in Mind

1.1 Introduction to Part I

Part I situates geoscience education as an indispensable facet of an educated public. Humans face tremendous challenges for survival, including water, energy, and mineral resources. Geologic hazards also represent an increasing threat, as our growing population occupies more and more of the earth's surface and demands more from it. Human interventions in the environment are also exacerbating the destructiveness of natural systems. In addition, professional geoscientists are starting to retire at a much higher rate than they are being replaced. This spells trouble for understanding and appropriately responding to issues that impact earth and its inhabitants. We need new geoscientists to help solve the problems we face, but we also need a geoscience literate populace to act in ways that are coherent with our well-being—including electing officials who will design and enact policies to mitigate these issues.

This requires teaching geoscience in a way that will engage *all* students. In other words, it demands transformative teaching. This means not just about incorporating some best practices into business-as-usual teaching, but rather taking a design approach, understanding the problem, delineating design constraints, developing and implementing teaching strategies, measuring outcomes, *and then* modifying the design in light of those outcomes. Part I of this book outlines this design process as it iteratively unfolded in an introductory geology course for non-science majors.

One constraint in this course is its setting: a large, lecture-style classroom with 200 to 300 non-science major students. The instructional design sought to create an atmosphere where all students felt they could be a part of the knowledge creation going on in the classroom by actually engaging in solving geoscience-related problems. They needed an environment where they could depend on others and knew that others depended on them; it had to be a space where they felt they could meaningfully contribute.

The transformative teaching strategies included team-based learning, problem-based learning, and inquiry-based learning, among others that provided the structural foundations for teaching geoscientific concepts within their historical context. In this way, students could experience the scientific process as they wrestled with the same problems their geoscience predecessors struggled—namely, the historical nature of the earth, the age of the earth, and the dynamics of the earth. These foundational concepts are a crucial starting point for understanding Earth in a broader context, developing foundational knowledge of geological concepts, and engaging with how scientific knowledge is generated.

By using history as a context, students experience the process of science and see geoscientists as regular people with a passion for solving natural world problems. This final point serves to invite all students into the experience of learning geology and acknowledging that perhaps they could be a geologist and solve such problems. These ends are foundational for creating a public better acquainted with geological processes and the role of professional geologists, and confident in their understanding of scientific knowledge. This, in turn, positions them to grapple with the severity of current environmental problems, natural hazards, and management of natural resources.

Part I provides the "cartography of teaching," a map for those who wish to teach the course in a fashion that is similar to the instruction presented. My hope is that readers get inspired to take a design approach to any course they teach, and then go on to practise transformative teaching. Ideally, readers will come to understand *the process* of successful and engaging curriculum design, a view that goes beyond just implementing bits and pieces of curriculum, or even the entire curriculum. That the next generation of students have multiple access points to a science education that centers humans in the creative process of knowledge development is crucial. This is opposed to an approach where they merely receive facts. In this way, students will have the opportunity to develop a more robust understanding of geoscience concepts along with insight into how that knowledge came to be developed and well-established.

Chapter 1
Earth Science Education: Rationales and Current State

After eighteen days of a space mission, I was convinced that all visible space—the black emptiness, the white, unblinking stars and planets— was lifeless. The thought that life and humankind might be unique in the endless universe depressed me and brought melancholy on me, and yet at the same time compelled me to evaluate everything differently. Nature has been limitlessly kind to us, having helped humankind appear, stand up, and grow stronger. She has generously given us everything she has amassed over the billions of years of inanimate development. We have grown strong and powerful, yet how have we answered this goodness?

Yuri Glazkov

If we intend to live on this planet, we truly need to understand how it works.

Thomas McGuire

1.1 Introduction: State of the Earth

This section outlines the aspects of Earth we will explore in this book, emphasizing appreciation for the earth, its scale and location, and its importance as a functioning system to all those living as part of it. It situates the earth in space, then situates life on it, existing at the interface between the solid (and very hot) interior of the earth and the emptiness of outer space. As we come to grips with the fact that humans are having an indelible (and in many cases, harmful) impact on the earth, and we look to other places, like Mars, to inhabit, we need to grapple with the possibility that we will never be able to fully disconnect from the earth (Wilson, 2002a). Since humans are the product of over four billion years of evolutionary "tweaking," trying to engineer a new environment in which we may thrive may appear to be a daunting challenge. If we can afford to make another planet habitable, surely, we can afford to make the earth habitable again.

G. Dolphin, *Teaching Geology Using the History and Philosophy of Science*, Science: Philosophy, History and Education,
https://doi.org/10.1007/978-3-031-68667-2_1

1.1.1 Our Home in Space

Imagine if you will, a yellow ball, one metre (m) in diameter, sitting in a large field. Also imagine another ball, a blue one, barely a centimetre (cm) in diameter, about 100 m from the yellow one. Picture this tiny blue ball revolving around the 1-m diameter yellow ball, maintaining the 100-m radius. You have just visualized a simple scale model of the earth–sun system. Within this 100-m radius circle is empty space, except for three other balls, smaller than the one representing the earth. One is also just less than a cm in diameter (Venus), the second is about 3 millimetres (mm) in diameter (Mercury), and the third one is about 2.5 mm in diameter (the moon). Over 3 kilometres (km) away from the sun is the orbit of Neptune, the outer edge of our eight-planet solar system. In your mind's eye, aim to situate these bodies within a huge circle with a 6-km diameter and virtually no other matter within. Outer space is spacious, as our planet travels within a small portion of this vast emptiness of the solar system.

Let's expand this mental model a bit by next imagining the tiny blue ball as 30.5 cm in diameter (that is, a standard 12-inch desktop globe). At this scale, you could barely perceive the most extreme topographic features, such as Mt. Everest and the Mariana Trench, as they would register approximately 0.1 mm above or below the globe's surface. Visualizing the actual 480-km thickness of the atmosphere as a cm-thick transparent layer blanketing this globe at this scale is not too difficult; however, 90% of the atmosphere exists within only the first 16 km, or 0.5 mm at the scale of our globe. Within this thin and delicate envelope separating the solid earth from the vast expanse of nothing is where all life we know of exists.

Despite having relatively little space to organize, life within this space is ubiquitous, opportunistic, and tenacious. Life, in its macroscopic form (mammals, birds, trees, flowers, and so forth) exists in most places on the earth, and it is diverse. However, the true scope of the extent of the biosphere comes into view when we investigate the realm of microbes. In his book, *The Future of Life*, Wilson (2002a) gave multiple examples of this opportunism and tenacity. There are whole communities of organisms (bacteria, fungi, algae, and so on) that live in the driest places of the world, experiencing no appreciable water for decades or even centuries at a time. Bacteria (extreme hyperthermophiles) survive in submarine environments near hydrothermal vents that get as hot as 149 °C (300 °F).

There are acidophiles, alkaliphiles (living at the very ends of their respective pH scales), halophiles, and barophiles, each named after the extreme environment they inhabit. For instance, researchers have found halophiles living in fluid inclusions within halite crystals that are thousands of years old (Lowenstein et al., 2011), and several species of barophiles (pressure lovers) live within the Marianna Trench, the deepest ocean on Earth at 10,900 m (35,750 ft), with a pressure 1000 times that at the earth's surface (Kato et al., 1998). Some bacteria can make their homes in environments with tremendously high radiation, having developed mechanisms that rebuild their radiation-damaged DNA (Pavlopoulou et al., 2016). There are even bacterial colonies living within the water trapped between interlocking crystalline rock situated deep below the sea floor (Suzuki et al., 2020). They are living in solid rock! This could have implications for finding life in the basalts on Mars.

The success of microbes, their capacity to flourish in such diverse and extreme conditions, has scientists reconsidering the status of these "primitive" organisms. Because of their diminutive size, most people ignore their existence, let alone their importance, until they experience a severe case of food poisoning or until the global COVID-19 pandemic occurred. However, the history of life demonstrates that humans, and all large animals, have evolved to fit into this bacterial world, to serve as hosts for these little microbes (Knoll, 2003). Their importance becomes more apparent when we count them. Imagine a human body with its 37 trillion cells. Now imagine a number 10 times that size: This is how many cells live on and in that body but do not carry that person's DNA. Nonetheless, they keep it alive. In contrast to bacteria, humans cannot survive extreme environments. Instead, we have evolved the ability to engineer our environments to make them livable for us; this, in turn, has enabled us to increase in population, expand our reach to all corners of the globe, and live in densely populated groups. This capability has also led to numerous problems, both for us and for the many other organisms with whom we share the planet.

Indications of these problems typically make daily headlines: "Natural hazards threaten 57% of US structures" (AGU, 2021), "Post-wildfire landslides becoming more frequent in Southern California" (USGS, 2021), "An earthquake in Japan speaks to a seawall in San Francisco" (Tada, 2021), and "St. Vincent hit by flooding, landslides following volcanic blasts" (Henry, 2021). We live on a dynamic planet—which puts the safety of human property and health at risk—but also affords us energy resources to power our societies, mineral resources to supply our consumer-based economies and transportation infrastructure, and soil and water resources to nourish our bodies. All this is wrapped in a delicately thin atmosphere that maintains a temperature conducive to the development and maintenance of life.

1.1.2 Natural Hazards

In their review of the literature on natural hazard risk assessment, Ward et al. (2020) highlighted the trends in natural hazard risks that included river, coastal, and pluvial flooding, tropical cyclones, severe convective storms, droughts, wildfires, earthquakes, tsunamis, volcanoes, and landslides. The risks are high and have been increasing in recent decades (Fig. 1.1, left). For instance, over the last 30 years, natural hazards have caused approximately 1.6 million deaths globally and $260–310 billion USD in damages annually (Fig. 1.1, right). This cost (both human and material) is expected to double by the year 2100, and double again when global warming is factored in. Ward et al. attributed this rise to several aspects, including a general increase in global population and the exacerbation of climate-related hazards (flooding, storms, droughts, fires, and landslides) via global climate change.

Fig. 1.1 Left: This shows the number of natural disaster events from 1900 to 2019. Right: This shows global cost of disasters from 1980 to 2019 in USD. (Note. Left: By The International Disaster Database (EM-DAT [graph]), 2020, Our World in Data, (https://ourworldindata.org/grapher/number-of-natural-disaster-events). CC BY 4.0; Right: By The International Disaster Database (EM-DAT [graph]), 2020, Our World in Data, (https://ourworldindata.org/grapher/damage-costs-from-natural-disasters) CC BY 4.0)

This second point, the impact of climate change, is especially important, as it highlights the complex interrelationships of the earth's systems. To mitigate the risk, we need to understand the hazard and risk within spatial scales (from household to global) and across time (from minutes to generations). For example, earthquakes result from the buildup of stress over decades to centuries, but last only seconds to minutes, and can cause damage that may take from years or decades to recover. They can cause small disruptions in households, knocking pictures off walls, plates off shelves, or as in the case of the 1906 San Francisco earthquake, they can cause a 200+ mile-long rip in the crust that devastated cities all along its path.

As the planet heats up, largely because of carbon dioxide added to the atmosphere from the combustion of fossil fuels, sea level will rise due to melting glacial ice but also from the thermal expansion of ocean water. This will lead to flooding of coastal areas, impacting approximately 40% of the global population that lives at near sea level and eight of the 10 largest cities that are located near a coast (National Oceanic and Atmospheric Administration, 2021). In addition to the hazards of higher sea level, tropical cyclones are more likely in these coastal locations and will increase in intensity due to the warming climate. In inland regions, global climate change also increases the threat of severe convective storms, with hail generating the most economic damage, but pluvial flooding and tornadic activity contributing to this.

Global climate change also contributes to increasing droughts and fires. At the end of June 2021, the village of Lytton, British Columbia experienced three consecutive days of record-breaking heat: 46.1 °C, 47.9 °C, and 49.6 °C (115 °F, 118.2 °F, and 121.3 °F, respectively). The following day, a fire swept through and burned the village beyond recognition. This is in Canada, a country with a normally temperate climate! The US has seen its wildfire season lengthen, with increasingly intense fires occurring all over the country. Despite this, people in the US are more often building in fire-prone areas, thereby increasing the hazard risk (Radeloff et al., 2018). This example demonstrates the scope, complexity, and interconnectedness of Earth phenomena and its effects. That said, as a species, we are facing numerous other natural hazards—among them, the impact of energy and mineral resource extraction, the endangerment of soils, and reduced water resources—which will all affect our quality of life.

1.1.3 Mineral Resources: Consumption and Extraction

If you can't grow it, you must mine it. Our lives are tied inextricably to this planet (Wilson, 2002a). Obviously, our food, a necessity, is grown, but a simple reflection will demonstrate the importance of the materials we mine. Maybe you are drinking a cup of tea or coffee while reading this book. You may well have heated the water for this hot beverage in an electric kettle. That kettle has several mineral components, as well as petroleum-based plastics. Let's look at just one of these materials.

Copper makes up portions of the kettle's electrical system. Copper processing takes the native element from the ground (mining) and processes it into the wire that is intrinsic to the kettle's electrical system. Mining involves technological equipment and earth-moving machines that all require their own mineral resources; these, in turn, also required mining, processing, and incorporation. That processing occurs in a factory made of more mineral resources and aggregate (concrete) that must be quarried and then transported and processed into concrete. Once manufactured, the processed copper wire travels to another factory, where it encounters other processed resources and gets fashioned into components of an electric kettle. Additionally, the ships, trucks, or trains that transport the copper also all comprise mineral and petroleum resources, made in other factories of minerals and aggregate.

The assembled kettle is then transported to a store for sale. The delivery mode (ship, truck, or train) and the store itself are similarly composed of minerals, petroleum resources, and/or aggregate. Happy consumers then make a trip to that store in a vehicle—a mode of transportation most likely originated within the earth too, as did all the pieces of technology that created that mode of transportation, and the technology that created the technology. Or perhaps they went on foot. What footwear did they where and how did it come into being? What kind of road or path did they travel on? These resources, of course, would all require multiple steps and materials to become service-ready.

Importantly, the copper wire is only ONE component of that teapot; many more components are necessary, each within its own web of creation. After the kettle, there is the water, the tea, and the teacup, and perhaps creamer and sweetener (whether natural or artificial). This mental inventory makes obvious that at every step of the way, from raw material to kitchen table, what we depend on, we dig out of the earth.

To put this dependency into another perspective, average Americans born in 2020, will *each* use these amounts of the following Earth-derived materials during their lifetime (Minerals Education Coalition, 2020):

- 0.04 kg gold
- 228 kg zinc
- 395 kg lead
- 431 kg copper
- 1466 kg phosphate rock
- 5757 kg bauxite (aluminum ore)
- 9818 kg iron ore
- 13,608 kg salt
- 128,114 kg coal
- 644,101 kg stone, sand, and gravel
- 26,656 kg other minerals
- 226,535 m^3 natural gas
- 284,337 L petroleum

Of course, Americans quite possibly are the largest consumers on the planet. At present, the earth holds almost 8 billion people, with most aspiring to a standard of living equivalent to that in the industrialized world. Clearly, we need to consider consumption and extraction through a long-term lens, one that accounts for the multiple generations to come.

A critical aspect here is the pace of growth. The global human population has doubled since the 1970s, and is projected to peak at approximately 10.4 billion by 2086 before steadying and slowly falling. To serve that growing population, resource extraction has increased by a factor of 3.5, from 27 billion tons in 1970 to 92 billion tons in 2017, and is expected to reach 190 billion tons by 2060 if we maintain a business-as-usual approach (Oberle et al., 2019). The International Resource Panel (IRP, Oberle et al., 2019) has documented the increased rate of change in extraction and consumption. The panel also stressed the need for complex modelling, especially regarding interactions between resource extraction and use, and society. Models must accurately represent complex systems and work within different scales of time and space.

Notably, resource extraction and processing contribute 50% of current greenhouse gas (GHG) emissions according to the IRP (Oberle et al., 2019), which must be mitigated if we hope to reach the Paris Climate Accord goals of 2015. These activities also increase water toxicity, especially due to heavy metal release. Since water extraction will also continue to rise, access to clean water will increasingly worsen says the IRP. Finally, the release of industrial emissions into the air, most markedly from coal electricity generation, contribute to diminishing biodiversity and respiratory illnesses, especially in developing countries.

1.1.4 Energy Resources: Consumption and Extraction

Some evidence points to coal use in East Asia over 4000 years ago (Barras, 2014), but it was not until the onset of the Industrial Revolution in Europe in the early 1800s that coal began generating electricity and powering factories on a comparatively large scale. However, a century later, energy generation jumped 100 times—from about 90 terawatt hours (TWh) to 8600 TWh (Fig. 1.2). At that time oil and then natural gas also became popular energy sources. From the early twentieth century to today, fossil fuel usage increased by yet another order of magnitude from about 9000 TWh to 135,000 TWh (Oxford Martin School, 2019), or about 8 billion tons of coal, 35 billion barrels of oil, and almost 4 trillion m^3 of natural gas per year, every year. Though estimates project that substantial carbon-based energy resources remain, the impact of GHG production from fossil fuel combustion mandates that we rethink how we meet our future energy needs.

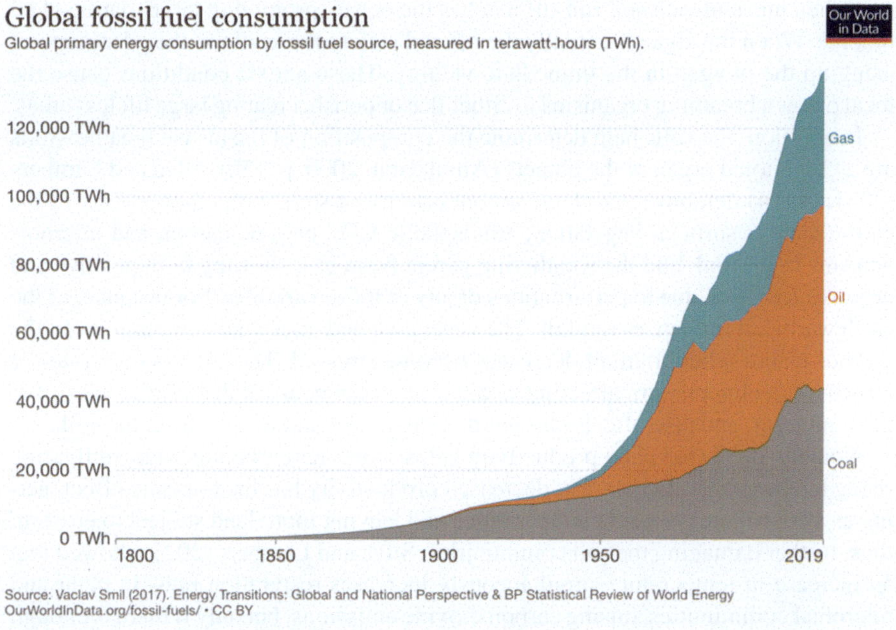

Fig. 1.2 Graph showing global consumption of fossil fuels in terawatt-hours from 1800–2019. (Note: From *Energy Transitions: Global and National Perspectives & BP Statistical Review of World Energy*, by Vaclav Smil, 2017, Our World in Data, (https://ourworldindata.org/grapher/global-fossil-fuel-consumption?country=~OWID_WRL), CC BY 4.0)

1.1.5 *Endangered Soils*

Another important resource lying right beneath our feet, one that often does not get much respect, is soil. Remember, if you can't grow it, you must mine it. But to grow it, you still need the solid Earth to assist. Although soil is comprised of much decayed organic material, the weathering of minerals from the bedrock provides the foundation and some nutrients for soil. Soil helps produce over 99% of the food we consume, but poor farming practices and other factors, such as deforestation and poor water use policies, are responsible for the erosion of soils at a rate that is currently unsustainable (Pimental & Sparks, 2000). The high rate of erosion will, of course, inhibit food production, but it also threatens the biodiversity of the planet. Additionally, eroding soils carry with them large amounts of herbicides and insecticides that harm aquatic ecosystems, and fertilizers that are responsible for creating dead zones in various portions of the world's oceans. Studies on waterways within the US show nutrient levels to be two to 10 times greater than recommended and well above natural background readings (Dubrovsky et al., 2010). Dead zones happen when nutrient-rich soil run-off reaches the ocean, promoting large marine algal blooms. When the algae eventually die off and sink to the seafloor, they decompose, using up the oxygen in the immediate vicinity. These anoxic conditions cause the local oxygen breathing organisms to either flee or perish, creating large lifeless areas.

In addition, the soils help determine the composition of the air we breathe. Soils are a "functional organ of the planet" (Amundson, 2000, p. 150). Silva and Lambers (2021, Fig. 4) demonstrated the interconnectivity of soils with other aspects of the earth (such as surficial vegetation, atmospheric CO_2, organic carbon and microorganisms in the soil, and the weathering parent bedrock), showing both positive and negative feedback due to perturbations of any of those variables. For instance, as the earth warms, it influences rainfall. The vegetative and microbial communities in the soils of places where rainfall decreases become stressed. This affects their rates of growth or decline, in turn affecting weathering of the bedrock that affects the soil in the long term, and also affects carbon dioxide uptake and release from the soil.

Notably, increased plant productivity helps sequester carbon dioxide, mitigating rising temperatures. In contrast, decreased productivity has the opposite effect, acting as a contributor to rising temperatures and leaving more land subject to erosion; thus, further damaging the soil communities. Silva and Lambers (2021) showed that an increase in temperature simultaneously increases respiration rates in plant and microbial communities, raising carbon dioxide emissions, but only if there is enough water to support the communities. Their work highlights that feedback mechanisms quickly become very complex. Significantly, soils cover most of the exposed land on Earth, and they interface between atmosphere, hydrosphere, biosphere, and geosphere, playing a major role in regulating each of these spheres via the others. Soils maintain a series of complex systems, but we have much we still need to learn about all this, to ensure our own survival (Silva & Lambers, 2021).

1.1.6 Diminishing Water Supply

Our planet was a distinctive "pale blue dot" when seen by the Voyager space craft looking back as it made its way to the edges of the solar system. That inviting blue colour comes in part from the reflection of sunlight off the water that covers about 70% of the planet's surface. Water is certainly necessary for life to exist on this planet, but that expansive layer of water is relatively thin even though it covers such a large portion of the earth's surface. Indeed, it averages only about 3700 m deep. If all the water on Earth were gathered into one big drop of water, it would be one-tenth the diameter of the earth—a 1.4 billion-km^3 volume, with a 1400-km diameter (United States Geological Survey, 2020). However, only a small portion of that total water is available for human consumption, with about 97% of it being salty marine water and just under 2%, frozen icecaps and glaciers. This leaves little more than 1% of ground water, which is also mostly inaccessible, and even less usable than surface water (Fig. 1.3). Thus, the full amount of freshwater would make a drop with only a 56-km diameter (compared to the 1400-km diameter sphere of total water on Earth), and approximately 93,000 km^3 (United States Geological Survey, 2020).

With these numbers in mind, that 47% of the global population currently lives in an area of water scarcity, an amount projected to rise to 52% by 2050, is unsurprising (Boretti & Rosa, 2019). Boretti and Rosa (2019) reported other alarming statistics. Globally, the demand for water, 70% of which is for agricultural purposes, has increased by about 600% in the last 100 years (Fig. 1.4) and will significantly increase for the foreseeable decades.

At the same time, water resources have been shrinking, with 30% of the largest aquifers under threat of being completely depleted. These drawdowns of fresh groundwater, especially along coastal regions, allow for the influx of saline waters from the oceans into aquifers, further limiting drinking water. The threat to water

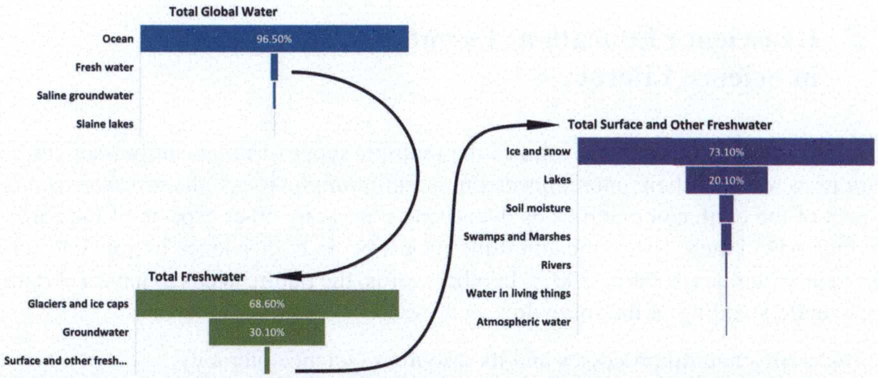

Fig. 1.3 Graphic showing the distribution of global water resources. (Note: After United States Geological Survey, Water Science School (2019))

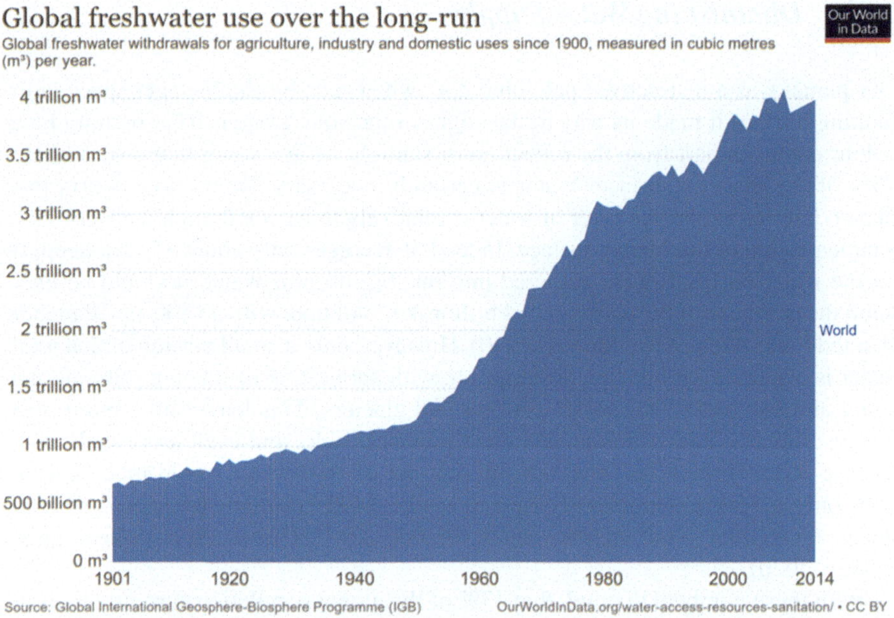

Fig. 1.4 Graph showing the increase in global freshwater usage from 1900 to 2014. (Note: By *Global International Geosphere-Biosphere Programme (IGB),* The International Disaster Database (EM-DAT) [graph]), n.d., Our World in Data, (https://ourworldindata.org/grapher/global-freshwater-use-over-the-long-run?country=~OWID_WRL). CC BY 4.0)

resources is especially felt in developing nations, where up to 80% of industrial effluent is released into the surface waters untreated (Boretti & Rosa, 2019). This trend will only increase as industrialization rises in these regions over the next several decades.

1.2 Geoscience Education: Essential, But Neglected in Science Literacy

The preceding information illustrates the multiple types of threats to human civilization, as well as their interconnectedness and complexity. A disturbance to one aspect of the earth can manifest as disturbances in many other aspects of the earth. To understand these issues meaningfully means being both science literate and scientifically literate (Roberts, 2007). In other words, the public needs to have a certain basic understanding of the following:

- the earth's natural processes and its resources (science literacy)
- the practices used to develop that knowledge (scientific literacy)
- how scientific knowledge fits into socio-cultural contexts (also scientific literacy)

Improving geoscience literacy and geoscientific literacy is central to the discussions in this book. Solving the issues outlined above will require people who can (a) think geologically (Frodeman, 1995), including how to do geology (Cleland, 2013; Dodick et al., 2009), (b) understand Earth's systems (Manduca & Kastens, 2012; Stillings, 2012) and (c) appreciate time scales over which these geological phenomena occur (Bjornerud, 2018; Cervato & Frodeman, 2012; Dodick & Orion, 2006).

Efforts to realize these goals have led to a decades-long call to action to enhance geoscience education globally. Many geoscientists and geoscience educators have been seeking ways to put geoscience on the map for students as a viable career option (Barstow & Geary, 2002; Corgan & Stearns, 2008; Hoffman & Barstow, 2007; Lewis, 2008). Adding to the sense of urgency is the growing shortage of geoscientists to replace expert retirees and fill new openings (American Geosciences Institute, 2008). 27% of the current geoscience workforce will reach retirement age by 2029 and 22,000 new jobs will emerge by then, resulting in a 130,000 geoscience workforce deficit (American Geosciences Institute, 2020).

1.2.1 Geoscience Literacy: Fundamental Ideas

Many disciplinary groups have responded to this call to action by creating "guiding standards" documents for geoscience literacy. Thus, overarching or core concepts form the framework for other related concepts within a domain (Wiggins & McTighe, 2006). For instance, the Earth Science Literacy Initiative (2010) outlines the following set of nine big ideas for understanding earth science:

1. Earth scientists use repeatable observations and testable ideas to understand and explain our planet.
2. Earth is 4.6 billion years old.
3. Earth is a complex system of interacting rock, water, air, and life.
4. Earth is continuously changing.
5. Earth is the water planet.
6. Life evolves on a dynamic Earth and continuously modifies Earth.
7. Humans depend on Earth for resources.
8. Natural hazards pose risks to humans.
9. Humans significantly alter the Earth.

Although these principles appear as propositional statements, they are big ideas because conceptually understanding each provides umbrella for many other important ideas. Each encompasses a breadth of information about the earth, its characteristics, and how we came to understand that information. In other words, geoscience literacy requires far more than memorizing these statements. Conceptual understanding is needed to comprehend their significance and impact on human welfare. Developing a robust understanding of such big ideas and their interconnections requires effort from teachers and students. The strategies described in this book can help facilitate the process.

Subdisciplines of the earth sciences, such as oceanography and climatology, have created more specific sets of literacy principles. For example, the seven principles of the Ocean Literacy Framework are (Ocean Literacy, 2015):

1. The earth has one big ocean with many features.
2. The ocean and life in the ocean shape the features of Earth.
3. The ocean is a major influence on weather and climate.
4. The ocean made Earth habitable.
5. The ocean supports a great diversity of life and ecosystems.
6. The oceans and humans are inextricably interconnected.
7. The ocean is largely unexplored.

Similarly, seven climate literacy principles have been established (US Global Change Research Program, 2009):

1. The sun is the primary source of energy for the earth's climate system.
2. Climate is regulated by complex interaction among components of the earth's systems.
3. Life on Earth depends on, is shaped by, and affects climate.
4. Climate varies over space and time through both natural and man-made processes.
5. Our understanding of the climate system is improved through observations, theoretical studies, and modelling.
6. Human activities are impacting the climate system.
7. Climate change will have consequences for the earth system and human lives.

Synthesizing these sets of principles results in four overarching themes. *First*, the environment is constantly changing. The notion of constant change also implies that the earth is an historical body. The future is contingent on the conditions of the present, with those conditions contingent on those in the past. This may seem obvious, but it is a relatively new idea (emerging in the eighteenth to nineteenth centuries), a realization on par with the Copernican revolution (Rudwick, 2014). That much of this change occurs very slowly is noteworthy: for example, erosion rates occur at fractions of a milimeter per year, continents move centimeters per year, and glaciers move meters per year. Thus, the record of many "former worlds" (Hutton, 1788) within the rock record highlights the *second* theme: the earth is very, very old. The concept of deep time is essential for understanding the nature of the earth and its processes. Comprehension of the vast age of the earth is also a relatively recent development, becoming accepted knowledge only in the early twentieth century.

The *third* theme is that earth phenomena operate within complex and dynamic systems. Viewing a particular phenomenon outside the context of a complex system gives, at best, an incomplete understanding of that phenomenon (Stillings, 2012). For instance, understanding the nature of a landslide (and related hazard mitigation) is much more robust if we look at that landslide within the context of the tectonics, climate, geology, and geomorphology of the area where the landslide took place. Here, noting the role humans play within the system is also important. A geologic phenomenon is only considered a hazard if it somehow poses a threat to human life or property. Also, as humans continue to modify their environments—cutting roads into mountainsides, drawing down aquifers, burning fossil fuels, and converting forest into farmlands—we affect the environment in ways that are detrimental to our well-being.

The *fourth* major theme addresses *how* geoscience is done. This aspect seldom makes it into classroom instruction, but it is crucial for understanding and accepting many geological concepts. I will address this aspect more fully when discussing the importance of teaching the history and nature of science in Sect. 1.1.3. The development of these geoscience literacy standards provides the foundation to develop geoscience teaching/learning standards—specifically, what we should teach students and when we should teach it.

1.2.2 Geoscience Teaching and Learning Standards

The geoscience, oceanic, and climate literacy standards are the foundation for the geoscience related standards appearing in national science education standards documents such as the U.S. Next Generation Science Standards (NGSS) (NGSS Lead States, 2013) and other national documents (Greco & Almberg, 2016). The NGSS standards represent three-dimensional curriculum design by providing grade K–12 teachers direction about the following:

- Important core disciplinary ideas—like the big ideas or literacy principles described above
- Crosscutting concepts—thematic concepts that transcend disciplinary boundaries, like the use of models and notions of scale
- Geoscience practices—such as aspects of the scientific inquiry process

Table 1.1 shows the components for secondary education in the US according to the NGSS. Importantly, national standards documents can be, and often are, modified at regional or local levels.

Table 1.1 highlights how the standards parallel the four themes described earlier—namely, the earth's fluctuating character, its historical nature, the importance of a systems approach, and the interactions between the earth and humans. Though not overtly stated in the standards descriptions, deep time is implicit in the notions of the origin of the universe and Earth, as well as the scales of time highlighted within systems interactions. The entire column of "Practices" also emphasizes the processes involved in doing earth science.

Despite the calls to action and efforts to transform the teaching of earth science, both in the US and globally, student exposure to geoscience education remains poor (Greco & Almberg, 2016; King, 2013). This is in large part due to a dearth of qualified earth science teachers (Lewis & Baker, 2010), but also a perception by many that geology is not a pure science. Rather, it is considered a derivative of physics (Dodick & Orion, 2003b; Kravitz, 2013) and therefore undeserving of the status of a science, one on par with physics and chemistry.

Table 1.1 NGSS Standards for Geoscience

Core Disciplinary Ideas	Crosscutting Concepts	Practices
HS-ESS1 Earth's Place in the Universe (origin of the universe, solar system, and earth; plate tectonics; Earth's history)	Patterns	Asking Questions and Defining Problems
HS-ESS2 Earth's Systems (cycling of matter and energy; structure of Earth; interactions and feedback mechanisms among components, and their expression at the surface; different scales of time and space for these interactions)	Cause and Effect	Developing and Using Models
HS-ESS3 Earth and Human Activity (human impacts on resources, hazards, environment, and climate	Scale Proportion, and Quantity	Planning and Carrying Out Investigations
	Systems and System Models	Analyzing and Interpreting Data
	Energy and Matter	Using Mathematics and Computational thinking
	Structure and Function	Constructing Explanations and designing Solutions
	Stability and Change	Engaging in Argument from Evidence
	Interdependence of Science, Engineering and Technology	Obtaining, Evaluating and Communicating Information
	Influence of Engineering, Technology, and Science on Society and the Natural World	

1.2.3 Geoscience Education Research Also Weighs In

The call for more and directed geoscience education research (GER) supplements the call for more comprehensive earth science instruction (Orion, 2019; Shipley et al., 2017). GER is an example of discipline-based STEM education research (DBER). DBER focuses on the following (Shipley et al., 2017, p. 354):

- Understanding how people learn concepts, practices, and ways of thinking within STEM disciplines
- Characterizing the nature and development of disciplinary expertise
- Measuring instructional strategies that advance student learning
- Guiding the incorporation of research findings into classroom practice
- Identifying inclusive approaches to STEM education

The American National Research Council (NRC, 2012) acknowledged that within every science and engineering discipline, students maintain conceptions that do not parallel the scientific consensus. These student perspectives are commonly referred to as naïve, alternative, or misconceptions. Identifying these misconceptions and how they came to be is important for focussing instruction to bring them into alignment with the consensus views. In the geosciences, misalignments exist primarily because many geoscience phenomena operate at scales of time and/or space that are well outside everyday human experience (Cheek, 2010; NRC, 2012; Niebert & Gropengiesser, 2015). Moreover, they may be inaccessible for direct observation. GER, therefore, has focused on untangling students' misconceptions and how to effectively teach concepts such as the dynamic nature of the earth (landforms, natural hazards, and more recently, climate change), and Earth's history and geologic time. GER inquiry has also focused on skills, such as spatial, temporal, and systems thinking (NRC, 2012).

Important research at the intersection of geology and cognitive science (for instance, developing visual-spatial skills) has targeted a range of topics such as: describing and interpreting geological objects (signs); identifying or classifying signs by shape; ascribing meaning to the shape, position, and/or orientation of an sign; discerning patterns or signals from background noise; synthesizing or visualizing one- or two-dimensional observations into a three-dimensional mental model; and using space as a proxy for time (Kastens & Ishikawa, 2006). Some research has turned to how and what students learn about (a) complex systems thinking (Herbert, 2006; Manduca & Kastens, 2012; Stillings, 2012), (b) discerning inference from observations in geoscientific processes (Ault Jr. & Dodick, 2010), and (c) geologic time—both relative time, by using signs in the rocks to logically sequence events from oldest to youngest (Dodick & Orion, 2006), and deep time (Cervato & Frodeman, 2012; Dodick & Orion, 2003a, c; Resnick et al., 2012).

In their review of the earth systems science education literature, Orion and Libarkin (2014) summarized key student learning outcomes in the earth sciences as the following:

1. Approaching scientific inquiry from a historical context, including retrospective thinking and retrodictive reasoning
2. Understanding the role of complex systems acting over the whole of the earth
3. Conceptualizing phenomena through time and space at scales from atomic to planetary, including integrating solutions to problems across multiple scales
4. Reasoning spatially about objects and phenomena

As noted earlier, many consider the geosciences a lesser science, when compared to physics or chemistry.[1] This is partly because the geosciences are viewed as historical and hermeneutic (Frodeman, 1995), as opposed to the physical sciences that are wrongly seen as entirely experimental. While experiments can be part of the geoscience endeavour, the geosciences primarily collect signs of past events and interpret them to tell the story of past events, trying to determine contingent causes. Part of the process, therefore, involves *retrospective thinking*, or projecting understandings of present-day phenomena into the past, as a way to explain the signs in the rocks; it also invokes *retrodictive reasoning*, or identifying a phenomenon as a causal factor of a past event whose effects are the signs observed in the rocks (Dolphin & Dodick, 2014).

The role of complex systems and their impact on the earth, noted earlier, is exemplified cyanobacteria when they began to populate the earth over two billion years ago. Their waste product (oxygen) began to concentrate in the oceans and the atmosphere, and it oxidized much of the dissolved oceanic iron that then precipitated out as iron oxide and settled to the seafloor to become banded iron formation, the largest

[1] In Alberta, where I teach at a research university, the provincial board of education does not consider geology to be a science—even though the provincial economy is largely based on the oil and gas industry.

source for today's iron. Higher oxygen levels in the atmosphere also fundamentally changed many of the chemical weathering processes happening on land, thereby affecting the landscape and the resulting sedimentary deposits. The higher oxygen concentrations also fundamentally influenced the subsequent evolution of life on the earth.

Conceptualizing natural phenomena (and systems) through time and space is at the heart of the geosciences. Always, we are trying to construct a story of the earth's past. That story may encompass the building or eroding of mountain ranges over millions of years, rising or falling sea level over hundreds of thousands of years, migrating rivers on a flood plain over hundreds of years, or a landslide occurring over minutes. Geoscientists interpret signs in the rocks in terms of causal processes. They interpret assemblages of signs in an outcrop as the collection of processes within the environment where the rock formed. They interpret changes of sign assemblages in both horizontal and vertical directions as an indication of changing environments through time and space.

Not only do we need to visualize these processes from the past, but we also need to imagine the structures the processes create (layers, folds, faults, etc.) and how they might look below the surface of the earth, where we have limited or no access. For instance, we may gain access to the subsurface via seismic waveform data (the squiggly lines on a seismogram, which records seismic energy travelling through the earth). How do we interpret those squiggles, and then what meaning can we make from that interpretation?

Despite all these informed perspectives on why it matters that people be geoscientifically literate, at all levels of education a general misalignment persists in the public's understandings compared with the scientific consensus (Francek, 2013; Kirkby, 2014; NRC, 2012). As noted, this is partly due to the lack of adequate earth science instruction, or even any, in primary and secondary education. However, even where such instruction takes place, geological concepts are difficult to learn. Libarkin and Anderson (2005) showed that many undergraduate students maintained their naïve conceptions about certain geological concepts even after direct instruction. Cheek (2010) reviewed 20 years of conceptual development literature to show that many students struggle with geological concepts because of the temporal and spatial scales involved, inaccessible phenomena, incomplete data, and the like.

For the most part, the public maintains ideas that are based in everyday perceptual experiences. Therefore, instructors need to provide learners with concrete experiences and/or embodied experiences (source knowledge) as building blocks for learning abstract concepts (the target knowledge) (Niebert & Gropengiesser, 2015). Niebert et al. (2012) described a hierarchy of strategies for introducing concrete embodied experiences to students to facilitate conceptual development. Specifically, instructors should (a) enable experiences in the domain of the target phenomenon. (b) enable experiences in an embodied source domain, (c) refer to an embodied source domain, and (d) reflect on an embodied source domain.

Dolphin and Benoit (2016) gave an example of enacting this series of strategies when teaching students about the mechanism of earthquakes, particularly elastic rebound theory (Reid, 1910). Because students do not readily attribute the property of elasticity to rocks, Dolphin and Beniot deployed the four-point approach described above, as follows:

1. *Enabled experience in that target domain*—using a sandstone drill core with a gap, sawed lengthwise two-thirds the way up the centre of the core. It pinches closed and then opens elastically like the marble tongs described by Hubenthal (2018).
2. *Enabled embodied experience by using the conceptual metaphor* (Lakoff & Johnson, 1980, 1999)—or a "bridging analogy" (National Research Council, 2012, p. 72). Here, that entailed the analogy that "the crust is the skin of the earth." All students have experience with their own skin and its elasticity, and we were able to draw the analogy of the deformation they can cause on their skin, using their hands, with the deformation that happens in the crust due to geophysical forces.
3. *Referred students to embodied experiences in the source domain*—by showing them molecular models with wood balls for atoms and springs as bonds holding the atoms together. Upon shearing the model, elastic deformation occurs in the springs (atomic bonds) and recovers once the stress is released. With too much shearing, the springs break from the balls and the model falls apart, a metaphoric earthquake.
4. *Students reflected on how these source domain examples mapped onto the target domain*—What did the models highlight, and more importantly, what did they hide? These strategies from Niebert et al. (2012) make clear that the role of the geology instructor is to engineer an environment with experiences that give students the best chance at developing the target understandings. This is different than just presenting content material in a transactional (lecture) format, where the instructors have the authority and students experience their role as passive consumers of knowledge.

1.3 Geoscience Education: The Role of History and Nature of Science

Geoscience education, as opposed to geoscience training, entails understanding both what we know (geoscience content) and how we know it (geoscience practices and the history and nature of geoscience). Without attention to both, students may understand what the scientific community claims about the natural world yet reject those claims because of misconceptions regarding how science knowledge is developed and becomes well established.

The past decades have witnessed a refinement of what it means to be science literate. As I noted above in Sect. 1.2, *science literacy* attends to the products and processes of science (Roberts, 2007); it is what mostly happens in science courses, representing the specific content material that is taught and tested. The geoscience literacy examples discussed in Sect. 1.2.1 demonstrate this. In contrast, *scientific literacy* focuses on the aspects of science that transcend any one discipline (Roberts, 2007), and link the science with the social—thereby contributing to the broader understanding of how science fits into the social environment. Such knowledge helps students to approach and solve science-related problems, but sadly does not usually make its way into instruction. One way to help students appreciate how and where science fits into the social environment is by recounting examples using the history and nature of science (HNOS; Allchin, 2013b; Matthews, 2015; McComas et al., 2020).

Broadly defined, the nature of science (NOS) addresses "how scientific knowledge is established and changes and how best to communicate about science as a way of knowing" within political, economic, sociological, or other contexts (Duschl, 2020, p. vii). NOS also considers various stances that inform scientific knowledge, such as epistemological, ontological, philosophical, and pedagogical. At the inception of the atomic age, James Bryant Conant (1947) pointed out the necessity for all students to understand science, its methods, and its societal connections. Though normally presented as a "final form" science (Duschl, 1990), Conant emphasized that even nonscientists should understand the "tactics and strategy of science" (p. 102). Conant's own answer to this was to develop a college science course using historical case studies, modelled after the success of case pedagogy in business and law schools.

Others followed this example of using the history of science as a vehicle for teaching content: for example, History of Science Cases (Klopfer & Cooley, 1963), and the Harvard Project Physics Course (Holton et al., 1970). Decades later, science education documents continue to emphasize the importance of learning *about* science, both in the US and globally (Asia-Pacific Economic Cooperation, 2012; Eurydice Network, 2011; Interamerican Network of the Academies of Sciences, 2011; McComas & Olson, 1998; NGSS Lead States, 2013; NRC, 1996; NSTA, 2003; Rutherford & Ahlgren, 1990). This emphasis has led to many discussions about how to define NOS, and what and how to teach it effectively. The history of geology illustrates how geologists have come to know what they know. Students see that geoscience knowledge development is not produced by "the scientific method"; further, controlled experiments and the notion of falsification are not always needed or appropriate for developing reliable knowledge. Reflecting on these conclusions, students are more likely to accept consensus science, where political or social pressures may have created controversy surrounding the consensus view (Herman et al., 2017; Höttecke & Allchin, 2020).

The following subsections review various educators' approaches to privileging scientific literacy over mere science literacy. They are organized chronologically to document the development of NOS research and approaches, and to illuminate the cross-pollination of ideas and strategies.

1.3.1 Characterizing NOS for Science Education

Defining NOS, or at least what aspects of NOS should be taught in schools has received considerable attention since Conant's early attempt. Most familiar to science educators are a list of seven NOS tenets that proponents maintain is a consensus about what characterizes science and the knowledge it generates (Lederman et al., 2002), as follows:

- Science has an empirical nature.
- We can distinguish between scientific theories and scientific laws.
- Science requires creativity and imagination.
- Science is theory-laden.
- Scientific knowledge is socially and culturally embedded.
- The notion of a single scientific method is a myth.
- Science is tentative.

These tenets have helped to generate much research literature on the pedagogy and learning of NOS in classrooms K–16 (Abd-El-Khalick & Lederman, 2000; Lederman, 2007), but as with all models, these seven tenets have limitations (Allchin, 2012; Clough, 2008, 2020a; Deng et al., 2011; Erduran & Dagher, 2014; Matthews, 2012). In general, they have too often become mere propositional statements devoid of the very historical and philosophical context that forms the basis of the propositions.

As an illustration, we can examine Lederman's et al. (2002) distinction between theories and laws—that is, explanatory claims and descriptive claims, respectively. At first glance, this is a very straightforward proposition, although it fails to address the nuanced difference between theories and laws. Thagard (2012) called explanation the goal of science, describing it as the most sophisticated way to think. But he further asserted that science both broadens explanatory models by making them apply to more situations and deepens them by exploring understanding in terms of their component parts. Those explanatory models in turn become descriptive knowledge for the next deeper level of explanation.

As an example, we can *describe what happens* during an earthquake (that is, the breaking and shaking of the crust of the earth) and *explain why* it happens (which is, the release of stored elastic potential energy). As we dig deeper into the phenomenon, the energy released becomes the description, which the breaking and motion of deformed crust explains. Then, the motion of the crust becomes the description, and the theory of plate tectonics explains that motion. Taking this further, plate tectonics (plate motions) is the description of events, which mantle convection (Isacks et al., 1968) or gravity dynamics explains (Anderson, 2006). And so it goes. The point is that the tenet brings up an important facet, but then just leaves it, without exploring nuanced iterations or related imbricated relationships, or why these or the phenomenon are important.

1.3.2 Ideas of Science

Osborne et al. (2003) stated the importance of teaching specific *ideas* of science, which they determined from the results of a Delphi study that included expert scientists, science educators, and historians, philosophers, and sociologists of science. Via participant consensus after three iterations of their survey, they discerned 10 important ideas of science they believe should be taught along with science content:

- Analysis and interpretation of data
- Science and certainty
- Hypothesis and prediction
- Cooperation and collaboration while developing scientific knowledge
- Scientific method and critical testing
- Science and technology
- Historical development of scientific knowledge
- Diversity in scientific thinking
- Questioning
- Creativity

The authors also noted the gap between summarizing these ideas and having them effectively implemented in science classrooms.

1.3.3 Knowledge of the Nature of Whole Science

In response to the limitations in the earlier approaches, Allchin (2011a, 2012) put forth a more holistic model—knowledge of the nature of whole science (KNOWS)—with the goal of "interpreting the reliability of scientific claims in personal and public decision making" (p. 521). Instead of teaching particular propositional statements about science, Allchin suggested teaching students the skills to evaluate scientific claims and evidence on their own, so they can understand what to look for to develop confidence in a scientific claim. Allchin also supplied a list of *dimensions* of science.

Though he acknowledged the list may seem unwieldly, as with Matthews (2012, 2015) and Osborne et al. (2003), but Allchin (2011a, 2012) does not expect teachers to address all dimensions. This formulation pivots around the theme of reliability; thus, it is most important to teach students how to discern reliability of evidence, claims, experts, and such. This entails that they learn what questions to ask to become better informed, including how to find the answers. For Allchin, the important dimensions of science revolve around the following:

- Observations and reasoning, methods of investigation
- History and creativity
- The human context
- Culture

- Social interactions among scientists
- Cognitive processes
- Economics/funding
- Instruments and experimental practices
- Communication and transmission of knowledge

Allchin suggested using case studies (Allchin, 2011b, 2013b, 2017) containing intermittent "thought questions," where students can explore such dimensions.

1.3.4 Features of Science

Matthews (2012) acknowledged the importance of the seven NOS aspects Lederman and his group were using but advocated instead to address *features of science* (FOS), essentially extending Lederman's list of tenets. He mainly critiqued *how* Lederman and his group emphasized their tenets; therefore, he suggested that teachers start with Lederman's tenets and explore their philosophical implications with students. Matthews addressed such FOS as the following, for instance:

- Experimentation, particularly what it is and how it is used
- Idealization and its function in theorizing
- The ubiquity and use of models in science
- Values and socio-scientific issues
- Mathematization
- Technology
- Explanation
- Worldviews and religion
- Theory choice and rationality
- Feminism
- Realism and constructivism

Matthews stressed that teachers need not address all FOS, but that they should facilitate conversations around them as appropriate, to help students develop a more complex understanding of science. This "move from universal conceptions of nature of science to an appreciation of particularities of its practice" has also been noted by Rudolph (2000, p. 416).

1.3.5 Family Resemblance Approach

More recently, Erduran and Dagher (2014) elaborated on the work started by Irzik and Nola (2011) in conceptualizing the Family Resemblance Approach (FRA) to NOS. Dagher and Erduran (2016) posited that this framework also carries implications for curriculum and instruction, in addition to being useful for developing a better understanding of NOS:

> [FRA] offers a fresh way of pointing out those overlapping features of science disciplines without overly generalizing them to all other sciences. It points to a wide range of shared and distinctive scientific practices, methodologies, aims and values, social norms and the very aspects that contextualize and frame scientific knowledge. (p. 150)

In general, this framework helps to identify a particular discipline of science *as a science*. Specifically, it displays a certain subset of characteristics from among all the characteristics of science. Thus, it is a metaphor derived from the notion that family members share certain characteristics, yet do not have to look exactly alike.

Erduran and Dagher (2014) went on to describe science as having two systems—cognitive-epistemic and social-institutional—with each system maintaining four subcategories. The cognitive-epistemic has these subcategories: processes of inquiry, aims and values, scientific methodology, and methodological rules. The social-institutional system has these subcategories: professional activities, scientific ethos, social certification and dissemination of scientific knowledge, and social values of science. In this way, different domains in science relate via certain common aspects, and yet are distinguished from others by differences in characteristics, without being classified as non-scientific altogether.

1.3.6 Questions Rather Than Tenets

In response to this limitation, Clough (2008 & 2020a) recommended an alternative to presenting NOS in the form of specific tenets. Teachers ought to reframe the tenets and address other NOS ideas/issues as questions that promote exploration of context and nuance. Some examples are (p. 273), but others appear in Table 1.2:

- Instead of stating that science has an empirical nature, ask, "To what extent is scientific knowledge based on and/or derived from observations of the natural world?"
- Instead of stating that science is tentative, yet durable, ask, "Regardless of how durable well-established science ideas may be, why is all science knowledge still potentially open to revision or even rejection by the scientific community?"

This approach forces students to think about the meaning behind the tenet, not just memorize an abstract proposition. It also emphasizes for instructors that tenets are not the end, but the means to the end. By revising the tenet to a question and facilitating discussion, instructors emphasize the subtleties hidden in the declarative statement. This prevents students from gravitating towards either pole of what is actually a continuum. This orientation counterbalances students' tendency to see the world in black and white terms. Having them reflect on the intricacies of NOS ideas will help them develop a more sophisticated understanding of NOS.

Table 1.2 Constituent aspects of diverse published approaches to NOS

Approach	Constituent aspects[a]
Seven NOS Tenets (Lederman et al., 2002)	Science has an empirical nature. We can distinguish between scientific theories and scientific laws. Science requires creativity and imagination. Science is theory-laden. Scientific knowledge is socially and culturally embedded. The notion of a single scientific method is a myth. Science is tentative.
Ideas of Science (Osborne et al., 2003)	Analysis and interpretation of data Science and certainty Hypothesis and prediction Cooperation and collaboration while developing scientific knowledge Scientific method and critical testing Science and technology Historical development of scientific knowledge Diversity in scientific thinking Questioning Creativity
Knowledge of whole science (KNOWS) (Allchin, 2011a)	Observations and reasoning Methods of investigation History and creativity The human context Culture Social interactions among scientists Cognitive processes Economics/funding Instruments and experimental practices And communication and transmission of knowledge
Features of Science (Matthews, 2012)	Experimentation—What it is and how it is used Idealization and its function in theorizing Models—Their ubiquity and use in science Values and socio-scientific issues Mathematization Technology Explanation Worldviews and religion Theory choice and rationality Feminism Realism and constructivism

(continued)

Table 1.2 (continued)

Approach	Constituent aspects[a]
Family resemblance approach (FRA) (Dagher & Erduran, 2016)	To what extent is scientific knowledge based on and/or derived from observations of the natural world? In what ways are scientific laws and theories different types of knowledge? How are they related to one another? What factors moderate imagination and creativity in the development and justification of science ideas? In what ways are particular aspects of scientists' work guided by existing knowledge and protocols? To what extent is scientific knowledge socially and culturally embedded? How does the notion of a universal step-by-step scientific method distort how scientists actually work? Regardless of how durable well-established science ideas may be, why is all science knowledge still potentially open to revision or even rejection by the scientific community?
Framing NOS Tenets as Questions (Clough, 2020a)	Cognitive-epistemic system Processes of inquiry Aims and values Scientific methodology Methodological rules Social-institutional system Professional activities Scientific ethos Social certification and dissemination of scientific knowledge Social values of science

[a] As with the geoscience literacy points noted above, the constituent aspects here are often given as propositional statements; however, they are more like topic headings, representing many dimensions and fuller ideas than what the propositional statement might indicate

Table 1.2 makes clear that despite the differences in these approaches, they share emphases and overlaps in the targeted NOS concepts. The point of this inventory is not an evaluation of which one is better or worse than the others; rather, the aim is to outline the existence of these diverse approaches, so that readers can consider the strengths and limitations of each for their own context and purpose.

Each has its limitations and affordances. As a beginning grade-9 earth science teacher several years ago, I found Lederman's seven tenets provided my springboard to teaching NOS. Eventually, I found the tenets to be very limiting and prescriptive, as I began to explore teaching the history of geoscience. I allowed myself the freedom to start with the history and develop NOS understandings from the ground up, contingent on the particular piece of history, with real examples of science in the making (Latour, 1987). Years later, when Allchin published his description of KNOWS (Allchin, 2011a), I found myself very much aligned with his description. Ultimately, Allchin's Whole Science description informed my approach to teaching NOS within the GLGY 209 context.

1.4 Incorporating the History and Nature of Geoscience: Helping Students Think Like a Geologist and Recognize Geological Expertise

The introductory geology course (GLGY 209) that is the focus of this book, extensively emphasizes the history and nature of geoscience, a particular context under the larger umbrella, HNOS. The history of geology demonstrates how to think like a geologist and helps students to better address such geological problems described above. This is an important skill for those who will be future geologists. While most students will not become professional geologists, as citizens, they will still be consumers of scientific information. They will continue to make decisions on a local, regional, and national level, where the ability to discern the reliability of scientific claims or the reliability of experts making those claims is vital (Norris, 1995, 1997). Reflecting on the implications of geologic thinking—how geologists come to know what they know—is necessary, as it can enable them to recognize such expertise and trust the knowledge coming from experts when making important decisions (Höttecke & Allchin, 2020).

Thinking like a geologist requires understanding how geology gets done—its methodologies and how the field moves from the development of ideas, through uncertainty and then modification, to eventually reject or accept knowledge and new practices. Most descriptions of how science gets done come from the very narrow perspectives of physics and chemistry. These disciplines have historically advanced, primarily, via direct observations from strictly controlled experimentation, along with making predictions based on a hypothesis and then testing to see if the prediction plays out according to the constraints put in place by the variables. Their data gathering methods more often include manipulating variables in an otherwise controlled setting, with the aim of generalizing the results. Not all physics and chemistry has been, or is, done through control-treatment experiments, but public perception of science wrongly thinks that good science must be experimental. In the geosciences, this is not how most research is done. Geology, like evolutionary biology, astronomy, and cosmology is about collecting signs of past events and trying to explain their origin.

Investigators in the historical sciences, however, are more interested in constructing a history, delineating a single contingent cause for the gathered signs. This often means starting out with multiple possible explanations—multiple working hypotheses (Chamberlin, 1965)—for the origin of the signs. This is followed by searching for a "smoking gun" (Cleland, 2013), evidence that will eventually support one explanation over all the others. Dodick et al. (2009) supported the divergent methodologies with evidence from discourse analysis of over 1600 peer-reviewed published journal articles from different science disciplines. Analysis of language used by scientists in disciplines categorized as experimental showed that such scientists write in more assertive language, focus on singular events or concepts, and describe testing hypothesis-based predictions. By comparison, those writing in historical scientific disciplines have a broader focus and describe multiple hypotheses. They qualify claims more often, allowing for the interpretation of signs of past events (Argamon et al., 2008).

Geologists mainly focus on phenomena that have happened in the distant past, like those who study evolution or the history of the universe. As such, arguing an explanation from a set of signs requires a different approach than when investigating the impact of a single variable. But for many questions in and out of geology, a control-treatment experimental design is not the appropriate approach. Instead, as alluded to above, a geologist might start abductively, with a couple of different possible explanations for a set of signs, then set about looking for evidence that would favour one explanation over another. However well-supported the best explanation is, it is still likely to present difficulties: in other words, the data may diverge from the explanation. When this happens, the geologist begins incorporating three research strategies (RS) to bolster their argument (Frankel, 2012):

- RS1: Lowering the number of difficulties in the explanation or solution, and therefore increase the problem-solving capabilities.
- RS2: Increasing the number of difficulties in competing explanations or solutions (in other words, "poking holes" in them), therefore lowering their problem-solving capabilities.
- RS3: Comparing problem-solving capabilities among different explanations or solutions.

Creating a difficulty-free explanation or solution may require the development and use of models because aspects of many geological phenomena are often inaccessible (within the earth) or are spatially or temporally outside the realm of everyday experience (Turner, 2013), as discussed earlier. Models are further necessary because geological phenomena take place within complex systems; trying to isolate one phenomenon from the rest of the system—as in a typical controlled experiment—would not yield accurate results, even if this were possible (Manduca & Kastens, 2012; Stillings, 2012). The prominence of models and modelling in the NGSS (2013) gives credence to its importance to the scientific process.

Geologists also need skills in the visual-spatial realm, such as discerning patterns or a signal within noisy data, inferring process from form, visualizing structure and shape in the subsurface, visualizing change through time, and substituting space for time. Further, speaking of time in terms of space, is important in geology. Certainly, in everyday life, such metaphoric language is quite common (Moore, 2014). We *look forward* to celebrating anniversaries: We know that birthdays *come* and *go;* that a whole lifetime might *pass you by* before you even know it; we *head into* summer, while at the same time knowing fall is *on its way*.

More explicitly, in geology, we conceptualize the rocks as a *record* of time. We can equate the thickness of a sedimentary sequence as a particular length of time. Unconformities record instances of erased or missing time. We see the passage of time occurring between the first appearance of a new fossil species and the last appearance when it goes extinct. This time and space occur in a very wide range of scales (eras, periods, epochs, and ages). For instance, cm-scale cross-stratification (a kind of sedimentary structure) is conceptualized in terms of the time the environmental process takes to produce it, through a unidirectional fluid flow. Suites of diverse sedimentary structures can then represent the time of an existing depositional environment. Variations among suites of sedimentary structures within a vertical rock section represent a change through time from one depositional environment to another. These spans of time are on the order of thousands, millions, or even billions of years. However, these magnitudes make them difficult to conceptualize because modern societies emphasize scales of weeks and months—as in election cycles, news cycles, and quarterly financial reports (Bjornerud, 2018).

Developing a sense of the passage of time within the context of geologic events recorded in a thickness of rock requires diachronic thinking, which means to ponder the development of something through time. Diachronic thinking requires knowledge of the processes taking place that create the signs, and of the spatial relationships that are usually quite complex due to both the interconnected nature of *influencing* processes, but also *intervening* processes—such as diagenesis, metamorphism, folding, and faulting, among others—that affect the rocks between their formation and subsequent observation (Dodick & Orion, 2003a, c). In addition to diachronic thinking, students need to be able to understand the immense age of the earth, relative to when certain events happened, like the evolution of humans relative to the origin of the earth. However, students often have trouble understanding relationships between thousands to millions to billions of years (Cheek, 2012).

Helping students to grasp such ways of thinking, including the related concepts that enable this thinking, requires a different approach than traditional, lecture-style teaching. Novices need to encounter instruction based in how the mind works and how learning happens. For instance, using history that is structured as interrupted case studies can help to introduce concepts in a concrete and descriptive fashion, which can develop into more abstract and explanatory understandings (Allchin, 2017; Dolphin & Benoit, 2016). At the same time, students can experience a sense of being a scientist as they make their way from novice towards expert.

1.4.1 Using History to Teach NOS

The history of science provides a means to assist students with understanding science and scientific thinking. By detailing the trajectory of scientific knowledge as it actually unfolded and developed, students can experience the process and reflect on the influences in play during the process. The benefits of using the history of science to teach science include (a) promoting better comprehension of science concepts and methods, (b) connecting students to the thinking involved in developing scientific knowledge, (c) understanding NOS better, (d) counteracting scientism and dogmatism, (e) humanizing science, and (f) connecting multiple disciplines within and outside of science (Matthews, 1994). In addition, using history as a tool to teach science helps to uncover students' alternative conceptions, celebrate the achievements of the scientific endeavour, promote scientific careers, develop inquiry skills, and emphasize the social and cultural contexts of science (Allchin, 2013b).

Teaching to develop students' NOS understanding often aims at intellectual independence where a student might be able to judge the reliability of a claim through their mastery of disciplinary knowledge and the data purported to support the claim. However, intellectual independence is not a reasonable goal for science education, not even for experts (Norris, 1995, 1997). Even scientists are rarely able to accurately judge the reliability of a claim made outside their narrow areas of expertise. Thus, students must learn to discriminate and then trust those who truly speak for science, especially now, when science information gets pushed out on different media and social platforms, along with misinformation and disinformation (Höttecke & Allchin, 2020). Invoking the history of science can help students successfully face this challenge. By highlighting aspects of knowledge as it develops, such as social, economic, and political pressures, conflicts of interest, and errors, students experience how to become critical and reflective in the face of claims they will be exposed to.

Although hurdles remain to incorporating the history and philosophy of science into science curricula (Clough et al., 2020; Höttecke & Silva, 2011), some implementation strategies have been successful. Storytelling is one such strategy that multiple studies support (Clough, 2011; Dahlstrom, 2014; Fawcett & Fawcett, 2011; Hadzigeorgiou & Schulz, 2019; Rudge et al., 2014; Stinner, McMillan, et al., 2003; b; Williams & Rudge, 2019; Wilson, 2002b). Evidence supports the notion that humans are hardwired for learning through storytelling (Zarkadakis, 2015). Therefore, telling a good story can grab students' attention, draw students emotionally into a problem, and offer a context for remembering content details (Stinner, McMillan, et al., 2003). The case study is one particular vehicle for storytelling (Herreid, 2007), which has been used as a pedagogical tool for decades in business schools, medical schools, and law schools. Conant (1947) was the first to try this method to teach science, later followed by Klopfer and Cooley (1963) with their Historical Cases of Science; Harvard's Project Physics Course followed (Holton et al., 1970). The historical case method met with some initial success but unfortunately never gained much popularity.

1.4.2 Case Studies as a Vehicle for History

Some within the science education community are developing and promoting cases as a pedagogical tool for teaching science. Active groups or locations include the National Center for Case Study Teaching,[2] The European-based HiPST (history, philosophy, and science teaching) group,[3] and the SHiPS Resource Center,[4] and the

[2] https://sciencecases.lib.buffalo.edu/

[3] https://atlaswiki.eled.auth.gr/

[4] http://www.shipseducation.net/

Tamaratt Teaching Professor in Geoscience website.[5] Reports generated from inquiries concerning the effects of using a "case method" for teaching, whether for science content or NOS understanding are usually anecdotal (Camill, 2006; Dinan, 2005), and descriptive (Dinan, 2005; Dunnivant et al., 2000; Fawcett & Fawcett, 2011; Hodges, 2005). They may rely on self-report data from students (Camill, 2006) or teachers (Yadav et al., 2007), or other instruments to measure the impact of using cases on students' critical thinking skills (Dori et al., 2003; Yadav & Beckerman, 2009) and conceptual understanding (Karuksis, 2003).

Using historical case studies has been effective at teaching students disciplinary content knowledge, as well as science and scientists more broadly (Klopfer & Cooley, 1963). More recently, instructors have reported that using cases to teach science helps them to stimulate interest by making science relevant to students (Dinan, 2005; Dunnivant et al., 2000). It has also enhanced critical thinking skills (Dori et al., 2003; Herreid, 2004; Hodges, 2005; Yadav & Beckerman, 2009), added alternatives to the instructor's pedagogical repertoire (Camill, 2006; Hodges, 2005), promoted more active participation from students (Camill, 2006; Dinan, 2005), and developed students' understandings of NOS (Allchin, 2011b; Clough, 2020b; Kapsala & Mavrikaki, 2020; Stinner, McMillan, et al., 2003). Vehicles for case teaching encompass current news articles (Dinan, 2005), historical figures or discoveries (Stinner & Teichmann, 2003), novels (Herreid, 2005), dramatizations (Stinner & Teichmann, 2003) and even photographs (Krauss et al., 2010). These studies mostly used single case studies intermittently in a class. In this book, however, I describe my effort at full immersion of historical cases throughout a semester geology course.

Herreid (2007) characterized effective cases as ones that do the following:

- Tell a story
- Focus on an interesting idea or issue but are also general enough to have applicability in other domains
- Are contemporary
- Create empathy for a central figure or figures
- Include original texts, are relevant to the reader
- Highlight conflicts and force a decision
- Are short and useful for teaching a targeted concept or concepts

The structure of cases can differ depending on their purpose and the comfort of the instructor with giving control of the learning to the students. Herreid (2007) also identified five main case formats:

- *Individual assignment*: Individual students take responsibility for solving the problem.
- *Lecture*: The instructor primarily disseminates pertinent case information or tells the story.

[5] https://live-ucalgary.ucalgary.ca/tamaratt-teaching-professorship/projects/tamaratt-case

- *Discussion*: The class works as a group to pose suggestions and solutions to the problem.
- *Small group*: Students in small, cooperative groups work through the issues posed in the case, similar to problem-based learning (Allchin, 2013a).
- *Interrupted*: The professor holds all the pertinent case information, releasing it in a piecemeal fashion, each time allowing students to work with the information so far.

Herreid asserted that the interrupted case was popular and often used by university professors, because much of the control (including a correct answer) remains with the instructor. Table 1.3 gives a more detailed account of each of these formats that may be used singularly or in concert.

The use of cases in science classes has a long history, although it certainly has not replaced the more traditional lecture-style teaching, especially at the university level (Stains et al., 2018). Where it has been used, instructors reported that the strategy seemed pedagogically sound, helping to draw students in by making the material more relevant and engaging. They reported that students' participation increased with the use of cases and that their conceptual understanding, problem-solving, and critical thinking skills all benefited from their experiences with cases. These results also largely aligned with student reports, where students said they preferred the case method to traditional lecture. Of the reports reviewed here, none were empirical investigations of the use of specifically historical case studies. Only a few researchers have collected data on student understandings of NOS in the context of HNOS instruction (Clough, 2020b).

A survey of three of the case repositories mentioned above (NCCSTS, SHiPS, and HiPST) shows that of the 500+ cases housed in these three sites, both contemporary and historical, only 24 were related to earth science, with six of those focused on global climate change. Only one of the 24 cases had any relevance to plate

Table 1.3 Different formats for teaching with case studies

Case study format	Characteristics
Individual assignment	Individuals read the case on their own and work through and solve the problem posed by the case.
Lecture	The instructor plays the major role in disseminating the story to the class as a narrative. A subset of this format is the interrupted case. Here the instructor presents the story in pieces, pausing on occasion to work on only what has been presented so far, then moving forward with more story.
Whole-class discussion	The instructor guides whole-class discussion concerning the case, in a Socratic method, as the entire class tries to solve the problem. Cases may be open or closed. Open means more than one possible solution exists.
Small group	Small groups of students work through the case collaboratively, solving the problems presented in the case. The group approaches the solution in an iterative fashion, receiving data and working on one or more hypotheses, then receiving more data and amending the possible solution(s) in light it the new data. Finally, they present a solution at the end of the case.

tectonics and even then, tectonics was treated as peripheral to the case. All eight of the cases studies on the Tamaratt Teaching Professor site focus on geoscience. Though they did not present their methods as formal cases, Dolphin (2009), Dolphin and Benoit (2016), and Paixao et al. (2004) focused on the historical aspects of the theory of plate tectonics and used many of the same strategies highlighted by the case study literature. Activities involved discussion and critique of models, and were very student-centered. Dolphin et al. (2018) reported on their development of geology-related historical cases that focused on understanding the seafloor, continental drift, the great Devonian controversy, the discovery of radium, global climate change, and the K-Pg extinction. My intent with this book is to promote developing and using more geology-specific case studies in the teaching and learning of geology.

1.5 The Structure of This Book

The remainder of this book will detail my process as design engineer of the learning ecosystem in an introductory non-majors geology course. It will communicate the design steps I took to develop, implement, and evaluate the course over five iterations.

Chapter 2 describes my initial thoughts about teaching the introductory geology course and teaming up with a colleague, Robert Kelly, from the Department of Creative and Performing Arts to help realize my vision for the course. Robert helped me through the design process by facilitating discussion around design constraints, ideation, and implementation of the teaching strategies.

Chapter 3 outlines three historical narratives targeting three fundamental geology ideas: the earth is a historical body, the earth is very old, and the earth is dynamic. I also describe a fourth unit that is an explicit and reflective summary of important aspects of NOS and particularly the nature of geoscience.

Chapter 4 will addresses literature and philosophical rationale for implementing specific teaching strategies, along with the evolution of some strategies over the five iterations of the course, three that were face-to-face and two that unfolded remotely.

Finally, Part II will look at the forms of evaluation employed to assess the learning happening in the course, both with respect to content knowledge as well as NOS. Chapter 5 takes advantage of a few different survey instruments, such as click-on diagrams used in an in-class student response system and items from the Geoscience Concept Inventory. These two surveys indicate that students did make significant gains in conceptual understanding as a result of the innovative teaching strategies implemented in the course. Finally, students answered multiple items from the Views of Science, Technology, and Society questionnaire. The results from these items corroborate the qualitative data in Chaps. 6 and 7 by showing a shift to a more robust understanding of science products and processes from pre-instruction to post-instruction.

Chapter 6 employs a phenomenographic theoretical framework to analyze qualitative data from students, both prior to and after course instruction. This analysis highlights a shift from pre-instruction to post-instruction in how students describe concepts such as science knowledge, science processes, and scientists. The analysis shows that students do develop a more nuanced understanding for both geoscience and more broadly, science and how it happens.

Chapter 7 incorporates conceptual metaphor theory to analyze the same qualitative data. In many cases, the interpretation of the metaphors used by the student participants reinforce the interpretations from Chap. 6.

As a whole, this book recounts my process of design and restructuring of a course traditionally referred to as "Rocks for Jocks" into one of jocks talking about rocks. It provides the reasons for making such a change as well as evidence-based rationales for implementing the various strategies central to the change. Finally, it models a way of evaluating student learning within the course, both in terms of the geology content as well as the nature of science content. The results of the evaluation support the notion that the instruction has been successful in shifting students' knowledge base in a direction more closely aligned with the intended learning outcomes. It has been a long and satisfying journey, and I hope the reader senses that enough to try some transformational teaching of their own.

References

Abd-El-Khalick, F., & Lederman, N. (2000). Improving science teachers' conceptions of nature of science: A critical review of the literature. *International Journal of Science Education, 22*(7), 665–701.

Allchin, D. (2011a). Evaluating knowledge of the nature of (whole) science. *Science Education, 95*(3), 518–542.

Allchin, D. (2011b). The Minnesota case study collection: New historical inquiry case studies for nature of science education. *Science & Education, 21*(9), 1263–1281.

Allchin, D. (2012). Towards clarity on whole science and KNOWS. *Science Education, 96*, 693–700.

Allchin, D. (2013a). Problem- and case-based learning in science: An introduction to distinctions, values, and outcomes. *CBE-Life Science Education, 12*, 364–372.

Allchin, D. (2013b). *Teaching the nature of science: Perspectives & resources*. SHiPS Education Press.

Allchin, D. (2017). Historical inquiry cases for nature of science learning. *Cadernos De História Da Ciência, 13*(2), 101–126.

American Geophysical Union. (2021, June 23). Natural hazards threaten 57% of US structures: Rising temperatures and risky development contribute to heightened exposure to earthquakes, flooding, tornadoes, hurricanes and wildfire. *Science Daily*. www.sciencedaily.com/releases/2021/06/210623091230.htm

American Geosciences Institute. (2008). Geoscience workforce age distritbution. *Geoscience Currents* (Data Brief 2008–013).

American Geosciences Institute. (2020). Geoscience workforce projections 2019–2029: Geoscience workforce expected to grow by 4.9%. *Geoscience Currents* (Data Brief 2020–025).

Amundson, R. (2000). Are soils endangered? In J. Schneiderman (Ed.), *The earth around us: Maintaining a liveable planet* (pp. 144–153). WH Freeman and Company.

Anderson, D. (2006). Plate tectonics; the general theory: Complex earth is simpler than you think. In C. A. Manduca & D. W. Mogk (Eds.), *Earth and mind: How geologists think and learn about the earth*. Special paper 413 (Vol. 413, pp. 29–38). Geological Society of America.

Argamon, S., Dodick, J., & Chase, P. (2008). Language use reflects scientific methodology: A corpus-based study of per-reviewed journal articles. *Scientometrics, 75*, 203–238.

Asia-Pacific Economic Cooperation. (2012). *APEC mathematics and science instruction (5th APEC Educational Ministerial Meeting*. Retrieved from http://aimp.apec.org/Documents/2012/MM/AEMM/12_aemm_005.pdf

Ault, C. R., Jr., & Dodick, J. (2010). Tracking the footprints puzzle: The problematic persistence of science-as-process in teaching the nature and culture of science. *Science Education, 94*(6), 1092–1122.

Barras, C. (2014). Coal fuelled China long before industrial revolution. *New Scientist* (2963, 5 April 2014). Retrieved from https://www.newscientist.com/article/mg22229632-600-coal-fuelled-china-long-before-industrial-revolution/

Barstow, D., & Geary, E. (2002). *Blueprint for change: Report from the national conference on the revolution in earth and space science education*. TERC. Retrieved from http://www.earthscienceedrevolution.org/

Bjornerud, M. (2018). *Timefulness: How thinking like a geologist can help save the world*. Princeton University Press.

Boretti, A., & Rosa, L. (2019). Reassessing the projections of the world water development report. *Npj Clean Water, 2*(1), 15. https://doi.org/10.1038/s41545-019-0039-9

Camill, P. (2006). Case studies add value to a diverse teaching portfolio in science courses. *Journal of College Science Teaching, 36*(2), 31–37.

Cervato, C., & Frodeman, R. (2012). The significance of geologic time: Cultural, educational, and economic frameworks. In K. A. Kastens & C. A. Manduca (Eds.), *Earth and mind II: A synthesis of research on thinking and learning in the geosciences* (pp. 19–29). The Geological Society of America.

Chamberlin, T. C. (1965). The method of multiple working hypotheses (reprint of the 1890 version). *Science, 148*, 754–759.

Cheek, K. (2010). Commentary: A summary and analysis of twenty-seven years of geoscience conceptions in research. *Journal of Geoscience Education, 58*(3), 122–134.

Cheek, K. (2012). Students' understanding of large numbers as a key factor in their understanding of geologic time. *International Journal of Science and Mathematics Education, 10*, 1047–1069.

Cleland, C. E. (2013). Common cause explanation and the search for smoking gun. In V. R. Baker (Ed.), *Rethinking the fabric of geology: Geologic Society of America Special Paper 502* (pp. 1–9). Geologic Society of America.

Clough, M. P. (2008). Teaching the nature of science to secondary and post-secondary students: Questions rather than tenets. *California Journal of Science Education, 8*(2), 31–40.

Clough, M. P. (2011). The story behind the science: Bringing science and scientists to life in post-secondary science education. *Science & Education, 20*(7–8), 701–717.

Clough, M. P. (2020a). Framing and teaching nature of science as questions. In W. F. McComas (Ed.), *Nature of science in science instruction: Rationales and strategies* (pp. 271–282). Springer.

Clough, M. P. (2020b). Using stories behind the science to improve understanding of nature of science, science content, and attitudes toward science. In W. F. McComas (Ed.), *Nature of science in science instruction: Rationales and strategies* (pp. 513–526). Springer.

Clough, M. P., Herman, B. C., & Olson, J. K. (2020). Preparing science teachers to overcome common obstacles and teach the nature of science. In *The nature of science: Rationales and strategies* (2nd ed., pp. 239–251). Springer.

Conant, J. B. (1947). *On understanding science; an historical approach*. Yale University Press.

Corgan, J., & Stearns, R. (2008). Rise and fall of geology in nineteenth century American secondary schools: History and textbook reviews. *Journal of Geoscience Education, 56*(1), 33–55.

Dagher, Z. R., & Erduran, S. (2016). Reconceptualizing the nature of science for science education: Why does it matter? *Science & Education, 25*, 147–164. https://doi.org/10.1007/s11191-015-9800-8

Dahlstrom, M. F. (2014). Using narratives and storytelling to communicate science with nonexpert audiences. *PNAS, 11*, 13614–13620.

Deng, F., Chen, D.-T., Tsai, C.-C., & Chai, C. S. (2011). Students' views of the nature of science: A critical review of research. *Science Education, 95*(6), 961–999.

Dinan, F. J. (2005). Laboratory based case studies: Closer to the real world. *Journal of College Science Teaching, 35*(2), 27–29.

Dodick, J., & Orion, N. (2003a). Cognitive factors affecting student understanding of geologic time. *Journal of Research in Science Teaching, 40*(4), 415–442.

Dodick, J., & Orion, N. (2003b). Geology as an historical science: It's perception within science and the education system. *Science & Education, 12*(2), 197–211.

Dodick, J., & Orion, N. (2003c). Measuring student understanding of geological time. *Science Education, 87*(5), 708–731.

Dodick, J., & Orion, N. (2006). Building an understanding of geologic time: A cognitive synthesis of the "macro" and "micro" scales of time. In C. A. Manduca & D. W. Mogk (Eds.), *Earth and mind: How geologists think and learn about the earth: Geological Society of America special paper 413* (pp. 77–93). Geological Society of America.

Dodick, J., Argamon, S., & Chase, P. (2009). Understanding scientific methodology in the historical and experimental sciences via language analysis. *Science & Education, 18*(8), 985–1004.

Dolphin, G. (2009). Evolution of the theory of the earth: A contextualized approach for teaching the history of the theory of plate tectonics to ninth grade students. *Science & Education, 18*(3), 425–441.

Dolphin, G., & Benoit, W. (2016). Students' mental model development during historically contextualized inquiry: How the 'tectonic plate' metaphor impeded the process. *International Journal of Science Education, 38*(2), 276–297.

Dolphin, G., & Dodick, J. (2014). Teaching controversies in earth science: The role of history and philosophy of science. In M. M. Matthews (Ed.), *International handbook of research in history, philosophy and science teaching* (Vol. 1). Springer.

Dolphin, G., Benoit, W., Burylo, J., Hurst, E., Petryshen, W., & Wiebe, S. (2018). Braiding history, inquiry, and model-based learning: A collection of open-source historical case studies for teaching both geology content and the nature of science. *Journal of Geoscience Education, 66*(3), 205–220. https://doi.org/10.1080/10899995.2018.1475821

Dori, Y. J., Tal, R. T., & Tsaushu, M. (2003). Teaching biotechnology through case studies—Can we improve higher order thinking skills of nonscience majors? *Science Education, 87*(6), 767–793.

Dubrovsky, N. M., Burrow, K. R., Clark, G. M., Gronberg, J. M., Hamilton, P. A., Hitt, K. J., et al. (2010). *The quality of our nation's waters: Nutirents in the nation's streams and groundwater, 1992–2004: U.S. Geological Survey circular 1350*. U. S. G. Survey.

Dunnivant, F., Moore, A., Alfano, M., Buckley, P., & Newman, M. (2000). Understanding the greenhouse effect: Is global warming real? An integrated lab-lecture case study for non-science majors. *Journal of Chemical Education, 77*(12), 1602–1603.

Duschl, R. A. (1990). *Restructuring science education: The importance of theories and their development*. Teachers College Press.

Duschl, R. (2020). Preface. In W. McComas (Ed.), *The nature of science in science instruction: Rationales and strategies* (pp. vii–xi). Springer.

Earth Science Literacy Initiative. (2010). Earth science literacy principles: The big ideas and supporting concepts of earth science. Earth Science Literacy Initiative. Retrieved July 2 from http://www.earthscienceliteracy.org/document.html

Erduran, S., & Dagher, Z. (2014). *Reconceptualizing the nature of science for science education* (Vol. 43). Springer.

Eurydice Network. (2011). *Science education in Europe: National policies, practices and research*. Education, Audiovisual and Culture Executive Agency. http://eacea.ec.europa.eu/education/eurydice/documents/thematic_reports/133EN.pdf

Fawcett, S., & Fawcett, A. (2011). The "living" case: Structuring storytelling to increase student interest, interaction, and learning. *Decision Sciences Journal of Innovative Education, 9*(2), 287–298.

Francek, M. (2013). Compilation and review of over 500 geoscience misconceptions. *International Journal of Science Education, 35*(1), 31–64.

Frankel, H. (2012). *Wegener and the early debate* (Vol. 1). Cambridge University Press.

Frodeman, R. (1995). Geological reasoning: Geology as an interpretive and historical science. *Geological Society of America Bulletin, 107*, 960–968.

Global International Geosphere-Biosphere Programme (IGB). n.d.. *Global freshwater use over the long-run* [graph]. The international disaster database (EM-DAT), our world in data. https://ourworldindata.org/grapher/global-freshwater-use-over-the-long-run?country=~OWID_WRL

Greco, R., & Almberg, L. (Eds.). (2016). *Earth science education: Global perspectives*. Pouso Alegre.

Hadzigeorgiou, Y., & Schulz, R. M. (2019). Engaging students in science: The potential role of "narrative thinking" and "romantic understanding" [Hypothesis and theory]. *Frontiers in Education, 4*(38). https://doi.org/10.3389/feduc.2019.00038

Henry, R. S. (2021, April 30). St. *Vincent hit by flooding, landslides following volcanic blasts*. Reuters. https://www.reuters.com/business/environment/st-vincent-hit-by-flooding-landslides-following-volcanic-blasts-2021-04-30/

Herbert, B. (2006). Student understanding of copmplex earth systems. Special paper 413. In C. A. Manduca & D. W. Mogk (Eds.), *Earth and mind: How geologists think and learn about the earth* (pp. 95–104). Geological Society of America.

Herman, B. C., Feldman, A., & Vernaza-Hernqndez, V. (2017). Florida and Puerto Rico secondary science teachers' knowledge and teaching of climate change science. *International Journal of Science and Mathematics Education, 15*(3), 451–471.

Herreid, C. F. (2004). Racism and all sorts of politically correct "isms" in case studies: What are we to do? *Journal of College Science Teaching, 34*(3), 10–11.

Herreid, C. F. (2005). Using novels as bases for case studies: Michael Crichton's state of fear and global warming. *Journal of College Science Teaching, 34*(7), 10–11.

Herreid, C. F. (2007). *Start with a story: The case study method of teaching college science*. NSTA Press.

Hodges, L. C. (2005). From problem-based learning to interrupted lecture: Using case-based teaching in different class formats. *Biochemistry and Molecular Biology Education, 33*(2), 101–104.

Hoffman, M., & Barstow, D. (2007). *Revolutionizing earth system science education for the 21st century: Report and recommendations from a 50- state analysis of earth science education*. TERC.

Holton, G., Watson, F., & Rutherford, J. (1970). *The project physics course*. Holt, Reinhart and Winston.

Höttecke, D., & Allchin, D. (2020). Reconceptualizing nature-of-science education in the age of social media. *Science Education, 104*, 641–666.

Höttecke, D., & Silva, C. C. (2011). Why implementing history and philosophy in school science education is a challenge: An analysis of obstacles. *Science & Education, 20*(3–4), 293–316.

Hubenthal, M. (2018). Exploring undergraduates' conceptions of elasticity, within a plate tectonics context, before and after experience with rock's elastic behavior. *Journal of Geoscience Education, 66*(4), 261–277.

Hutton, J. (1788). Theory of the earth. *Transactions of the Royal Society of Edinburgh, 1*(2), 209–304.

Interamerican Network of the Academies of Sciences. (2011). *About IANAS*. Interamerican Network of the Academies of Sciences. Retrieved 15 November 2015 from www.ianas.org

Irzik, G., & Nola, R. (2011). A family resemblance approach to the nature of science for science education. *Science & Education, 20*(7), 591–607. https://doi.org/10.1007/s11191-010-9293-4

Isacks, B., Oliver, J., & Sykes, L. R. (1968). Seismology and the new global tectonics. *Journal of Geophysical Research, 73*(18), 5855–5899.

Kapsala, N., & Mavrikaki, E. (2020). Storytelling as a pedagogical tool in nature of science instruction. In W. F. McComas (Ed.), *Nature of science in science instruction: Strategies and rationales* (pp. 485–512). Springer.

Karuksis, K. (2003). Examining technology's impact on society: Using case studies to introduce envirnmental and economic considerations. *Journal of College Science Teaching, 33*(2), 36–40.

Kastens, K. A., & Ishikawa, T. (2006). Spatial thinking in the geosciences and cognitive sciences: A cross-disciplinary look at the intersection of the two fields. In C. A. Manduca & D. W. Mogk (Eds.), *Earth and mind: How geologists think and learn: Special paper 413* (pp. 53–76). Geological Society of America.

Kato, C., Li, L., Nogi, Y., Nakamura, Y., Tamaoka, J., & Horikoshi, K. (1998). Extreme barophilic bacteria isolated from the Mariana trench, challenger deep at depth of 11,000 meters. *Applied Environmental Microbiology, 64*(4), 1510–1513.

King, C. (2013). Geoscience education across the globe–results of the IUGS/IGEO survey. *Episodes, 36*(1), 19–30.

Kirkby, K. (2014). *Easier to address earth science misconceptions.* SERC. Retrieved 12 August 2014 from http://serc.carleton.edu/NAGTWorkshops/intro/misconception_list.html

Klopfer, L. E., & Cooley, W. W. (1963). The history of science cases for high schools in the development of student understanding of science and scientists: A report on the HOSG instruction project [https://doi.org/10.1002/tea.3660010112]. *Journal of Research in Science Teaching, 1*(1), 33–47. https://doi.org/10.1002/tea.3660010112

Knoll, A. H. (2003). *Life on a young planet: The first three billion years of evolution on earth.* Princeton University Press.

Krauss, D. A., Salame, I. I., & Goodwyn, L. N. (2010). Using photographs as case studies to promote active learning in biology. *Journal of College Science Teaching, 40*(1), 72–76.

Kravitz, G. (2013). The thermodynamics time arrow ad the logical function of the uniformity principle in geohistorical explanation. In V. R. Baker (Ed.), *Rethinking the fabric of geology: Geological Society of America special paper 502* (pp. 19–40). Geological Society of America.

Lakoff, G., & Johnson, M. (1980). *Metaphors we live by*. University of Chicago Press.

Lakoff, G., & Johnson, M. (1999). *Philosophy in the flesh: The embodied mind and its challenge to western thought.* Basic Books.

Latour, B. (1987). *Science in action: How to follow scientists and engineers through society.* Harvard University Press.

Lederman, N. (2007). Nature of science: Past, present and future. In S. Abell & N. Ledderman (Eds.), *Handbook of research in science education* (pp. 831–880). Lawrence Erlbaum Associates.

Lederman, N., Abd-El-Khalick, F., Bell, R., & Schwartz, R. (2002). Views of nature of science questionnaire (VNOS): Toward valid and meaningful assessment of learners' conceptions of nature of science. *Journal for Research in Mathematics Education, 39*(6), 497–521.

Lewis, E. (2008). Content is not enough: A history of secondary earth science teacher preparation with recommendations for today. *Journal of Geoscience Education, 56*(5), 445–464.

Lewis, E. B., & Baker, D. R. (2010). A call for a new geoscience education research agenda. *Journal of Geoscience Education, 47*(2), 121–129.

Libarkin, J. C., & Anderson, S. W. (2005). Assessment of learning in entry-level geoscience courses: Results from the geoscience concept inventory. *Journal of Geoscience Education, 53*(4), 394–401.

Lowenstein, T. K., Schubert, B. A., & Timofeeff, M. N. (2011). Microbial communities in fluid inclusions and long-term survival in halite. *GSA Today, 21*(1), 4–9.

Manduca, C. A., & Kastens, K. A. (2012). Mapping the domain of complex earth systems in the geosciences. In K. A. Kastens & C. A. Manduca (Eds.), *Earth and mind II: A synthesis of research on thinking and learning in the geoscience: Geological Society of America special paper 486* (pp. 91–96). Geological Society of America.

Matthews, M. (1994). *Science teaching: The role of history and philosophy of science*. Routledge.

Matthews, M. (2012). Changing the focus: From nature of science (NOS) to features of science (FOS). In M. S. Khine (Ed.), *Advances in nature of science research: Concepts and methodologies* (pp. 3–26). Springer.

Matthews, M. M. (2015). *Science teaching: The contribution of history and philosophy of science.*, 20th anniversary revised and, *Expanded Edition*. Routledge.

McComas, W., & Olson, J. (1998). The nature of science in international science education standards documents. In W. McComas (Ed.), *The nature of science in science education: Rationales and strategies* (pp. 41–52). Kluwer.

McComas, W. F., Clough, M. P., & Nouri, N. (2020). The nature of science and classroom practice: A review of the literature with implications for effective NOS instruction. In W. F. McComas (Ed.), *Nature of science in science instruction: Rationales and strategies* (pp. 67–111). Springer.

Minerals Education Coalition. (2020). *2020 MEC mineral baby*. Minerals Education Coalition. Retrieved 22 February 2022 from https://mineralseducationcoalition.org/mining-mineral-statistics

Moore, K. E. (2014). *The spatial language of time* (Vol. 42). John Benjamins Publishing Company.

National Oceanic and Atmospheric Administration. (2021). Is sea level rising? Yes, sea level is rising at an increasing rate. National Oceanic and Atmospheric Administration. Retrieved July 12 2021 from https://oceanservice.noaa.gov/facts/sealevel.html

National Research Council. (1996). *National Science Education Standards*. National Academy Press.

National Research Council. (2012). *Discipline-based education research: Understanding and improving learning in undergraduate science and engineering*. N. A. Press. https://doi.org/10.17226/13362

NGSS Lead States. (2013). *Next generation science standards: For states, by states*. National Academies Press. Retrieved July 2 2021 from https://www.nextgenscience.org/

Niebert, K., & Gropengiesser, H. (2015). Understanding starts in the mesocosm: Conceptual metaphor as a framework for external representations in science teaching. *International Journal of Science Education, 37*(5–6), 903–933.

Niebert, K., Marsch, S., & Treagust, D. F. (2012). Understanding needs embodiment: A theory-guided reanalysis of the role of metaphors and analogies in understanding. *Science Education, 96*(5), 849–877.

Norris, S. P. (1995). Learning to live with scientific expertise: Toward a theory of intellectual communalism guiding science teaching. *Science Education, 79*(2), 201–217.

Norris, S. P. (1997). Intellectual independence for nonscientists and other content transcendent goals of science education. *Science Education, 81*(2), 239–258.

NSTA. (2003). Standards for science teacher preparation. In (pp. 42). NSTA Presshttp://www.nsta.org/pdfs/NSTAstandards2003.pdf: .

Oberle, B., Bringezu, S., Hatfield-Dodds, S., Hellweg, S., Schandl, H., & Clement, J. (2019). *Global resources outlook 2019: Natural resources for the future we want*. U. N. I. R. Panel.

Ocean Literacy. (2015). *Ocean literacy framework*. Ocean Literacy. Retrieved July 2 2021from http://oceanliteracy.wp2.coexploration.org/

Orion, N. (2019). The future challenge of earth science education research. *Disciplinary and Interdisciplinary Science Education Research, 1*(3), 1–8.

Orion, N., & Libarkin, J. (2014). Earth systems science education. In N. G. Lederman & S. K. Abell (Eds.), *Handbook of research on science education* (Vol. II, pp. 481–496). Routledge.

Osborne, J., Collins, S., Ratcliffe, M., Millar, R., & Duschl, R. (2003). What "ideas-about-science" should be taught in school science? A Delphi study of the expert community. *Journal of Research in Science Teaching, 40*(7), 692–720.

Oxford Martin School. (2019). *Our world in data: Global fossil fuel consumption*. Oxford Martin School. Retrieved July 16, 2021 from https://ourworldindata.org/grapher/global-fossil-fuel-consumption?country=~OWID_WRL

Paixao, I., Calado, S., Ferreira, S., Salves, V., & Smorais, A. M. (2004). Continental drift: A discussion strategy for secondary School. *Science & Education, 13*(3), 201–221.

Pavlopoulou, A., Savva, G. D., Louka, M., Bagos, P. G., Vorgias, C. E., Michalopoulos, I., & Georgakilas, A. G. (2016). Unraveling the mechanisms of extreme radioresistance in prokaryotes: Lessons from nature. *Mutation Research-Reviews in Mutation Research, 767*, 92–107. https://doi.org/10.1016/j.mrrev.2015.10.001

Pimental, D., & Sparks, D. L. (2000). Soil as an endangered ecosystem. *Bioscience, 50*(11), 947–947.

Radeloff, V. C., Helmers, D. P., Kramer, H. A., Mockrin, M. H., Alexandre, P. M., Bar-Massada, A., et al. (2018). Rapid growth of the US wildland-urban interface raises wildfire risk. *Proceedings of the National Academy of Sciences, 115*(13), 3314–3319. https://doi.org/10.1073/pnas.1718850115

Reid, H. F. (1910). *The California earthquake of April 18, 1906: The mechanics of the earthquake* (Report of the state earthquake investigation commission in two volumes and atlas, Issue. N. Press.

Resnick, I., Atit, K., & Shipley, T. F. (2012). Teaching geologic events to understand geologic time. In K. A. Kastens & C. A. Manduca (Eds.), *Earth and mind II: A synthesis of research on thinking and learning in the geosciences: GSA special paper 486* (pp. 41–43). Geological Society of America.

Roberts, D. A. (2007). Scientific literacy/science literacy. In S. K. Abell & N. G. Lederman (Eds.), *Handbook of research on science education* (pp. 729–780). Lawrence Earlbaum Associates, Inc..

Rudge, D. W., Cassidy, D. P., Fulford, J. M., & Howe, E. M. (2014). Changes observed in views of nature of science during a historically based unit. *Science & Education, 23*(9), 1879–1909.

Rudolph, J. L. (2000). Reconsidering the 'Nature of Science' as a curriculum component. *Journal of Curriculum Studies, 32*(3), 403–419.

Rudwick, M. J. S. (2014). *Earth's deep history: How it was discovered and why it matters*. Chicago University Press.

Rutherford, F. J., & Ahlgren, A. (1990). *Science for all Americans*. Oxford University Press.

Shipley, T. F., McConnell, D., McNeal, K. S., Petcovic, H. L., & St. John, K. E. (2017). Transdisciplinary science education research and practice: Opportunities for GER in a developing STEM discipline-based education research alliance (DBER-A). *Journal of Geoscience Education, 65*(4), 354–362.

Silva, L. C. R., & Lambers, H. (2021). Soil-plant-atmosphere interactions: Structure, function, and predictive scaling for climate change mitigation. *Plant and Soil, 461*(1), 5–27. https://doi.org/10.1007/s11104-020-04427-1

Smil, V. (2017). Global fossil fuel consumption: Global primary energy consumption by fossil fuel source, measure in terawatt-*hours (TWh)* [graph]. In *Energy Transitions: Global and National Perspectives & BP Statistical Review of World Energy*. Our World in Data. https://ourworldindata.org/grapher/global-fossil-fuel-consumption?country=~OWID_WRL

Stains, M., Harshman, J., Barker, M. K., Chasteen, S. V., Cole, R., DeChenne-Peters, S. E., et al. (2018). Anatomy of STEM teaching in north American universities. *Science, 359*(6383), 1468–1470.

Stillings, N. (2012). Complex systems in the geosciences and in geoscience learning. In K. A. Kastens & C. A. Manduca (Eds.), *Earth and mind II: A synthesis of research on thinking and learning in the geosciences: Geological Society of America special paper 486* (pp. 97–111). Geological Society of America.

Stinner, A., & Teichmann, J. (2003). Lord Kelvin and the age-of-the-earth debate: A dramatization. *Science Education, 12*(2), 213–228.

Stinner, A., McMillan, B. A., Metz, D., Jana, M., & Klassen, S. (2003). The renewal of case studies in science education. *Science & Education, 12*, 617–643.

Suzuki, Y., Yamashita, S., Kouduka, M., Ao, Y., Mukai, H., Mitsunobu, S., et al. (2020). Deep microbial proliferation at the basalt interface in 33.5–104 million-year-old oceanic crust. *Communications Biology, 3*(1), 136. https://doi.org/10.1038/s42003-020-0860-1

Tada, G. M. (2021, June 22). An earthquake in Japan speaks to a sea wall in San Francisco. *The New York Times*. https://www.nytimes.com/2021/06/22/climate/san-francisco-sea-wall-earthquake.html

Thagard, P. (2012). *The cognitive science of science: Explanation, discovery, and conceptual change*. MIT Press.

Turner, D. (2013). Hisotrical geology: Methodology and metaphysics. In V. R. Baker (Ed.), *Rethinking the fabric of geology: Geological Society of America special paper 502* (pp. 11–18). Geological Society of America.

United States Geological Survey. (2020). *How much water is there on earth?* United States Geological Survey. Retrieved14 July 2021 from https://www.usgs.gov/special-topic/water-science-school/science/how-much-water-there-earth?qt-science_center_objects=0#qt-science_center_objects

United States Geological Survey. (2021, February 25). *Post-wildfire landslides becoming more frequent in Southern California.* https://www.usgs.gov/news/post-wildfire-landslides-becoming-more-frequent-southern-california

United States Geological Survey, Water Science School. (2019). *The distribution of waster on, in and above the earth.* United States Geological Survey. Retrieved 5 January 2022 from https://www.usgs.gov/media/images/distribution-water-and-above-earth

US Global Change Research Program. (2009). *Climate literacy: The essential principles of climate science.* US Global Change Research Program. Retrieved July 2 2021 from https://www.climate.gov/teaching/essential-principles-climate-literacy/essential-principles-climate-literacy

Ward, P. J., Blauhut, V., Bloemendaal, N., Daniell, J. E., de Ruiter, M. C., Duncan, M. J., et al. (2020). Review article: Natural hazard risk assessments at the global scale. *Natural Hazards and Earth System Sciences, 20*(4), 1069–1096. https://doi.org/10.5194/nhess-20-1069-2020

Wiggins, G. P., & McTighe, J. (2006). *Understanding by design*. (Expanded 2nd ed. Pearson Education, Inc.

Williams, C. T., & Rudge, D. W. (2019). Effects of historical story telling on student understanding of nature of science. *Science & Education, 28*, 1105–1133.

Wilson, E. O. (2002a). *The future of life/Edward O. Wilson*. Alfred A. Knopf.

Wilson, E. O. (2002b). The power of story. *American Educator, 26*(1), 8–11.

Yadav, A., & Beckerman, J. L. (2009). Implementing case studies in a plant pathology course: Impact on student learning and engagement. *Journal of Natural Resources and Life Sciences Education, 38*, 50–55.

Yadav, A., Lundeberg, M., DeSchryver, M., Dirkin, K., Schiller, N. A., Maier, K., & Herreid, C. F. (2007). Teaching science with case studies: A national survey of faculty perceptions of the benefits and challenges of using cases. *Journal of College Science Teaching, 37*(1), 34–38.

Zarkadakis, G. (2015). *In our own image: Savior or destroyer? The history and future of artificial intelligence*. Pegasus Books.

Chapter 2
Transformative Teaching: From Educator to Design Engineer

Glenn Dolphin and Robert Kelly

There is a strong temptation to assume that presenting subject matter in its perfected form provides a royal road to learning. What [is] more natural than to suppose that the immature can be saved time and energy, and be protected from needless error by commencing where competent inquiries have left off? The outcome is written large in the history of education. Pupils begin their study…with texts in which the subject is organized into topics according to the order of the specialist. Technical concepts and their definitions are introduced at the outset. Laws are introduced at an early stage, with at best a few indications of the way in which they were arrived at…The pupil learns symbols without the key to their meaning. He acquires a technical body of information without ability to trace its connections [to what] is familiar—often he acquires simply a vocabulary.

John Dewey (1916)

(Democracy and Education, 1916, Chap. 18, http://xroads.virginia.edu/~Hyper2/Dewey/ch17.html)

A new type of thinking is essential if [humankind] is to survive and move to higher levels.

Albert Einstein (1946)

With Dr. Robert Kelly

G. Dolphin (✉)
Department of Earth, Energy, and the Environment, University of Calgary, Calgary, AB, Canada
e-mail: glenn.dolphin@ucalgary.ca

R. Kelly
University of Calgary, Calgary, AB, Canada

G. Dolphin, *Teaching Geology Using the History and Philosophy of Science*, Science: Philosophy, History and Education,
https://doi.org/10.1007/978-3-031-68667-2_2

2.1 Introduction

In the winter of 2016, I was assigned to teach Introduction to Geology, a course for non-science majors open to all students across campus. Perhaps you are familiar with such a course. It draws 300–400 students from all over campus into a large-lecture environment, lacks a laboratory component, and is affectionately known as "Rocks for Jocks." It is mainly taken by students needing to fulfill a science requirement—because it is NOT chemistry or physics. My approach to teaching has always been quite different from the more traditional mode of teaching. I had been teaching our introductory geology course for geoscience majors and saw my assignment as chance to explore some new tactics and strategies for engaging non-science majors in a journey of discovery of geology as a human endeavour, with all the implications that come from that.

This chapter recounts a bottom-up design challenge to transform the dynamics of the educational ecosystem from one of knowledge consumption to knowledge creation, from passive learning to active learning, from instruction that is didactic and authoritative, to instruction that is dialogic and empowering. I envision this chapter as an example of thought processes, decisions, and considerations that may be drawn upon to design or revamp courses at other universities. Think of this as a demonstration of my approach to invigorate your own design process. It is not *the* way to transformative teaching, but it documents my way.

Geoscience is a key to many societal concerns, such as natural hazards, mineral and energy resource exploitation, clean air and water, global warming, and the like that I introduced in Chap. 1. Hence, it must be well-positioned in society as a platform to help meet these challenges. However, despite calls for enhanced geoscience education (Barstow & Geary, 2002; Hoffman & Barstow, 2007; Orion, 2019; Orion & Libarkin, 2014), reports focused on geoscience education around the globe show a great disparity between what is ideal and what is reality (Greco & Almberg, 2016; King, 2008, 2013; Orion, 2019).

Geology is typically offered at the secondary school level as an elective course and is often taught as disconnected topics that rarely inspire interest and conceptual understanding (Guffey et al., 2016). Post-secondary environments are where most students get their first comprehensive exposure to geoscience. Despite the literature demonstrating that student-centred, active engagement in STEM courses is best for effective student learning (Fairweather, 2008; Freeman et al., 2014), most geoscience courses still reflect traditional, lecture-based instruction (Macdonald et al., 2005). Research on students' conceptual change in traditional, lecture-based classes shows that many of their original conceptions, which normally do not match the consensus view, remain unchanged in these traditional learning environments (Clark et al., 2011; Libarkin, 2005; Libarkin & Anderson, 2005; Wandersee et al., 2003). This traditional approach reflected what existed at my university; therefore, I sought to transform business-as-usual into an opportunity for students to experience the excitement of science-in-the-making (Latour, 1987), to make *them* the scientists developing knowledge that was new to them, even if not new to the world. This strategy was driven by my view that when students take on the role of knowledge creator, they better understand the content and develop a more nuanced understanding for the process of science.

Helping students to become *science* literate, but also *scientifically* literate has been a decades-long goal for teachers of science (Conant, 1947; Holton et al., 1970; Lederman, 2007; Matthews, 2012). As I argued in Chap. 1, students should be knowledgeable about the products and processes of science (science literate) that are traditionally considered "the content," and knowledgeable about where and how science fits in with society (scientifically literate) (Roberts, 2007). Globally, educational documents have thus emphasized scientific practices and inquiry to help students develop a more nuanced understanding of science as a human undertaking (Asia-Pacific Economic Cooperation, 2012; Eurydice Network, 2011; Interamerican Network of the Academies of Sciences, 2011; McComas & Olson, 1998; National Research Council, 1996; National Science Teachers Association, 2003; States, 2013).

Both types of literacy call for understanding the nature of science (NOS), with various positions positing how to best inculcate this knowledge (Allchin, 2011a; Erduran & Dagher, 2014; Lederman et al., 2002; Osborne et al., 2003; Rudolph, 2000; see also Chap. 1). Inquiry methods (Bybee, 2006) and teaching the history of science (Allchin, 1997; Matthews, 2015), particularly via historical case studies (Allchin, 2011b, 2020; Dolphin et al., 2018; Höttecke, 2012) and narratives (Clough, 2011; Metz et al., 2007; Stinner et al., 2003) can be very effective. Still, most teachers are reluctant to do so because of the prevailing teaching culture, the perception of irrelevance within the already content-heavy curriculum, and because textbooks rarely treat the history of science in any meaningful way (Höttecke & Silva, 2011).

In my design process, I sought a fresh way forward through several transformative pedagogical lenses (Kelly, 2020). I aimed to enable an educational culture where students learned to understand science through their own engaged experiences like that of scientists. This depends on creating an amenable and meaningful context where disciplinary content knowledge is developed through highly collaborative engagement, leveraging the strengths of all involved, and encouraging active, investigative research to mobilize diverse resources, near and far. Ultimately, this approach will create collaborative innovation networks to support creative practice (Gloor, 2017).

Central to collaborative and creative development is students who communicate with extensive interactivity—characterized by interpretive, empathetic, and generative listening. Such a process supports the proliferation of ideas and feedback among learners, within a culture of trust. However, this transformative practice crucially depends on students perceiving the learning environment as safe for them to take risks with their learning.

This chapter describes my journey from educator to design engineer, as I built this new course from the ground up: I started with establishing design constraints and followed this with ideation, goal setting, experimentation, and evolution of both my teaching and the learning environment. The steps along the journey focus on these questions:

- Can I use historically contextualized geoscience instruction to promote both science literacy and scientific literacy?
- *More specifically:* Can students experience useful geoscience conceptual development via instruction that emphasizes the historical trajectory of those geological concepts?
- How do students describe science before and after instruction that emphasizes the historical trajectory of geological concepts?

2.2 Surveying the Educational Landscape: Design Constraints

To facilitate the kind of learning I sought, I had to be very purposeful in all aspects of the course. When talking about the educational landscape, I am referring to the many aspects of the teaching/learning context that shape the overall learning environment. This includes *the physical space*: a large-lecture classroom that can hold up to 450 students, and the furniture placed in long rows and bolted to the floor. It also includes *the students*: their background, where their interests lie, their prior experiences, and the number of students in the class. Further, this includes *the course content*: the "big ideas," how abstract they are, and what their concrete elements are. In later iterations of the course (Fall of 2020 and Winter 2021), it also included the *remote aspect of teaching* the course, because no classes were held on campus due to the COVID-19 pandemic.

In considering all these aspects of how I would teach GLGY 209—to facilitate robust student understanding—I transformed from *an educator to a design engineer*, meaning, I was not just putting activities together because they seemed like they would work. I was being deliberate and systematic with what I was implementing, how I was implementing it and when it was happening, with a definite plane to evaluate this new structure. For readers considering this endeavour, it entails a major mental shift. This course regularly consists of between 200 and 400 students, and was held in a large-lecture theatre, twice a week for 75 minutes per class, with no laboratory component. In such an environment, students may feel anonymous, and believe that the instructor expects them to take notes and perform well on multiple choice exams. However, I took a *constructivist perspective* on learning—which is to say, one in which I understand that students are responsible for developing their own knowledge by actively engaging with the content and each other in an environment I have engineered to give the best possible chance that student s' knowledge would align with the learning goals. In other words, they would need to invest in the course work so that they are empowered by their learning. Moreover, from a social constructivist perspective, a dialogical component would be necessary (Nesari, 2015; Redman, 2007), both *between* the instructor and the students, and *among* students. Unfortunately, large-lecture environments are not easily conducive, physically or socially, to facilitate either one of these conditions.

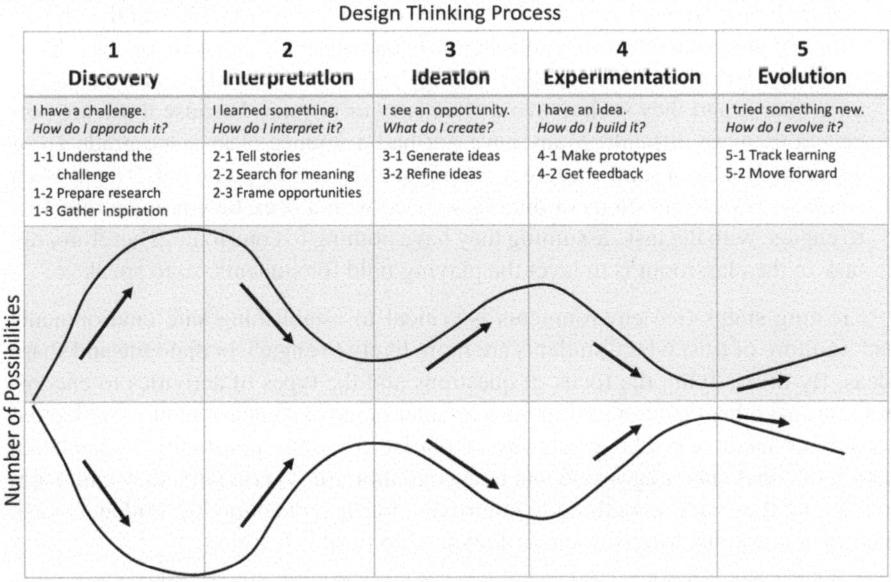

Fig. 2.1 Design Process as described by IDEO. (Note. A diagrammatic representation of the design process as described by IDEO as it progresses through the 5 stages of design (IDEO & Riverdale Country School, 2017). The shape of the graph represents divergent idea generation (wider spots) and convergence of ideas (narrower spots) to solve the problem under investigation)

These foundational considerations were the impetus for me to approach this transformative initiative from a design perspective. Indeed, this educational challenge *required* that I approach this as a designer, along with my mutable roles of facilitator, mentor, and collaborator. My approach to this was significantly informed by the design process described by IDEO, an international design company, which moves through five identifiable stages (Fig. 2.1):

1. ***Discovery:*** How I approach the design challenge, via research to understand the challenge more deeply
2. ***Interpretation:*** How I interpret this educational challenge, to specifically frame design opportunities
3. ***Ideation:*** What I design through numerous generative iterations of ideas, and what alternative courses of action are available
4. ***Experimentation:*** How I bring ideas into practice and the form they take
5. ***Evolution:*** How I evolve the whole design initiative, through recurrent iterations of ideation and experimentation over time, to improve its educational efficacy

As I engaged in the discovery and interpretation stages of the IDEO design process, I identified four impediments to facilitating the kind of student engagement needed—the constructivist model of learning—to achieve the intended learning outcomes.

1. ***Non-relevant background.*** Students in the course come from all over the university, bringing with them a multiplicity of backgrounds and experiences. Most students register for a non-majors geology course because they must take a science course and they seek to avoid chemistry or physics, because those are perceived as more difficult. Many have not had a science class since grade 10 or 11 in high school, and they feel intimidated by science content. Thus, when asked to answer questions or discuss science with a peer, they may be reluctant to engage with the task, assuming they have nothing to contribute. Therefore, my task in the classroom is to level the playing field for students, so to speak

Creating status-free environments is critical to establishing safe environments and a culture of trust where students are more likely to engage in dialogue and share ideas. By diversifying the focus of questions and the types of activities to encompass various other disciplines, I invite a broader range of students to take intellectual risks. This tactic acts like a gateway or border-crossing opportunity (Aikenhead, 2006), one that encourages students to engage more fully; crucially, they can begin to feel that they have something to contribute. Further, encouraging students to ask their own questions fosters a sense of ownership for this learning.

2. ***Large-lecture classes.*** These classes have their own unwritten rules that make deeper levels of engagement more difficult. For instance, in such courses, students often feel overwhelmed by the environment, and often are reluctant to answer the instructor's questions or ask questions because they fear being wrong or appearing ignorant in front of peers. Also, the physical design of the lecture hall makes the instructor seem inaccessible, while promoting the idea of the student as passive listener (Geske, 1992). Placing students in smaller, more intimate settings is critical to facilitating the deep student engagement necessary for effective learning (Hommes et al., 2014).
3. ***Anonymity.*** In large classes, students are more likely to feel anonymous, which undermines their engagement since they are generally less invested in the class. In turn, they feel less responsibility or accountability towards the class and attendance slips (Cooper & Robinson, 2000). These learning circumstances further emphasize students' passive role in the classroom and contribute to little interaction among students. Since interaction is essential to the learning process, I seek to create a level of comfort in class so that students will talk, discuss, and even argue their points with others in the class. I especially want them to attend class and this is more likely when they feel part of the class, even if only a small part, and feel that others are counting on them to show up and engage.
4. ***Little latent interest in topic.*** Finally, many of the students take this course because they must take a science course. It has held the label, "Rocks for Jocks," lowering students' expectations for the kind of work required for them to meaningfully learn material. Iaria and Hubball (2008) asserted that the large class format inhibits students from discussions with other students. To enhance student learning, students in this course will need to engage in numerous fertile discussions about the course content, which is easier when they become more familiar with one another.

2.3 Cultivating High Levels of Interactivity: Interpretation

To address these design constraints, I needed to change the perceived rules of the course *and* have students buy-in to the novel nature of the course I created. These identifiable impediments informed specific design opportunities necessary to transform this learning environment, and thus the course itself. During the "Interpretive" stage of the design process, the four design opportunities encompassed the following:

- "Level the classroom"
- Make the large class feel small
- Have students lose the feeling of anonymity
- Get students to talk about geology without realizing they are doing so (Fig. 2.2)

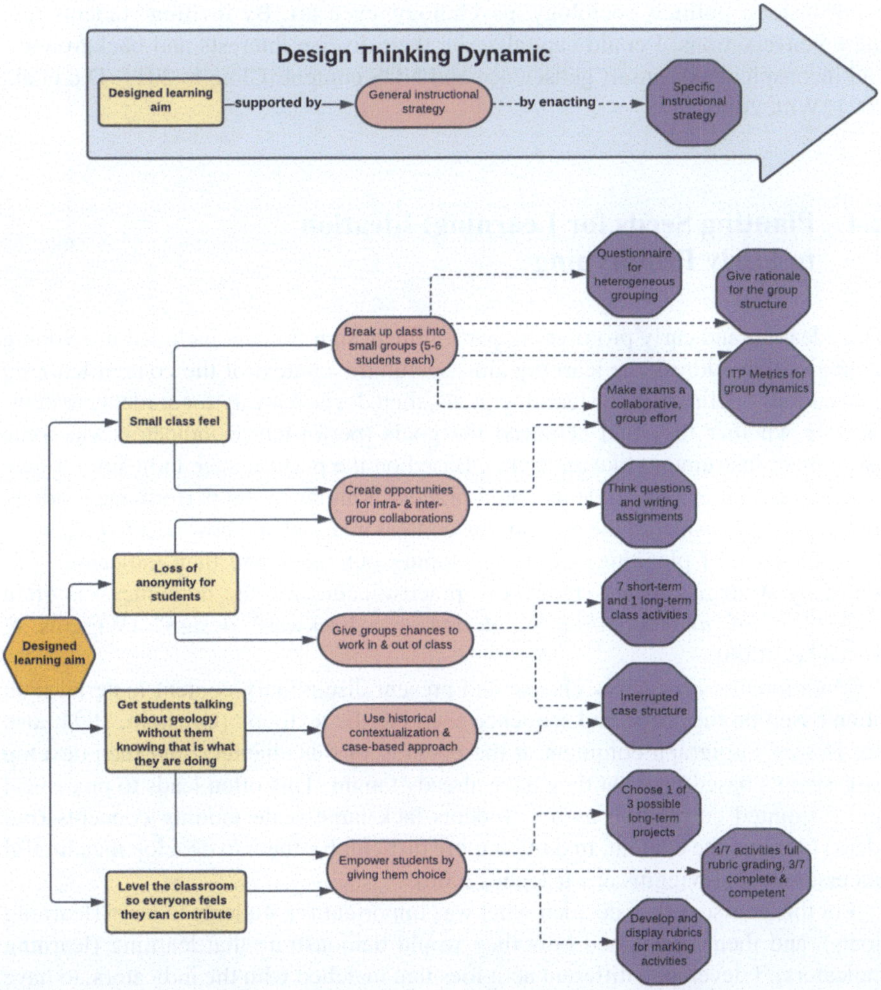

Fig. 2.2 Interpretation of preliminary design constraints. (Note. Diagram illustrating learning-ecosystem goals, and general and specific instructional strategies for achieving the goals)

Figure 2.2 shows these steps in graphic form. All four required high levels of student interactivity that is conspicuously absent in traditional learning environments.

To achieve these four goals, I broke the large class into small groups of five to six students, each heterogeneous in composition, that remain stable throughout the entire term, and work together on all learning and assessment activities within the course (short- and long-term assignments and all exams). Students would also have opportunities for inter-group work. I created multiple opportunities for students to make choices about what they did and how that work was graded. This opportunity to choose helps empower students and facilitate authentic learning (Rule, 2006). Finally, I chose to present content material within the context of interrupted historical narratives (Allchin, 2017). Because the human brain is wired for storytelling (Zarkadakis, 2015), recounting scientific discoveries in a narrative form would draw students in (Dahlstrom, 2014), engaging them to discuss the stories I was telling using history, politics, sociology, psychology, even art. By inviting students into these conversations, I could capitalize on their diverse interests and backgrounds, and inconspicuously insert geoscience and NOS content (Clough, 2011; Dai et al., 2021; Will, 2006).

2.4 Planting Seeds for Learning: Ideation to Early Prototyping

This ideation and early prototyping part of the design process included developing strategies that address the learning aims within the context of the content learning goals. First, I delineated the learning goals, then devised a way for students to demonstrate whether they had achieved the goals (performance indicators via some assessment instrument, like an exam). Based on the performance indicators, I then designed activities that students could practise, similar to the performance indicators. Lastly, I considered the appropriate content material and how best to present it. This sequence of planning—from developing outcomes and then indicators, followed by strategies to get students to practise achieving the outcomes—is often referred to as *understanding by design*, or "backward design" (Wiggins & McTighe, 2006).

Traditionally, instructors choose and present disciplinary content material first, often based on the scope and sequence found in the textbook (Ornstein, 1994; also see Dewey's epigraph comment at the opening of this chapter), and then develop assessments based on what they have already taught. This often leads to piecemeal and disjointed curriculum as the students lack large-scale guiding concepts (big ideas) to frame the content, making it more difficult for them to develop meaningful understandings (Wiggins & McTighe, 2006).

For this course, I thus decided what was important for students to learn (learning goals), and then determined how they would demonstrate that learning (learning indicators). I developed different activities that matched with the indicators, to have students demonstrate their learning. Because I expected more time would be required having students actively participate in learning, presenting the amount of content characteristic of most traditional, lecture-based courses would be impossible. Therefore, I settled on three important and foundational learning goals that were geology content–related, and a fourth goal focused on the practices and processes in geology research:

- First goal: *Students will understand the historical nature of the earth.* Rudwick (2005, 2014) equated the importance of this understanding to that of the Copernican Revolution. That the earth is historical means change through time is directional. The conditions existing during the present are the direct result of conditions and contingencies of the past. This means that during any time in history a unique set of conditions exists resulting from the progression of conditions from the past and irreversibly progressing to the present and leading to brand new conditions in the future (Rudwick, 2014; Şengör, 2001).
- Second goal: *Students will appreciate that the history of the earth is very, very long.* Understanding the scale of time encompassing the age of the earth is a cornerstone of geological thinking, but it is also a very difficult concept for most people to understand (Cervato & Frodeman, 2012; Dodick & Orion, 2003, 2006). This difficulty in understanding is often the root of students' alternative conceptions (Cheek, 2010). However, it is a crucial concept to grasp for solving the many geoscience-related problems facing humanity (Bjornerud, 2018).
- Third goal: *Students will appreciate the dynamic nature of the earth.* The focus here is on the solid portion of the earth that functions as a complex system (Anderson, 2006) and that we currently explain using the theory of plate tectonics (Dewey & Bird, 1970; Isacks et al., 1968).
- Fourth goal: *Students will appreciate the practices and process of geology and how to judge the reliability of scientific claims and experts.* This is the important science literacy aspect, helping student to develop a better understanding of NOS. Recent decades have seen increasing emphasis on achieving this goal in science education standards (National Research Council, 1996; National Science Teachers Association, 2003; NGSS Lead States, 2013). Many students learn that only one way exists to do science—the scientific method—when, as noted earlier, this is not how much science and most geology is done. Geology is a historic and hermeneutic science (Dodick et al., 2009; Frodeman, 1995) that relies much more on the grouping and interpretation of signs in the rock to interpret causation (Cleland, 2013; Turner, 2013).

Concerning learning goals, I encourage readers to consider the foundational ideas they wish to inculcate in their own course, whatever the subject matter, and treat my account here as an example of how they might proceed. Of course, each discipline will have its own focus, its own foundational concepts, but the general understanding-by-design model is valid and will hold across diverse subject areas. The design structure presented in Fig. 2.3 maps out how to move from learning goals, to indicators, to activities, bearing in mind the need to construct a learning environment to support these ends.

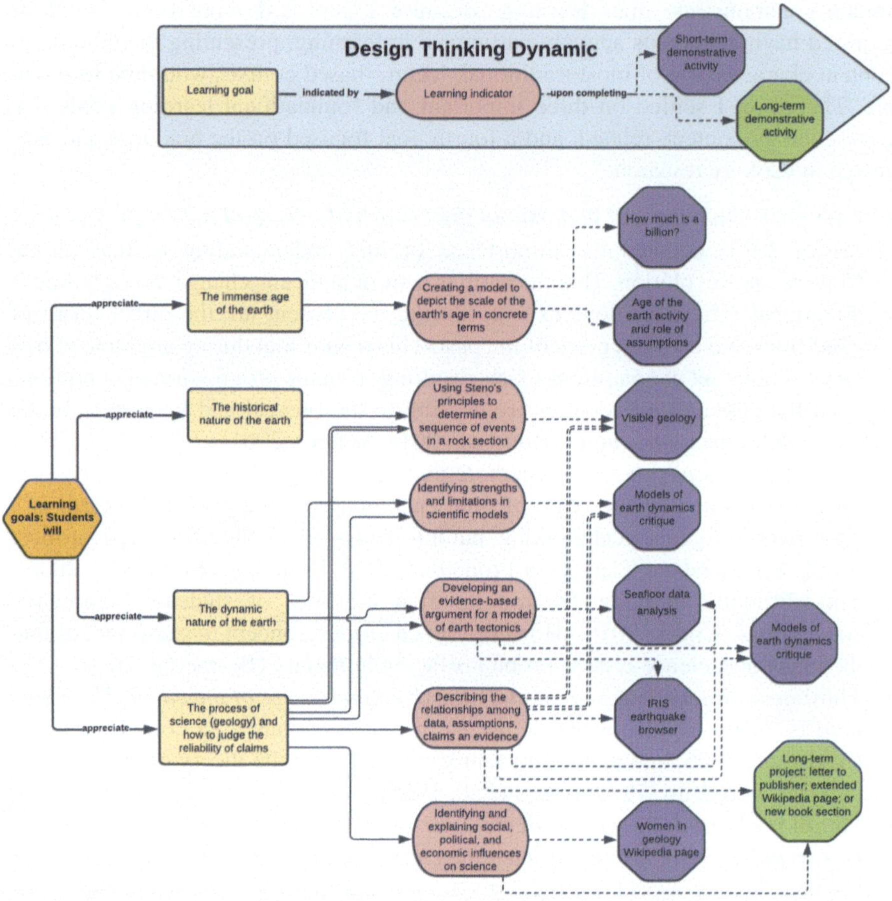

Fig. 2.3 Preliminary learning goals. (Note. Diagram showing the design of GLGY 209, from learning goals, to learning indicators, to student demonstrative activities)

After identifying the general learning goals (course-scale learning; see Fig. 2.3), I thought specifically of the target concepts for students to learn—both at the unit and at the course level. Target concepts are those that are "developmentally and academically appropriate…and are scientifically sound" (Rae-Ramirez, 2008, p. 47). Helping students develop all the facets of a target concept works best by breaking it into hierarchical levels. Thus, Fig. 2.4 provides examples of the integrated target concepts, using the first three content-domain learning goals: *Students will appreciate the historical nature of the earth*; *Students will appreciate the immense age of the earth;* and *Students will appreciate the dynamic nature of the earth*. I have broken each integrated target concept into the primary elements required to understand it. I then broke each of these primary element into component parts (secondary elements), which I further deconstructed to tertiary elements,

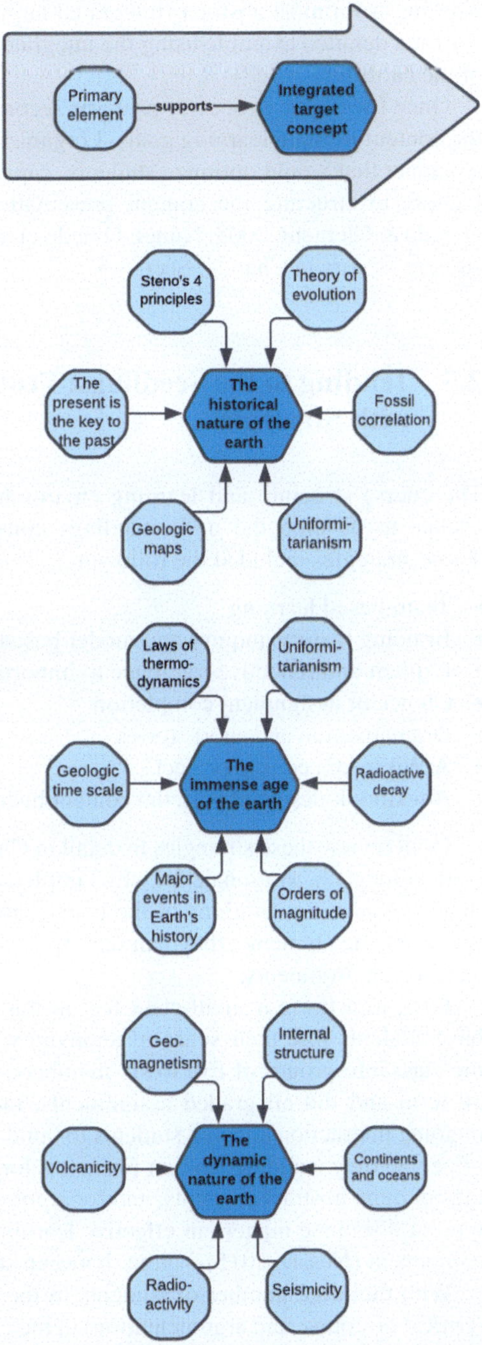

Fig. 2.4 Integrated target concept. (Note. Three integrated target concepts (dark blue hexagons) supported by primary elements (light blue octagons): *The historical nature of the earth (top), The immense age of the earth (middle), and the dynamic nature of the earth (bottom)*)

working from most abstract (integrated target concept) to more concrete (see Fig. 4.1 for a detailed example using the integrated target concept, "The dynamic nature of the earth").

Once I was able to specify primary, secondary, and tertiary elements for each of the content-domain learning goals, I organized them for presentation to the class in a manner that would optimize students' capacity to develop the learning outcomes. I chose to structure the content presentation in a model-based format (Núñez-Oviedo & Clement, 2008; Núñez-Oviedo et al., 2008). An in-depth treatment of this strategy occurs in Chap. 4, Sect. 4.5.

2.5 Tending to the Seedlings: Prototypes and Experiments with Strategies

The course structure and learning environment were created to provide the best chance to develop the understandings consistent with the scientific consensus. These strategies included the following:

- Team-based learning
- Braiding history, inquiry and model-based learning
- Explicit and reflective reference to important science-process themes
- Choice of assignment completion
- Dramatization and short stories
- A Wikipedia editing project
- A textbook designed with the content historically contextualized

I will review these strategies in detail in Chap. 4, but below is an introduction and context for each, a rationale for why I implemented them, and how that implementation evolved over iterations of the course. Readers will be able identify from this inventory the strategies they can deploy in their own classrooms to enhance their learning environments.

First, to achieve a small-class feel in this large class of 200–300 students, and have students lose their sense of anonymity, I used team-based learning–dividing the class into groups of five to six members. Students remained in the same group all term and did all graded assignments, including exams in these groups. This ongoing interaction allowed students to build a relationship with a small core group of people they could depend on for notes, for studying, for developing understandings of more abstract concepts, and for arguing out answers to exam questions. This type of discourse matters in effective learning, as it provides a necessary dialogic component (Nesari, 2015)—either between student and teacher or among students.

With the large number of students in the course, cultivating dialogic activities (critical discourse and argumentation) in the class requires students to communicate among themselves. Certainly, they are not experts, but with five or six different interpretations of the information, my intent was that the proper environmental conditions, engaging activities, thoughtful instruction, combined with intra-group discussions would tend towards the scientific consensus view.

I taught the course via this structure for two terms, during which a few students expressed concerns about particular group members who were not seemingly pulling their weight. I dealt with some team issues personally, but eventually, I implemented Individual, Team Performance (ITP) Metrics (www.itpmetrics.com) (O'Neill et al., 2017, 2019, 2020a, b). This team building platform, developed by the Psychology Department at the University of Calgary, involves automated self- and team-evaluations. I incorporated the evaluations into individual student grades to maintain fairness for different levels of effort.

Second, to portray science as it happened, I used historical case studies in line with the learning goals (itemized in Sect. 2.4, above) to structure and trace the trajectory of geological understandings as three larger, historical narratives (Allchin, 1997, 2020; Hodges, 2005): *How can we understand the history of the earth? How can we understand the age of the earth? What is the origin of continents and oceans, or why are there still mountains?* Each narrative is interrupted by various inquiry activities that lead to students' model-based learning. The narrative then continues, building to the next inquiry activity (see Figs. 3.1, 3.2, and 3.3 for examples).

Third, to promote an accurate understanding of NOS, I incorporated explicit and reflective questions and activities throughout instruction. I initially placed an overt NOS focused unit at the beginning of the course. My reasoning was that it would introduce the various facets of NOS students should be sensitive to as we walked through the narratives. But this approach did exactly what I had been critical of in traditional teaching (please revisit Dewey's epigraph quote as a reminder). I wanted students to identify and develop the concepts *based on* the data, rather than rationalize the data based on the concepts. I especially wanted them to use instances from the narratives to discern patterns and then generate possible explanations. This changed the schedule of events within the course and made some activities more difficult since they occurred far earlier in the course. Importantly though, I believe it made the NOS unit more effective as students were already grounded in the narratives, better preparing them for the abstract NOS conceptions since they'd already experienced the concrete examples in the stories.

Fourth, to empower students and increase engagement, I gave students choices in different aspects of activities and assignments. Aside from three exams—two midterms and one final—the course has seven short-term writing assignments and one long-term writing assignment. While students must complete all the short-term assignments, they can choose to have three marked as complete and competent (rubric grade 0 or 1), while the other four get graded in detail according to a more comprehensive rubric (rubric grade 0–12).

Part of giving students choice is to establish transparency regarding how their work will be evaluated. For this, I relied on rubrics I made available at the beginning of term (see Fig. 2.5), so students were aware of the expectations for each assignment. To create fair and effective rubrics, I met with two learning and instructional design specialists from my university teaching centre, the Taylor Institute for Teaching and Learning. We used my learning goals and performance indicators to

Extended Seismicity Activity

Due: by end of day Thursday 21 March

Points	Criteria	Comments
	No demonstrated use of IRIS Earthquake Browser Did not develop a question that can be answered by the Earthquake Browser Answer to question does not follow from data No evidence presented or evidence does not support claim	
	Demonstrated remedial use of IRIS Earthquake Browser Question is simplistic, not requiring much investigation Developed an accurate answer (claim) to the question Supplied little, or weak evidence to support claim	
	Demonstrated good use of IRIS Earthquake Browser Developed an insightful question that uses the capabilities of the Earthquake browser Developed an accurate and insightful answer (claim) to the question Supplied strong evidence to support claim	

Fig. 2.5 Sample rubric for short-term activity. (Note. This sample makes visible the different criteria and indicators for the extended seismicity activity)

develop the rubrics for the seven short-term activities and the long-term writing assignment (Fig. 2.5). Also, for the long-term assignment, students could choose one of three different possible projects:

- Create an advanced woman-in-geology Wikipedia page (see Chap. 4, Sect. 4.6), where they can choose the female geologist they wish to highlight.
- Author a short section to add to the course textbook (see Chap. 4, Sect. 4.7), as a topic not dealt with but which would contribute to the purpose of the book.
- Write a letter to a geology textbook author critiquing the presentation of NOS within their book. Students choose one of six textbook samples to address this problem.

Fifth, to engage students, especially those less inclined toward science, I included story telling. Having a storyline that includes a beginning, middle, and end helps structure content presentation and engage students in developing coherent understandings (Dahlstrom, 2014). For this course, it included a student-acted dramatization of a debate concerning the age of the earth[1] (Stinner & Teichmann, 2003), and two science fiction short stories (Harrison, 1992; Preuss, 1992). These various story forms are both novel additions to most science classes and offer a more human

[1] See https://vimeo.com/217090549 to view one class's presentation of "Age of the earth debate."

context to the science content and process, garnering student interest and critical thinking when students are asked to reflect on the science as part of the whole story.

Sixth, to improve students' digital literacy, develop collaborative ability, and enhance student motivation and satisfaction (McDowell, 2017), I have them create or edit a Wikipedia page of a famous woman geologist. Wiki Education helps set the schedule and provides all the support and training modules for students to navigate editing in the Wiki space. I constrained the topics to a list of about 100 important female geologists. Students do background research to learn about a famous woman geologist and edit or create a site for her. After completing the activity for five different course offerings, the students have added about 250,000 words, edited 294 different Wiki pages, and created 78 pages, garnering almost two million individual views. Students are very conscious that their work is going out to the whole world for others to see, as opposed to just the desk of an instructor.

Lastly, to reinforce the course structure and emphases, I wrote the course's companion textbook that will be detailed in Chap. 4 Sect. 4.7). Typical introductory geoscience textbooks begin with a short nature of science section followed by chapters that describe the universe and the creation of the solar system and Earth, onto a brief description of plate tectonics, then right into the submicroscopic chemistry, describing how atoms build into minerals. Since my presentation of the geology content is vastly different than the typical textbook presentation, writing a textbook that emphasizes a sounder pedagogical landscape (Dolphin, 2016) became necessary. The book traces the historical trajectory of three main ideas that align with the learning goals described above, by addressing these three questions: *How can we understand the history of the earth? How can we find out the age of the earth? Why are there still mountains?* The text presents the stories of these major concepts, the characters involved, and the social, political, and economic milieux that affects and are affected by the developing science. All through the text are opportunities for students to put themselves in the place of the scientists and reason out what they would think or do in each scenario.

2.6 Summary

This chapter describes the approach I took to develop a course that engages students in collaborative and creative experiences to have them experience geology-in-the-making. The aim is to help students learn both *what we know in* geology, but also *understand the contexts* (social, political, economic, and methodological) in which we came to know such things. Personally, this was a transformative change, from educator to design engineer. When I first contemplated diverging from the status quo, I knew I would need help and sought the assistance of this chapter's co-author, Robert Kelly. His research focuses on creativity and design thinking in educational practice.

We knew that students were well socialized into understanding education within a transactional context by this point in their studies—a process largely focused on the transfer and memorization of disciplinary content knowledge. Changing this context would be unsettling for many students who were used to the expectations of large classes that I have described: anonymity, little one-on-one communication, an atmosphere of little accountability or sense of responsibility to anyone else in the class, and so forth. The significant changes to what students expected raised the challenge of getting them to understand and buy into the changes. Students would need to feel they had something to offer in the class, and further, that they had a responsibility to others in the class to offer this. Ideally, the course structure would enable them to learn that the discipline of science is conducive to creative design practice. As such, their learning involved creativity, a necessarily collaborative effort. I aimed for them to understand that all scientists are humans with passion and training, and that any human with passion and training—them, for instance—could also be a scientist.

Robert and I set about understanding what my learning aims were: to create a small-class feel, to replace anonymity with high interactivity, to level the classroom so everyone feels safe contributing to discussions, and to have students talking about geology without realizing that is what they were doing. Ultimately, we would need to address these aims and the educational landscape—this included the physical space, the students' expectations, and the teaching strategies—to have the best chance at reaching those goals (IDEO's interpretation stage).

We generated several strategies (IDEO's ideation stage) that we thought would work well for reaching these aims (Table 2.1), and then it was up to me to prototype these strategies andd experiment with different aspects and different approaches, while being sensitive to students' responses. I used activity evaluations, formal assessment tools (midterm and final exams), and interpretation of pre- and post-instruction survey responses to guide my understanding of students' learning. This information informed the implementation of the strategies in subsequent iterations (IDEO's evolution stage). As a result, students have been able to demonstrate

Table 2.1 Alignment of instructional strategies with designed learning aims

Instructional strategy	Small-class feel	Loss of anonymity	Level the classroom	Students talking about geology
5.1 Team-based learning	X	X	X	X
5.2 Historical, Inquiry, MBL			X	X
5.3 Explicit/reflective NOS instruction			X	X
5.4 Student choice	X	X	X	X
5.5 Dramatization and short stories		X	X	X
5.6 Wikipedia project	X	X	X	X
5.7 Parallel structured textbook			X	X

learning in the disciplinary content domain, as well as more robust understandings of how science works and how we know what we know (See Chaps. 5 and 6 for more on this).

The next chapter will detail the three historical narratives used in the class, the key vehicles for presenting content information and also for drawing students' attention to various aspects of NOS. Chapter 4 will provide thorough descriptions of the strategies I introduced above and specifically how I implemented them in the course.

References

Aikenhead, G. S. (2006). *Science education for everyday life: Evidence-based practice*. Teachers College Press.

Allchin, D. (1997). The power of history as a tool for teaching science. In A. Dally, T. Nielsen, & F. Reiß (Eds.), *History and philosophy of science: A means to better scientific literacy?* (pp. 70–98). Evangelische Akadamie Loccum.

Allchin, D. (2011a). Evaluating knowledge of the nature of (whole) science. *Science Education, 95*(3), 518–542.

Allchin, D. (2011b). The Minnesota case study collection: New historical inquiry case studies for nature of science education. *Science & Education*, 1–19. Published on-line http://www.springerlink.com/content/v42561276210585q/. Accessed 1 Mar 2012.

Allchin, D. (2017). Historical inquiry cases for nature of science learning. *Cadernos De História Da Ciência, 13*(2), 101–126.

Allchin, D. (2020). Historical inquiry cases for teaching the nature of science and analytical skills. In W. F. McComas (Ed.), *Nature of science in science instruction: Rationales and strategies* (pp. 595–606). Springer.

Anderson, D. (2006). Plate tectonics; the general theory: Complex earth is simpler than you think. In C. A. Manduca, D. W. Mogk, & A. Geological Society of (Eds.), *Earth and mind: how geologists think and learn about the earth*. Special paper 413 (Vol. 413, pp. 29–38). Geological Society of America.

Asia-Pacific Economic Cooperation. (2012). *APEC mathematics and science instruction* (5th APEC Educational Ministerial Meeting), Issue.

Barstow, D., & Geary, E. (2002). Blueprint for change: Report from the national conference on the revolution in earth and space science education. TERC. Retrieved from http://www.earthscienceedrevolution.org/ 5 September 2011.

Bjornerud, M. (2018). *Timefulness: How thinking like a geologist can help save the world.* Princeton University Press.

Bybee, R. W. (2006). Scientific inquiry and science teaching. In L. B. Flick & N. G. Lederman (Eds.), *Scientific inquiry and nature of science: Implications for teaching, learning, and teacher education* (pp. 1–14). Springer.

Cervato, C., & Frodeman, R. (2012). The significance of geologic time: Cultural, educational, and economic frameworks. In K. A. Kastens & C. A. Manduca (Eds.), *Earth and mind II: A synthesis of research on thinking and learning in the geosciences* (pp. 19–29). The Geological Society of America.

Cheek, K. (2010). Commentary: A summary and analysis of twenty-seven years of geoscience conceptions in research. *Journal of Geoscience Education, 58*(3), 122–134.

Clark, S., Libarkin, J., Kortz, K., & Jordan, S. (2011). Alternate conceptions of plate tectonics held by non-science undergraduates. *Journal of Geoscience Eductation, 59*, 251–262.

Cleland, C. E. (2013). Common cause explanation and the search for smoking gun. In V. R. Baker (Ed.), *Rethinking the fabric of geology* (Geologic Society of America Special Paper 502) (pp. 1–9). Geologic Society of America.

Clough, M. P. (2011). The story behind the science: Bringing science and scientists to life in post-secondary science education. *Science & Education, 20*(7–8), 701–717.

Conant, J. B. (1947). *On understanding science; an historical approach.* Yale University Press.

Cooper, J. L., & Robinson, P. (2000). The argument for making large classes seem small. *New Directions for Teaching and Learning, 81*(Spring 2000), 5–16.

Dahlstrom, M. F. (2014). Using narratives and storytelling to communicate science with nonexpert audiences. *PNAS, 11*, 13614–13620.

Dai, P., Williams, C. T., Witucki, A. M., & Ridge, D. W. (2021). Rosalind Franklin and the discovery of the structure of DNA. *Science & Education, 30*, 659–692.

Dewey, J. F., & Bird, J. M. (1970). Mountain belts and the new global tectonics. *Journal of Geophysical Research, 75*(14), 2625–2647.

Dodick, J., & Orion, N. (2003). Cognitive factors affecting student understanding of geologic time. *Journal of Research in Science Teaching, 40*(4), 415–442.

Dodick, J., & Orion, N. (2006). Building an understanding of geologic time: A cognitive synthesis of the "macro" and "micro" scales of time. In C. A. Manduca & D. W. Mogk (Eds.), *Earth and mind: How geologists think and learn about the earth: geological society of america special paper 413* (pp. 77–93). Geological Society of America.

Dodick, J., Argamon, S., & Chase, P. (2009). Understanding scientific methodology in the historical and experimental sciences via language analysis. *Science & Education, 18*(8), 985–1004.

Dolphin, G. (2016). *Stories in geology: What we know and how we figured it out.* Kendall-Hunt Publishing.

Dolphin, G., Benoit, W., Burylo, J., Hurst, E., Petryshen, W., & Wiebe, S. (2018). Braiding history, inquiry, and model-based learning: A collection of open-source historical case studies for teaching both geology content and the nature of science. *Journal of Geoscience Education, 66*(3), 205–220. https://doi.org/10.1080/10899995.2018.1475821

Erduran, S., & Dagher, Z. (2014). *Reconceptualizing the nature of science for science education* (Vol. 43). Springer.

Eurydice Network. (2011). *Science education in Europe: National policies, practices and research.* C. E. A. Education. http://eacea.ec.europa.eu/education/eurydice/documents/thematic_reports/133EN.pdf

Fairweather, J. (2008). *Linking evidence and promising practices in science, technology, engineering, and mathematics (STEM) undergraduate education: A status report.* A. G. Institute.

Freeman, S., Eddy Sarah, L., McDonough, M., Smith Michelle, K., Okoroafor, N., Jordt, H., & Wenderoth Mary, P. (2014). Active learning increases student performance in science, engineering, and mathematics. *Proceedings of the National Academy of Sciences, 111*(23), 8410–8415. https://doi.org/10.1073/pnas.1319030111

Frodeman, R. (1995). Geological reasoning: Geology as an interpretive and historical science. *Geological Society of America Bulletin, 107*, 960–968.

Geske, J. (1992). Overcoming the drawbacks of the large lecture class. *College Teaching, 40*(4), 151–154.

Gloor, P. A. (2017). *Swarm leadership and the collective mind: Using collaborative innovation to build a better business.* Emerald Publishing.

Greco, R., & Almberg, L. (Eds.). (2016). *Earth science education: Global perspectives.* Pouso Alegre.

Guffey, S. K., Slater, S. J., Schleigh, S. P., Slater, T. F., & Heyer, I. (2016). geology concepts in education standards for a rapidly changing global context. *Contemporary Issues in Education Research, 9*(4), 167–188.

Harrison, H. (1992). Dawn of the endless night. In B. Preiss & R. Silverberg (Eds.), *The ultimate dinosaur* (pp. 264–273). Bantam Books.

Hodges, L. C. (2005). From problem-based learning to interrupted lecture: Using case-based teaching in different class formats. *Biochemistry and Molecular Biology Education, 33*(2), 101–104.
Hoffman, M., & Barstow, D. (2007). *Revolutionizing earth system science education for the 21st century: Report and recommendations from a 50-state analysis of earth science education*. TERC.
Holton, G., Watson, F., & Rutherford, J. (1970). *The project physics course*. Holt, Reinhart and Winston.
Hommes, J., Arah, O. A., de Grave, W., Schuwirth, L. W., Scherpbier, A. J. J. A., & Bos, G. M. (2014). Medical students perceive better group learning processes when large classes are made to seem small. *PLoS One, 9*(1), e93328. https://doi.org/10.91371/journal.pone.0093328
Höttecke, D. (2012). HIPST–History and philosophy in science teaching: A European project. *Science & Education, 21*(9), 1229–1232.
Höttecke, D., & Silva, C. C. (2011). Why implementing history and philosophy in school science education is a challenge: An analysis of obstacles. *Science & Education, 20*(3–4), 293–316.
Iaria, G., & Hubball, H. (2008). Assessing student engagement in small and large classes. *Transformative Dialogues: Teaching & Learning Journal, 2*(1), 1–8.
IDEO, & Riverdale Country School. (2017). *Design thinking for educators* (2nd ed.).
Interamerican Network of the Academies of Sciences. (2011). *About IANAS. www.ianas.org*. Interamerican Network of the Academies of Sciences. Retrieved 15 November from www.ianas.org
Isacks, B., Oliver, J., & Sykes, L. R. (1968). Seismology and the new global tectonics. *Journal of Geophysical Research, 73*(18), 5855–5899.
Kelly, R. (2020). *Collaborative creativity: Educating for creative development, innovation and entrepreneurship*. Brush Education, Inc..
King, C. (2008). Geoscience education: an overview. *Studies in Science Education, 44*(2), 187–222.
King, C. (2013). Third international geoscience education survey – 2012 with 2013 updates.
Latour, B. (1987). *Science in action: How to follow scientists and engineers through society*. Harvard University Press.
Lederman, N. (2007). Nature of science: Past, present and future. In S. Abell & N. Ledderman (Eds.), *Handbook of research in science education* (pp. 831–880). Lawrence Erlbaum Associates.
Lederman, N., Abd-El-Khalick, F., Bell, R., & Schwartz, R. (2002). Views of nature of science questionnaire (VNOS): Toward valid and meaningful assessment of learners' conceptions of nature of science. *Journal for Research in Mathematics Education, 39*(6), 497–521.
Libarkin, J. C. (2005). Conceptions, cognition, and change: Student thinking about the earth. *Journal of Geoscience Education, 53*(4), 2.
Libarkin, J. C., & Anderson, S. W. (2005). Assessment of learning in entry-level geoscience courses: Results from the geoscience concept inventory. *Journal of Geoscience Education, 53*(4), 394–401.
Macdonald, H., Manduca, C. A., Mogk, D. W., & Tewksbury, B. J. (2005). Teaching methods in undergraduate geoscience courses: Results of the 2004 On the Cutting Edge Survey of U.S. Faculty. *Journal of Geoscience Education, 53*(3), 237–252.
Matthews, M. (2012). Changing the focus: From nature of science (NOS) to features of science (FOS). In M. S. Khine (Ed.), *Advances in nature of science research: Concepts and methodologies* (pp. 3–26). Springer.
Matthews, M. M. (2015). *Science teaching: The contribution of history and philosophy of science* (20th anniversary revised and expanded edition). Routledge.
McComas, W., & Olson, J. (1998). The nature of science in international science education standards documents. In W. McComas (Ed.), *The nature of science in science education: Rationales and strategies* (pp. 41–52). Kluwer.
McDowell, Z. (2017). *Student learning outcomes using Wikipedia-based assignments: Fall 2016: Research report*. W. E. Foundation. https://commons.wikimedia.org/wiki/File:Student_Learning_Outcomes_using_Wikipedia-based_Assignments_Fall_2016_Research_Report.pdf

Metz, D., Klassen, S., McMillan, B., Clough, M., & Olson, J. (2007). Building a foundation for the use of historical narratives. *Science Education, 16*(3–5), 313–334.

National Research Council. (1996). *National science education standards*. National Academy Press. https://doi.org/10.17226/4962

National Science Teachers Association. (2003). *Standards for science teacher preparation*. NSTA Press.

Nesari, A. J. (2015). Dialogism versus monologism: A Bakhtinian approach to teaching. *Social and Behavioral Science, 205*, 642–647.

NGSS Lead States. (2013). *Next generation science standards: For states, by states*. National Academies Press. Retrieved July 2 2021 from https://www.nextgenscience.org/

Núñez-Oviedo, M., & Clement, J. (2008). A competition strategy and other modes for developing mental models in large group discussion. In J. Clement & M. A. Rae-Ramirez (Eds.), *Model based learning and instruction in science* (Vol. 2, pp. 117–138). Springer.

Núñez-Oviedo, M. C., Clement, J., & Rae-Ramirez, M. A. (2008). Developing complex mental models in biology through model evolution. In J. Clement & M. A. Rae-Ramirez (Eds.), *Model based learning and instruction in science* (Vol. 2, pp. 173–194). Springer.

O'Neill, T. A., Deacon, A., Gibbard, K., Larson, N., Hoffart, G., Smith, J., & Donia, M. (2017). Team dynamics feedback for post-secondary student learning teams. *Assessment and Evaluation in Higher Education, 43*, 571–585.

O'Neill, T. A., Larson, N., Smith, J., Deng, C., Donia, M., Rosehart, W., & Brennan, R. (2019). Introducing a scalable peer feedback system for learning teams. *Assessment and Evaluation in Higher Education, 44*, 848–862.

O'Neill, T. A., Boyce, M. A., & McLarnon, M. J. W. (2020a). Team health is improved when peer evaluation scores affect grades on team projects. *Frontiers in Education, 5*, 1–10. https://doi.org/10.3389/feduc.2020.00049

O'Neill, T. A., Pezer, L., Solis, L., Larson, N., Maynard, N., Dolphin, G., et al. (2020b). Team dynamics feedback for post-secondary student learning teams: Introducing the "Bare CARE" assessment and report. *Assessment and Evaluation in Higher Education, 45*, 1121–1135. https://doi.org/10.1080/02602938.2020.1727412

Orion, N. (2019). The future challenge of earth science education research. *Disciplinary and Interdisciplinary Science Education Research, 1*(3), 1–8.

Orion, N., & Libarkin, J. (2014). Earth systems science education. In N. G. Lederman & S. K. Abell (Eds.), *Handbook of research on science education* (Vol. II, pp. 481–496). Routledge.

Ornstein, A. C. (1994). The Textbook-Driven Curriculum. *Peabody Journal of Education, 69*(3), 70–85.

Osborne, J., Collins, S., Ratcliffe, M., Millar, R., & Duschl, R. (2003). What "ideas-about-science" should be taught in school science? A Delphi study of the expert community. *Journal of Research in Science Teaching, 40*(7), 692–720.

Preuss, P. (1992). Rhea's time. In B. Preiss & R. Silverberg (Eds.), *The ultimate dinosaur* (pp. 88–99). Bantam Books.

Rae-Ramirez, M. A. (2008). Determining target models and effective learning pathways for developing understanding in biological topics. In J. Clement & M. A. Rae-Ramirez (Eds.), *Model based learning and instruction in science* (pp. 45–58). Springer Science.

Redman, C. (2007). *Towards a dialogical perspective on agency in student learning*. Australian Association for Research in Education.

Roberts, D. A. (2007). Scientific literacy/science literacy. In S. K. Abell & N. G. Lederman (Eds.), *Handbook of research on science education* (pp. 729–780). Lawrence Earlbaum Associates, Inc.

Rudolph, J. L. (2000). Reconsidering the 'nature of science' as a curriculum component. *Journal of Curriculum Studies, 32*(3), 403–419.

Rudwick, M. J. S. (2005). *Bursting the limits of time: The reconstruction of geohistory in the age of revolution*. University of Chicago Press.

Rudwick, M. J. S. (2014). *Earth's deep history: How it was discovered and why it matters*. Chicago University Press.

Rule, A. C. (2006). Editorial: The components of authentic learning. *Journal of Authentic Learning, 3*(1), 1–10.

Şengör, A. M. C. (2001). *Is the present the key to the past or is the past the key to the present? :James Hutton and Adam Smith versus Abraham Gottlob Werner and Karl Marx in interpreting history* (Vol. 355). Geological Society of America.

States, N. L. (2013). *Next generation science standards: For states*. By States.

Stinner, A., & Teichmann, J. Â. (2003). Lord Kelvin and the age-of-the-earth debate: A dramatization. *Science Education, 12*(2), 213–228.

Stinner, A., McMillan, B. A., Metz, D., Jana, M., & Klassen, S. (2003). The renewal of case studies in science education. *Science & Education, 12*, 617–643.

Turner, D. (2013). Hisotrical geology: Methodology and metaphysics. In V. R. Baker (Ed.), *Rethinking the fabric of geology* (Geological Society of America Special Paper 502) (pp. 11–18). Geological Society of America.

Wandersee, J. H., Clary, R. M., Anderson, S. W., & Libarkin, J. (2003). *The retention of geologic misconceptions: Alternative ideas that persist after instruction*. American Geophysical Union. Fall Meeting, 2003.

Wiggins, G. P., & McTighe, J. (2006). *Understanding by design* (Expanded 2nd ed.). Pearson Education, Inc.

Will, W. (2006). Science as story: Communicating the nature of science through historical perspectives on science. *The American Biology Teacher, 68*(4), 200–205.

Zarkadakis, G. (2015). *In our own image: Savior or destroyer? The history and future of artificial intelligence*. Pegasus Books.

Chapter 3
Biographies of Three Different Concepts

A noteworthy and often-remarked similarity exists between the facts and methods of geology and those of linguistic study. The science of language is, as it were, the geology of the most modern period, the Age of the Man, having for its task to construct the history of development of the earth and its inhabitants from the time when the proper geological record remains silent…The remains of ancient speech are like strata deposited in bygone ages, telling of the forms of life then existing, and of the circumstances which determined or affected them; while words are as rolled pebbles, relics of yet more ancient formations, or as fossils, whose grade indicates the progress of organic life, and whose resemblances and relations show the correspondence or sequence of the different strata; while, everywhere, extensive denudation has marred the completeness of the record, and rendered impossible a detailed exhibition of the whole course of development.

William Dwight Witney

Language and the Study of Language (1867), p. 47

As geology is essentially a historical science, the working method of the geologist resembles that of the historian. This makes the personality of the geologist of essential importance in the way he [or she] analyzes the past.

Reinout Willem van Bemmelen

"The Scientific Character of Geology," The Journal of Geology (Jul 1961), 69(4), p. 453

G. Dolphin, *Teaching Geology Using the History and Philosophy of Science*, Science: Philosophy, History and Education,
https://doi.org/10.1007/978-3-031-68667-2_3

3.1 Introduction: The Structure of the Historical Narratives

This chapter recounts the three narratives that constitute the backbone of the conceptual development promoted within the course. The narratives reflect the teaching I do during the entire course. The term is approximately 13 weeks long with an 75-min class twice a week. All but the final three classes are spent telling the narratives via lectures, demonstrations, activities, and open-ended questioning, in the fashion of interrupted case studies. I have come to look at these narratives as three separate *biographies* of the relevant key concepts: the historical nature of the earth, the age of the earth, and the dynamics (or tectonics) of the earth. In turn, an essential question frames each of these three narratives:

- How can we understand the historical nature of the earth? (narrative I: The earth's biography)
- How can we determine the age of the earth? (narrative II: The age of the earth)
- Why are there still mountains? (narrative III: Developing a theory of the earth)

The answers to these essential questions live at the core of the discipline. They cannot be answered by short and simple statements; moreover, they point towards the discipline's big ideas (Wiggins & McTighe, 2006). Within this larger frame, these narratives braid multiple pedagogical strands, such as storytelling, inquiry, and model-based learning (MBL), in an iterative fashion that has students experience open inquiry—or science in the making (Latour, 1987). Open inquiry encourages students to expand their paths in any direction, even while they must arrive at explicit learning goals or endpoints.

The iterative structure of each narrative begins with the related question (as listed above), but then moves to a key subsidiary question. As shown in Fig. 3.1, in the first instance, this becomes "How can we understand the historical nature of the earth?" This secondary question, though more limited in scope than the opening essential question framing the entire narrative, is still broad and open-ended, pointing towards big ideas necessary for understanding the integrated target concept (see Fig. 2.4 and Sect. 4.5). It thus generates several more subsidiary questions, also demonstrated in Fig. 3.1.

The essential question frames the historical interlude, with the purpose of providing temporal contextual for the essential question by telling the story related to the question. It also builds the concept of inquiry by presenting specific observations as well as competing interpretations of those observations. The historical interludes are therefore important as they act as constraints that keep students from veering too far from the historical path of discovery (Allchin, 2017, 2020). In other words, students offered a historical problem have the freedom of inquiry to explore as they see fit; however, the narrative reveals the historical direction of the inquiry, clarifying for students why investigation or rationalization went this way, instead of that way.

The interlude also then sets up the next opportunity for inquiry, where students make sense of the observations and the competing interpretations. They may have to develop their own alternative or weigh the explanatory power of one set of

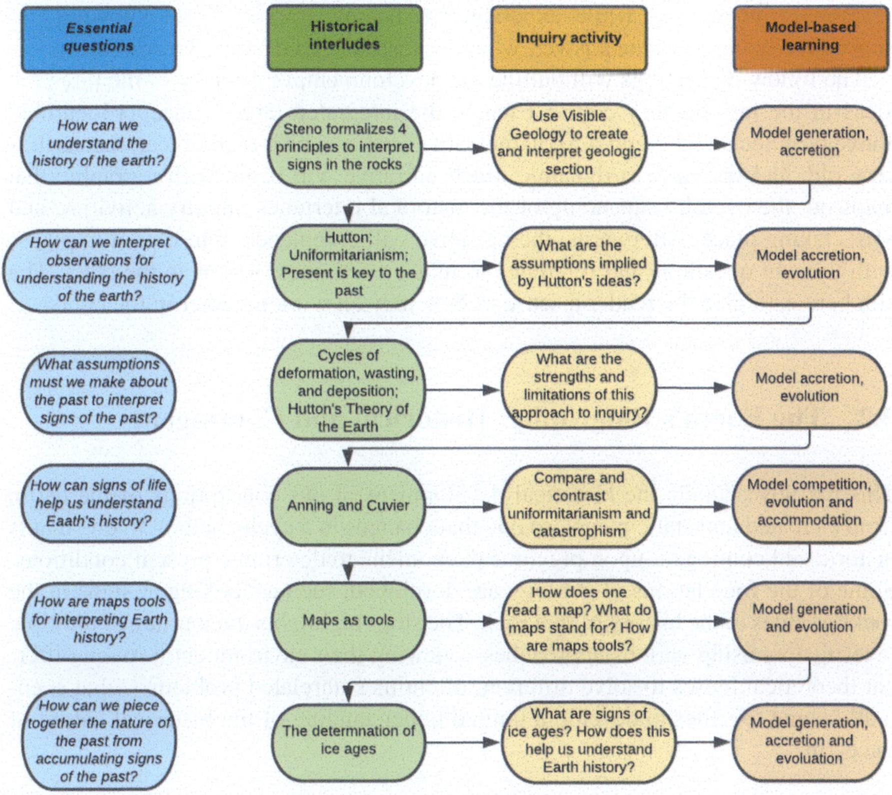

Fig. 3.1 How can we understand the historical nature of the earth?

interpretations with those of the competing interpretations. This puts them in the shoes of the scientists of the day, making decisions about data and interpretations without the benefit of hindsight and already knowing the "right" answer. Within the historical interludes are many names and dates that are not for students to memorize, but rather to give context to the story.

As a result of these inquiry activities, students are performing the work of MBL. Specifically, they are generating, modifying, accreting, and deleting from their own mental models of the target concepts. As this is happening, a new iteration starts with the next historical interlude that wraps up the loose ends from the previous interlude and then carries the inquiry forward. This phase typically introduces new data that students can incorporate within their evolving mental model. The iterations continue until a point is reached where the history culminates in what is largely the current consensus view of the target concept.

Students become invested in finding the answer as they wrestle through the process to arrive at the integrated target concepts. This, in turns means they engage in the process of science and understand the relationship between data and interpretation, and how that ends up as scientific knowledge. Students therefore invest in developing an answer, while experiencing the uncertainty of not knowing—as opposed to being given an answer, which is abstract and devoid of any context.

The following sections will outline the developmental trajectory of the three big ideas in the introductory course I teach, the integrated target concepts identified above (see Sects. 2.5.2 and 4.5): namely, the earth is a historical object, the earth is very old, and the earth is dynamic. Each narrative will begin with a graphic that maps out the overall sequencing of the historical interludes, inquiry activities, and MBL taking place. Afterwards, the narrative will commence, interrupted in places with thought questions and reference to inquiry activities, just as in the class. The aim here is to give the reader a sense of how they all work together in the course.

3.2 The Earth's Biography: Historical and Contingent

This narrative details the historical development of our conceptions of the earth, from a created and static planet, to one that changes in a cyclic nature, to one that is historic and contingent upon previous phenomena to determine present conditions. Some of the foundations of geology are developed, such as looking at signs in the rocks as clues to the history of that rock. The story highlights the journey of multiple investigators using various approaches—viewing their environment through different theoretical lenses to solve different, oftentimes unrelated problems—that eventually come together to develop a unified understanding of the historical nature of the earth.

3.2.1 *Signs of Earth's History: Learning to Read the Rocks*

This narrative starts at the beginning of the Enlightenment with Danish-born anatomist, ***Niels Stensen (1638–1687)***. Stensen would later Catholicize his name to Nicholas Steno and will henceforth be referred to as Steno. As a child, Steno was quite sickly and therefore isolated from many childhood activities. This allowed him to develop interests in both natural philosophy and religion. I want to stress here that the modern portrayal of rational thought (or scientific thought) versus religion is inaccurate when dealing with the earth's antiquity or any other concepts where modern-day tensions exist; indeed, it is a myth, as no controversy existed. For Steno, as with others at the time, engaging with natural philosophy was an endeavour to reconcile observations with biblical verse (Rudwick, 2014).

As Steno's father was a jeweller, Steno was exposed to and became proficient with the use of delicate tools for performing intricate tasks. With this background, he went on to study medicine at the University of Leiden and became a famous anatomist. He discovered that it was tear ducts that were responsible for human tears. Also, he found that it was muscle contraction that moved the bones, rather than the movement of bones that compressed muscles, the prevailing thought at the time (Cutler, 2003). Because of his insights and the corresponding fame, he secured a position in the court of King Ferdinand of Florence. Steno dissected the head of a great white shark, brought to the king by local fishermen. During the dissection, he noticed the similarities between the teeth of the shark and locally found "tongue stones" he had pulled out of the rocks around Tuscany, where he lived. He surmised that the tongue stones (named for their linguoid shape) were sharks' teeth embedded into the rocks. But how did they get into the rocks, and so far above sea level? These were questions that occupied much of Steno's thinking.

While exploring, Steno also observed different geometrical relationships in the rocks of the mountains in and around Tuscany. He began to interpret these relationships in terms of the relative age or time when events could have or would have occurred with respect to other events. Part of this process was formalizing four principles:

- *Superposition:* For a stack of layered rocks, the oldest layers are at the bottom and the youngest at the top
- *Original horizontality:* The rocks in a layered sequence were originally laid down horizontally
- *Lateral continuity:* The rocks in a layered sequence were originally laid down in continuous layers
- *Crosscutting relationships:* Where layers of rock are crosscut by other rocks, the rocks that are crosscut are older than what is crosscutting them.

Activity Alert
Visible Geology

Students use a computer app to create a geological cross section that they then trade with another group that will try to interpret the sequence of events for that cross section. This activity puts students in the position of Steno to look at the signs in a geological cross-section and through logical argument, develop a sequence of events history for it. See Chap. 4, Sect. 4.4.1 for a full description.

He did not develop these principles, as they were obvious to most people in that period (Rudwick, 2014). The significance of these principles is that Steno used them to logically reason out a history of the earth. Steno saw the rocks as a book—the signs in the rock were a language he could interpret to tell a story, but the language needed to be deciphered. Notably, Steno's work put forth some very important ideas for understanding the history of the earth:

- Interpreting the signs in rocks can provide information about the history of the earth. For instance, if layered rocks are tilted, folded, faulted, or otherwise deformed, then something must have happened to those rocks to change them from originally horizontal and laterally continuous.

- The formed stones (like tongue stones and other examples that resemble living organisms) *are* actually organisms preserved in the stone through petrifaction (mineral replacement).
- The signs in the rocks support a story of the earth that aligns with biblical description. Though many people in the mid-1600s considered the earth to have experienced no major changes through time—to essentially be in the same state as it was when it was created by God—Steno used direct quotes from scripture to support his interpretations of change through time.

Steno eventually withdrew from his naturalistic studies of the earth to concentrate more on religious studies. He had trouble reconciling his observations of the natural world with a literal reading of the Bible, but he came to see his science as a way to glorify God's creation. Eventually he was ordained as a priest and took a position in a small church in Germany where he encountered very difficult living conditions, became ill, and died at the age of 48. His work had no immediate impact in how people perceived the earth; however, over the next hundred years, this would change.

3.2.2 Eternalism: The Present Becomes the Key to the Past

The narrative continues into the early to mid-eighteenth century, with class discussion about ***James Hutton (1726–1797)***. Hutton was born in Scotland, one of five children. Hutton's father worked in the mercantile, affording the family a higher socio-economic status than many of his peers. At 14, Hutton entered university to become a lawyer, but after a brief apprenticeship realized that this line of study and work were not to his liking. Instead, he went back to school at the University of Leiden to study medicine. Hutton made this choice not because he was especially interested in becoming a doctor but because it enabled him to get exposure to what really interested him — chemistry. Sir Isaac Newton's mechanistic world view and descriptions of cyclical planetary motions greatly influenced Hutton's thinking. He graduated in 1749 with a medical degree after writing his dissertation on circulation of the blood in the human body (Repcheck, 2003).

Ultimately, Hutton settled on a farm formerly owned by his uncle and became a gentleman farmer. There, he implemented multiple innovate farming practices he learned while studying in continental Europe. He also became keenly aware of the small changes taking place on his land over time. For instance, he noticed that after heavy rains, water washed bits of dirt and soil into the rivers, which eventually travelled to the sea to be deposited on the ocean floor. He was familiar with Steno's work and so assumed this deposition to be in thin continuous layers that eventually turned to rock. Over time, this rock would experience pressure from the earth's internal heat engine; this would deform the rocks into mountains that would become exposed to wasting processes bringing sediments back to the oceans once again (Hutton, 1788).

He supposed that these processes of wasting, deposition, and deformation happened very slowly. He also observed evidence in different locations that suggested multiple such cycles had taken place in the past. Invoking his experiences with Newton's ideas of cycles, he surmised the earth as constantly rejuvenating itself in an endless round of this three-step cycle—thereby maintaining a dynamic equilibrium. The implication of this proposition was that the earth had "no vestige of a beginning, no prospect of any end." Hutton did not see the earth as a historic entity, but one that is eternally cyclical (Rudwick, 2014). An interesting apocryphal story has Hutton viewing an unconformity at Siccar Point, which led him to a flash of insight concerning this cyclic nature of the earth. However, Hutton had given much thought to the idea and had written about it and presented to the Geological Society of London years before observing the unconformity at Siccar Point (and the less well-known unconformity at Jedburgh). In fact, he searched for these examples specifically to support his ideas.

A significant implication arises from the cyclical quality and dynamic equilibrium at the foundation of Hutton's idea: namely, the notion that the processes involved in the wasting, deposition, and deformation of the surface of the earth have remained constant throughout its history. These repetitive processes produce the same signs in the past, which get preserved in the rocks. When we observe signs and structures in the rocks that resemble, say, modern-day river deposits, we can safely assume an ancient river was responsible for those signs. Thus, we can interpret the past in terms of present-day processes; conversely, the present is the key to the past (Şengör, 2001). This understanding would later by identified as uniformitarianism by William Whewell (1794–1866) and be popularized by Charles Lyell (1797–1895).

Another important contribution by Hutton was the fortification of the "Plutonist" understanding of the earth, a developing view at the time. Plutonists saw the earth as having started out as a molten ball of material that cooled and solidified at the surface, even though the interior remained hot and dynamic. This constituted the internal heat engine that Hutton saw generating and deforming rocks. Another view, "Neptunism," was also popular then, championed by the famous ***Abraham Gottlob Werner (1749–1817)***. Neptunism held that the earth started out covered by water, and as that water evaporated, dissolved minerals precipitated, creating a bottom-most layer of crystalline *primary* rocks. On top of those, intermediate and secondary rocks occurred, composed of the broken pieces of the primary rocks (sedimentary). Intermediate rocks had no signs of life (fossils), while secondary rocks contained signs of previous life (fossils).

Again, Hutton searched for evidence of the Plutonist view over the Neptunist perspective. He found examples where crystalline (primary) rocks had intruded into sedimentary (secondary) rocks, supporting the perspective that crystalline rocks did not necessarily come from precipitation, and they were not necessarily the oldest rocks. These observations also supported the idea of pressures from within the earth causing the intrusion of hot, liquid rock into surrounding secondary rocks. These last ideas will become more important in the third narrative describing the dynamics of the earth.

3.2.3 Constructing Earth's History: Giving Fossils Meaning

Mary Anning (1799–1847) enters the narrative next. Her father, a cabinet maker, also searched for, excavated, cleaned, and sold fossil shells he found in the sea cliffs near his home in Lyme Regis, England. He died when Mary was only 10, but not before he had taught her, the eldest of his two children, this hobby. With her father dead, selling these fossils became the only income for the family. Anning had a very keen eye for finding fine fossils and put a lot of effort into this endeavour (Rudwick, 2014). One day, she and her younger brother found the first Ichthyosaur, and later a full Plesiosaur, which were far more complex specimens than had been found to date. Some people, like noted anatomist ***Georges Cuvier (1769–1832)***, at first considered them to be fraudulent. However, having read other expert opinions, he finally acknowledged their veracity.

Anning became quite proficient in finding, cleaning, and selling these artifacts of former worlds, with her finds becoming part of many fossil collections throughout Europe. Of course, being both poor and a woman, her name and credit for the finds were omitted from the specimen identification cards. Her importance was nearly forgotten until her "story" was penned by English songwriter, Terry Sullivan in the famous ditty "She sells seashells at the seashore…" The accumulation and diversity of fossils being found and catalogued generated more interest in finding and describing them; soon they became tools for interpreting a history of the earth, a history that appeared to be quite different from present-day reality (eg., De la Beche, 1830). Collecting and studying these curious artifacts became a popular pastime for wealthier gentlemen. As more fossils were found, it became increasingly obvious that fossils from shallower rocks—and therefore younger ones—bore greater resemblance to modern life forms. Conversely, those fossils found deeper, from older zones, offered up forms increasingly different from modern life forms.

Georges Cuvier would contribute a possible explanation to these observations. Cuvier earned a medical degree, specializing in animal anatomy, and became famous for his ability to reconstruct entire animals with only minimal remains. He saw the diversity of life on Earth resulting from the diversity of environments. He determined that animals were born already adapted to their environment and if the environment changed, they would not survive. He did not, however think that animals *evolved* to fit into the environment, eschewing the notion that they changed over time due to environmental pressures. In his mind, "kinds" of animals, what we now call "species," were stable and unchanging. Significantly, Charles Darwin had not yet published *On the Origin of Species* (Darwin & Kebler, 1859).

Cuvier used his ideas about animals, and the constancy of their form and their environments, to explain the growing difference between modern life forms and the accumulation older and less familiar fossils. His ideas opposed some of his contemporaries' ideas of species dynamics, or evolution. ***Jean Baptiste de Lamarck (1744–1829),*** for example, asserted that animal forms were quite dynamic, changing with the changing environment. The changes incurred by a particular animal during its life would be passed down to the next generation. The concept of species was arbitrary because it depended on when, during the slow constant change, the observer demarcated the classification.

Darwinian evolution, published in *On the Origin of Species* in 1859, blended these two ideas. Darwin emphasized the importance of variation of offspring at birth and the selective pressures influencing which individuals would survive long enough to reproduce. These selective pressures would lead, eventually, to the generation of new species. Animal kinds were static within the lifetime of any organism, as Cuvier claimed, but the environmental pressures would influence survival over the long term, making animal kinds more dynamic, more sympathetic to Lamarck and contrary to Cuvier. Noteworthy, the resolutions of many scientific controversies were not the result of "either-or" propositions. Because everyone was working from mostly the same data, that various explanations overlapped is unsurprising, and resolutions often used facets from originally competing ideas.

In addition to his ideas about species, Cuvier started to interpret fossils in the rocks as indicators of depositional environments. Marine-looking fossils would mean the rocks were most likely deposited in a marine setting. He mapped the sedimentary rocks of the Paris Basin this way. Cuvier focused his attention on fossil mammals coming from the topmost unconsolidated sediments. These big mammals were very different from the mammals observed in his day. He postulated that the organisms had suffered a mass extinction. Cuvier was the first to put forth the idea of past mass extinctions. He saw quick, massive environmental change, or "violent revolutions," as the cause of these massive die-off events. He posited that a large global flood would have caused such massive and seemingly quick extinctions.

The theory of violent revolutions became the foundation of "catastrophism" that was in direct opposition to uniformitarianism, an increasingly popular view in that period. Perhaps Cuvier, having lived through the French Revolution, used his life experiences as a metaphor (Indurkhya, 1992; Kuhn, 1993; Lakoff & Johnson, 1980; Taylor & Dewsbury, 2018) for the violent revolution he talked about as the cause of mass extinctions (Rudwick, 2014). Of note here is that the global flood invoked by Cuvier to explain some mass extinction events was a way to reconcile observations with the biblical story of Noah, as opposed to using the Bible to constrain understanding of events on Earth.

Catastrophism was not religion's explanation for change in the past; indeed, what we currently experience as a "science versus religion" argument when considering the earth's past is really a very recent phenomenon—an early twentieth century reaction to the popularization of Darwinian evolution (Rudwick, 2005, 2014). Moreover, Cuvier pointed out other evidence for catastrophic events. For instance, the only conceivable explanation for the emplacement of very large boulders that rest in low lying topographies, with compositions quite different from the surrounding bedrock, was that they were pushed into place by huge walls of flowing water or were dropped out of melting ice floating in much higher water. These boulders would later become known as glacial erratics.

Charles Lyell (1797–1895) restricted the definition of uniformitarianism by asserting that any cause for a sign in the past had to be observable in the present. Further, whatever led to signs in the past were of no greater magnitude or intensity than what was currently observed. He could not refute the observation of large-scale die-offs. However, he did view it through the lens of uniformitarianism by explaining that locations displaying evidence of mass extinctions were locations where little to no deposition had taken place for a very long time, so the slow change from one animal type to another *appeared* to be rapid. This exemplfies how a scientist's theoretical framework impacts how they see and interpret the world; what is data and what is noise. Cuvier and Lyell had the same data but developed completely different explanations.

Through the eighteenth and nineteenth centuries, tensions developed among different schools of thought regarding interpretations of the signs of Earth's past. Those promoting Neptunism imagined a massive original flood to be responsible for the signs in the rocks. Evaporation and disappearance of that water led to the deposition of various rock types (primary, intermediate, secondary, and so on). Aside from that, the earth was static. On the other hand, Plutonism supposed a hot early Earth, with that heat trapped in the centre of the earth and driving change through time at the surface. Hutton and Lyell saw the earth as eternal, always remaining in a kind of dynamic equilibrium. For instance, Lyell supposed that a return to environmental conditions that supported dinosaur existence in the past would bring the dinosaurs back (Rudwick, 2014). This "eternalism" saw Earth amid endless cycles of change, in contrast to "historicism," which posited a unique and contingent past that determined the present.

Uniformitarianism set the framework for that change through time; it meant slow steady change over vast periods of time. This contrasted with catastrophism where large-scale and quick processes were at least partially responsible for the change recorded in the rocks. Those that adhered to catastrophism were not trying to promote a young Earth. The age of the earth did not matter, here. What mattered was that the observations could not be explained by slow and steady changes over long periods of time.

3.2.4 Using Fossils to Make Maps

The story next turns to ***William Smith (1769–1839)***. Smith had humble beginnings as the son of a blacksmith. Growing up in a working-class context, Smith did not have the benefit of the best education or social opportunities. Fortunately, he did navigate his way into engineering amid the Industrial Revolution, where he was able to capitalize on the need to transport coal from rural mines to urban centres. The steam train had not been invented yet, so the only reasonable manner to transport coal was via canals. Smith became widely known for his abilities designing and building them. Pertinent to this story is that his growing success afforded him opportunity to work across Great Britain. Because of the nature of his work, he gained access to the land's subsurface (Winchester, 2001).

Smith became fascinated by the little formed stones, many that looked like seashells that he found, collected, cleaned, and meticulously catalogued. As he collected more and more—making notes on where he found them and the kind of rocks in which he discovered them—he discerned patterns in their placement. Certain fossils seemed to come out of particular kinds of rocks. As well, he observed a trend: particular kinds of fossils were always found above a different, but specific, kind of fossil. Based on the kind of rock and the fossils located within a particular area, Smith learned to predict where he would find another kind of fossil. He knew which fossils were associated with the coal layers; a crucial insight given this period of the Industrial Revolution.

He also believed that he could connect layers of rock in the south of Britain to equivalent layers in the north, strictly by using the fossils they contained. His new "stratigraphy" was not concerned with what the fossils might mean or whether they might have some temporal significance (Rudwick, 2014). They were merely signs to him, markers of the strata that he could now correlate across the entire country. He eventually correlated rocks across all of Great Britain and drew the first geologic map. His method of correlating rocks based on fossils instead of rock-type was considered a controversial approach because it was so theoretical (Rudwick, 1985). The prevailing view then was to enter a field area as objectively as possible, with no guiding theoretical framework that could bias observations. This, of course, was an unreasonable ideal that many espoused but no one ever really fulfilled. For example, identifying the kind of rock (such as primary, intermediate, and secondary) based on its composition (crystalline, stratified with no fossils, stratified with fossils, respectively), had been the accepted practice to that point, but nevertheless invokes the same kind of theoretical starting point.

3.2.5 *Discerning Ice Ages Fortifies the Notion of a Historic Earth*

Jean de Charpentier (1786–1855), a geologist who studied glaciers in the Swiss Alps, noted on some of his travels to the lowlands of Europe many similar signs of glaciation he had documented at higher elevations. These included deposits of highly unsorted materials, scratched and polished bedrock, and very large boulders sitting on flat land with no nearby hill or mountain to slide down from. From a catastrophist perspective, many of these signs were then attributed to a large flooding event, what today we would call a tsunami; this was the only way it was considered possible to move such large rocks or to have emplaced so much sediment. However, Charpentier thought it was unlikely that sea level could have been so high as to emplace the large boulders. In addition, scraped and polished rocks occurred on higher ground where one would not expect to find the kind of running water linked to such signs. He therefore posited, based on the similarities to contemporaneous Alpine glacial signs, a time of far more extensive glaciation, where ice covered large portions of Europe and North America. This view also aligned the same kinds of signs reported elsewhere in these regions.

The proposition of far more extensive glaciation faced two challenges from the geologic community. First, the idea of continental glaciation covering large portions of Europe and North America did not agree with the uniformitarian framework. Claiming such widespread glaciation implied a far greater magnitude than had ever been observed. The second challenge rested with the increasing popularity of the "nebular hypothesis of planet formation." This hypothesis explained how gravitational collapse of space debris led to the earth's formation. Such gravitational collapse would result in a massive release of heat, possibly even melting the growing globe in space. This initially very hot planet would then cool over time. However, if the current temperature of the earth could not support massive glaciation as suggested by Charpentier, then how could a theoretically warmer, earlier Earth support such glaciation? Signs in the rocks, such as tropical-looking fossil plants and the massive coal seams that were driving the Industrial Revolution, appeared to support the notion of a much warmer Earth in the past.

Because Charpentier was not an eloquent communicator, his ideas about glaciation did not gain much immediate traction. At one point, he did engage famed geologist ***Louis Agassiz (1807–1873)*** in the discussion. Agassiz had by then made quite a name for himself through his work on fossil fish. Charpentier shared his idea with Agassiz, who was dismissive until Charpentier took him around to show him the evidence he had amassed. Eventually, Charpentier won over Agassiz, who then became a great proponent of the idea. A far better communicator than Charpentier, Agassiz persuaded many people through multiple field trips to become sympathetic to the idea. However, not everyone was convinced, especially because it clashed with the prevailing uniformitarian framework. Trying to create an explanation that remained within the framework of uniformitarianism, Agassiz suggested that instead of cooling in a uniform and steady manner, the earth cooled in a stepwise fashion, longer periods of stability punctuated by quick drops in temperature. During these quick drops in temperature, Agassiz presumed multiple ice ages happened.

By the late nineteenth century, ***T. C. Chamberlain (1843–1928)*** delineated the extent of North American glacial ice (Geikie, 2012). Continued observations from the field offered supporting evidence so that eventually the notion of multiple sequential ice ages became irrefutable, along with the realization that each one took a substantial amount of time to happen. However, no one could yet explain how a global ice age would be triggered. Human remains and other common modern animals had only been found in the most recent (uppermost) glacial deposits. Though the glacial periods were thought to be quite lengthy, they still only represented a tiny portion of the time recorded in the rocks. This understanding led to core new insights in how we thought about the history of the earth. First, the earth's history must be very long (Sect. 3.3), with humans only a minute part of that history. Second, accepting the occurrence of global ice ages meant Lyell's very restrictive definition of uniformitarianism had to be broadened to incorporate such global events, though they had not actually been witnessed.

3.3 The Age of the Earth: The Discovery of Deep Time

This second narrative (Fig. 3.2) pulls us back in time from the end of the previous narrative. The process of trying to determine the age of the earth has, itself, a very long history. Most cultures around the world have a creation story, explaining the origin of the universe or the earth within that universe. Many of the stories emphasize a cycle, where the earth is born, develops, and subsequently dies to be reborn once again. As the previous narrative demonstrated, geologists came to understand the earth as a historical entity, with the present determined by contingent causes of the past. Investigations into the age of the earth were happening even while investigations into the historical nature of the earth were also underway, as described above; in some cases, the same investigators were involved. Another theme that presents itself is the notion of developing scientific knowledge going hand-in-hand with developing technology: in some cases, the idea of a process was only hampered by the lack of technology to accomplish it, while in others, it was the development of new science knowledge that influenced new technologies.

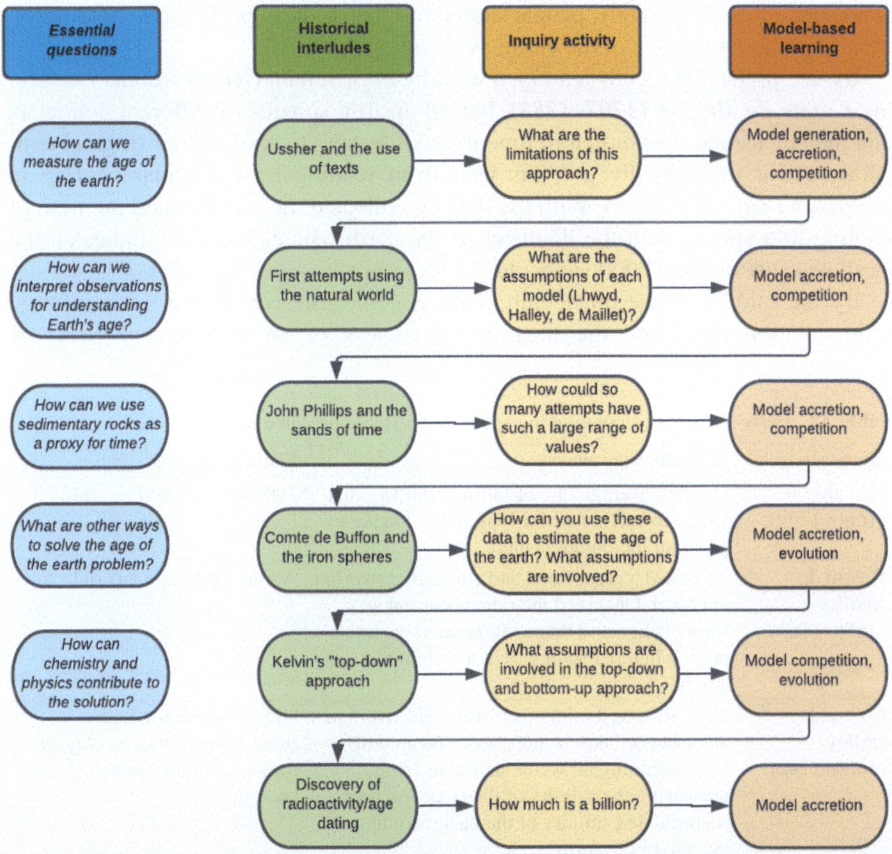

Fig. 3.2 How can we determine the age of the earth?

3.3.1 *Breaking with Tradition: Empiricism and a "Bottom-Up" Approach*

One of the most famous (or infamous) attempts at dating the earth was conducted by ***Archbishop James Ussher (1580–1656)***. Ussher attempted to determine the moment of creation though biblical scripture, coupled with other important texts to fill holes in scripture. Importantly, this was not an attempt by Ussher to usurp knowledge of the earth's beginning from science. In fact, science as we think of it today did not even exist yet. Ussher's attempt at pinpointing the origin of the earth in time was a purely scholarly endeavour. He tried to calculate an age, not to constrain what he thought that age should be.

However, Ussher's attempt is often derided by science-minded folk as an effort to establish biblical supremacy. Really though, like hundreds of others at the same time, he was trying to answer a question using the most reliable information available at the time: How old is the earth? ***Johannes Kepler (1571–1630)*** and ***Isaac Newton (1642–1727)*** also notably developed their own biblically based chronologies for the earth, each establishing an age of just over 6000 years, similar to Ussher (Jackson, 2006). Eventually, people started looking at the signs in nature to develop their ideas about the age of the earth, as shown in Table 3.1.

By late in the eighteenth century, a wealthy Frenchman, ***Georges-Louis Leclerc, the Comte du Buffon (1707–1788)***, heated up iron spheres of different diameters and then measured the amount of time it took for them to cool to their original temperature. He undoubtedly got this idea from reading about secular cooling in Newton's *Principia* (1846). With the data he collected, he extrapolated the time of cooling to a sphere with the diameter of the earth. His calculations indicated the earth was just under 87,000 years old.

By the middle of the nineteenth century, geologists considered using average sedimentation rates and the measured thickness of all sedimentary rocks as

Table 3.1 Early ideas for calculating the age of the earth using calculated estimates

Name	Procedure/rationale	Estimated age of earth
Edward Lhwyd (1660–1709)	Used an average rate of falling boulders in a valley to calculate how long it would take to place all the boulders there	200,000–300,000 year old
Benoit de Maillet (1656–1783)	Assumed a great flood and currently lowering sea level. Observed the rate of current sea level change and then extrapolated back to a point when sea level was at a certain assumed height	About 2 billion years old
Edmund Halley (1656–1742)	New dissolved minerals travelled in rivers to an ocean or lake. If there was a basin with no exit, assuming all water started as fresh, measuring the salinity of the river and the accumulated salinity of the basin would give the age of the earth	Did not attempt a figure because there was no way to measure salinity precisely enough

variables to calculate an age of the earth when sedimentation started. Those researching this problem would have assumed an average sedimentation rate and then a total measured thickness of sedimentary rocks. Even with just these two variables, estimates of the age of the earth ranged from about three million years old to about 1.5 billion years old, although most settled on many 10 s to about a 100 million years old. No consensus emerged, however.

What does stand out in all these diverse attempts at calculating the age of the earth is the number and kinds of assumptions that each approach must make for their calculations. Assumptions are crucial in science because they help to simplify otherwise complex systems and complex calculations. Some of the following questions would have also informed these calculations:

- Does the earth behave mostly like a cooling iron ball?
- Is the sedimentation rate stable through time?
- Do the measured sections of sedimentary rocks equal the measurement of the sediments as they were originally deposited?

As all models incorporate assumptions, there are certain affordances as well as limitations. Identifying the assumptions that go into a particular model or equation is crucial for judging the reliability of that model or equation.

At the turn of the twentieth century, a young geologist, ***John Joly (1857–1933)*** published a paper that would become famous as the "Joly sodium method." In essence, Joly's paper was very similar to Edmund Halley's from a century earlier, though he had never read Halley's paper. Joly posited that if one measured the amount of sodium in the oceans and divided that number by the rate that dissolved sodium enters the oceans through rivers, one could calculate the age of that ocean—this would be a proxy for the age of the earth. While Halley's paper garnered almost no attention, Joly's paper would be his most important. Using this method, Joly calculated the earth's age at between 90 and 100 million years (Jackson, 2006).

3.3.2 Enter Thermodynamics: A Top-Down Approach

Towards the end of the nineteenth century, the problem of determining the age of the earth was getting attention from chemistry and physics in response to the advent of the laws of thermodynamics. For the preceding few centuries, those we would today identify as chemists and physicists had worked to take individual observations of phenomena and consolidate them into the general dictates of behaviour in the universe. Laws, however, have been considered the design of God, dictating the way things are. When we impose such so-called universal truths on our observations as a means to explain those observations, this is a "top-down" approach to solving a problem (Schindler, 2007). This contrasts with the way many geologists worked: starting with observations and using abductive reasoning to reach a generalization or explanation. This also contrasts with the way scientists often think outside of geology. There is a persistent tension in science research where theory does and ought to guide observations. Scientists are not blank slates, but at times that can be problematic.

William Thomson, Lord Kelvin (1824–1907), showed a great aptitude for math that was supported by his father, a mathematics professor at the University of Glasgow. The younger Thomson Glasgow as a student at age of 10 and later taught mathematics and physics with his father. Through the progression of his career, Thomson became very interested in the nature of heat flow and was instrumental in developing the second law of thermodynamic–that heat always flows from a region of higher temperature to lower temperature–an example of the laws of nature that he believed were divinely authored. By using this second law and assuming certain physical characteristics and generalizations of the earth—heat flow, composition, historical change in rotational velocity, and so on—Thomson calculated an age of the earth to be between 20 and 400 million years. Over time, he settled for a maximum of about 20 million. However, this figure was unacceptably low to some geologists, like Charles Lyell. Based on their observations of the natural world—how slow change happens to the land and how slow change happens in animals—they were more comfortable with an estimate in the billions or even 10 s of billions of years. This set the stage for a controversy that would last through the end of the nineteenth century. Nonetheless Thomson asserted that the earth was giving off too much heat to the rest of the universe for it to be very old. According to his calculations, the earth would be much colder if it was much older.

Thompson, himself, used multiple modes to derive an answer. For instance, while Thomson used the second law of thermodynamics to assess the earth's age, he calculated the age of the sun in the same manner, arriving at a value of 20 million years. He also noted that the tidal-induced friction on the earth would slow its rotation and that the actual shape of the earth is an oblate spheroid. It is wider by about 50 km at the equator than it is through the poles, as the spinning of the earth causes the poles to flatten and the equator to bulge. Thomson figured that at the origin of the earth, it was spinning faster than at his present but has since slowed down due to the tidal friction.

He postulated that the original rotational velocity would be responsible for the earth's oblateness. He also observed that the current rotational velocity of the earth was not much slower than the rotational velocity of the earth when the earth solidified, and thus the earth could not be that old; it has not slowed appreciably from the original velocity responsible for the earth's observed oblateness. His multiple approaches produced results of an age from 10 s to 100 s of millions of years old. The estimates of the age of the earth may appear to be divided between the physicists and chemists on one hand, and the geologists and biologists on the other, but this was not really the case. Thomson's calculation very much aligned with the range of ages calculated by many geologists who were using sedimentary rock thicknesses and deposition rates to calculate the earth's age, strengthening the reliability of Thomson's results. Thomson was also not a fan of Darwinian evolution; to him the mechanism of randomness in evolution was not consistent with his ideas of design in nature (Stinner & Teichmann, 2003). Throughout his life, Thomson consistently maintained the age of the earth was 20 million years.

Activity Alert
Age of the Earth Controversy: The Role of Assumptions

Students watch the dramatization by Stinner and Teichmann (2003), then identify the assumptions made on each side of the argument and determine the most reliable argument based on the identified assumptions. This activity emphasizes the role and importance of initial assumptions when solving a problem. Students must distinguish between observation, inference, and assumptions, and then decide who has the stronger argument based on those variables. See Chap. 4, Sect. 4.4 for a full description.

3.3.3 How the Discovery of Radioactivity Changed the Dating Game

Near the beginning of the twentieth century, ***Marie Curie (née Sklodowska; 1867–1934)*** and ***Pierre Curie (1859–1906)***, two graduate students working in a French laboratory, were given a unique task by their supervisor ***Antoine Becquerel (1852–1908)***. Some days earlier while tidying his office, Becquerel had placed a sample of pitchblende (a rock related to uranium mining) on some photosensitive glass plates that were inside a cabinet. Days later, upon retrieving the rock sample, he noticed that the photosensitive plates had been activated where the sample had been sitting. The rock was apparently emitting energy. He thus set Marie and Pierre to figure out the nature of the energy in the rock (Mullner, 1999).

Their laboratory was not much more than a large garage with holes in the roof and little heat. Still, they processed tons of pitchblende and were able to isolate fractions of a gram of both polonium and radium. Marie had named the former after her home country of Poland, and the latter for its ability to radiate energy. Pierre kept some of this first sample of radium in a small glass vial in his shirt's breast pocket. At the end of the day, he noticed his skin, just opposite the vile was blistered and inflamed, which led him to test the energy emissions of the sample. He taped the vile to his upper arm for varying lengths of time and observed progressively more skin reaction as the time increased (Mullner, 1999). Not long after this, doctors developed a way to focus the energies emitted from this strange material and used it to destroy cancerous tissue and eradicate tumours (Lederman, 1981). This usage of radium helped shape the public perception of the element as a "health giving" material, as opposed to a "tissue destroying" material.

Becquerel and the Curies shared the Nobel Prize in physics in 1903 for their discovery of these radioactive elements. Marie would eventually win a second Nobel Prize in Chemistry in 1911 with her descriptions of the behaviour of these elements. She accomplished this amazing feat as a single mother raising two daughters, as Pierre was killed in 1906 when he was hit by a truck. Both Becquerel and Curie would eventually die of radiation poisoning, the former in 1908, and the latter in 1934.

As news started to circulate about this material and its seemingly mysterious powers, radium became a very popular ingredient in a surprising array of products that included bottled water, milk, cooking utensils, shoe polish, and makeup, among others. The public saw it as a way to achieve better health. Women would be glowingly beautiful, while men would gain energy and virility. By the early 1920s, radium became the key ingredient in a glow-in-the-dark paint used on watch dials to help in the war effort. The dials could be used during the night without the need of an external light source that could give someone's location away.

The dials were painted by young women, usually aged 16–24. As they were paid per piece painted, speed and precision were the goals. Notably, the women would incidentally ingest small portions of radium, since they repeatedly used their lips to bring their brushes to a point. As radium behaves chemically like calcium, it found its way into the women's bones and teeth, resulting in the development of carcinomas, unhealing wounds, lost teeth, rotting jaws, and finally, death. Once the news got out about the connection between radium and the plight of the dial painters, public opinion about radium immediately shifted from a health-inducing agent to a poison (Mullner, 1999).

How is this story about the discovery of radioactivity relevant to the age of the earth controversy? First, it challenges the reasoning behind Thomson's equations, as it was a source of heat not previously known. In other words, Thomson said that the earth was releasing too much heat to the universe for it to be more than a few 10 s of millions of years old. But this newly understood radioactive phenomena meant that the earth is internally generating heat due to the decay of radium and other radioactive elements, making his assertion no longer valid. We believe today, however, that the concentration of these elements within the earth are insufficient to have really had much influence on the earth's internal temperature. Instead, it seems that a main source of Earth's internal energy is the latent heat of fusion, which is the heat released as the liquid outer core solidifies and becomes part of the solid inner core; this supplements the heat generated during the earth's formation.

The second reason why radioactivity becomes pertinent to this historical trajectory is that ***Ernest Rutherford (1871–1937)*** determined that radioactive decay happens at a constant rate, regardless of any environmental conditions. This means that a measurement of the ratio between parent and daughter elements within a rock sample indicates how long the decay process had been underway after the rock solidified. Rutherford's research would earn him a Nobel Prize as well. It also led to age-dating of different rock samples bearing radioactive elements. In 1907, ***Bertram Boltwood (1870–1927)*** calculated the first age-dates of uranium-bearing rocks, using the uranium-to-lead decay rate. His results revealed rocks of more than a few billion years old, far greater than Thomson's calculations using the laws of thermodynamics (Jackson, 2006). Of course, these methods have been refined, resulting in increases in both precision and accuracy. Our current understanding puts the earth at approximately 4.6 billion years old.

Activity Alert
How Big is a Billion?

Students attempt to make concrete the abstract size of a billion by using everyday object or experiences as analogies. The goal is to have students at least begin to internalize the magnitude of such numbers of a million or a billion of something.

3.4 Developing a Theory of the Earth: Why Are There Still Mountains?

This third and longest narrative (Fig. 3.3) begins, as did the previous two narratives, with efforts to understand the earth. As with the biographies of the other two concepts, the initial biblical explanations change due to empirical observations. This narrative illustrates how many different explanations for the same set of data can exist. As with the previous narratives, advancing technologies helped the science advance. This narrative also showcases how the many subdisciplines in geology, including the burgeoning field of geophysics, each sought to solve diverse problems central to their discipline that then contributed evidence synthesized into a grand theory of the earth. Ultimate acceptance of this grand theory was a struggle for many reasons; this story helps elucidate the scientific process of idea generation, exploration, validation, and acceptance.

3.4.1 Initial Thoughts Concerning the Origin of Continents and Oceans

This narrative begins with a rather straightforward question. Given that weathering and erosion continue to break down the surface of the earth at a rate of about 0.005 mm/year (Protenga & Bierman, 2011), and the earth is well over 4 billion years old —resulting in approximately 2000 km of weathering and erosion — why do mountains remain as part of the landscape? Throughout history many mythologies have sought to explain the existence of mountains, as well as continents and oceans. Ther Sumerian deluge myth from about 1535–1700 BC, Akkadian Gilgamesh myth at about 1634 BC, the Judaic Noachian myth at the sixth century BC, and the Grecian Deucalian myth of the early fifth century all provided stories where water rises and falls over a stable ground and the wind is responsible for the movement of the waters (Şengör, 2003).

A fundamental shift came when ***Aristotle (384–322 BC)*** developed a model that accounted for mountains and oceans by invoking vertical movements of the land. Aristotle lived in Greece, where he experienced a karstic landscape (limestone bedrock that has dissolved into caves and caverns) and volcanic eruptions. He came to

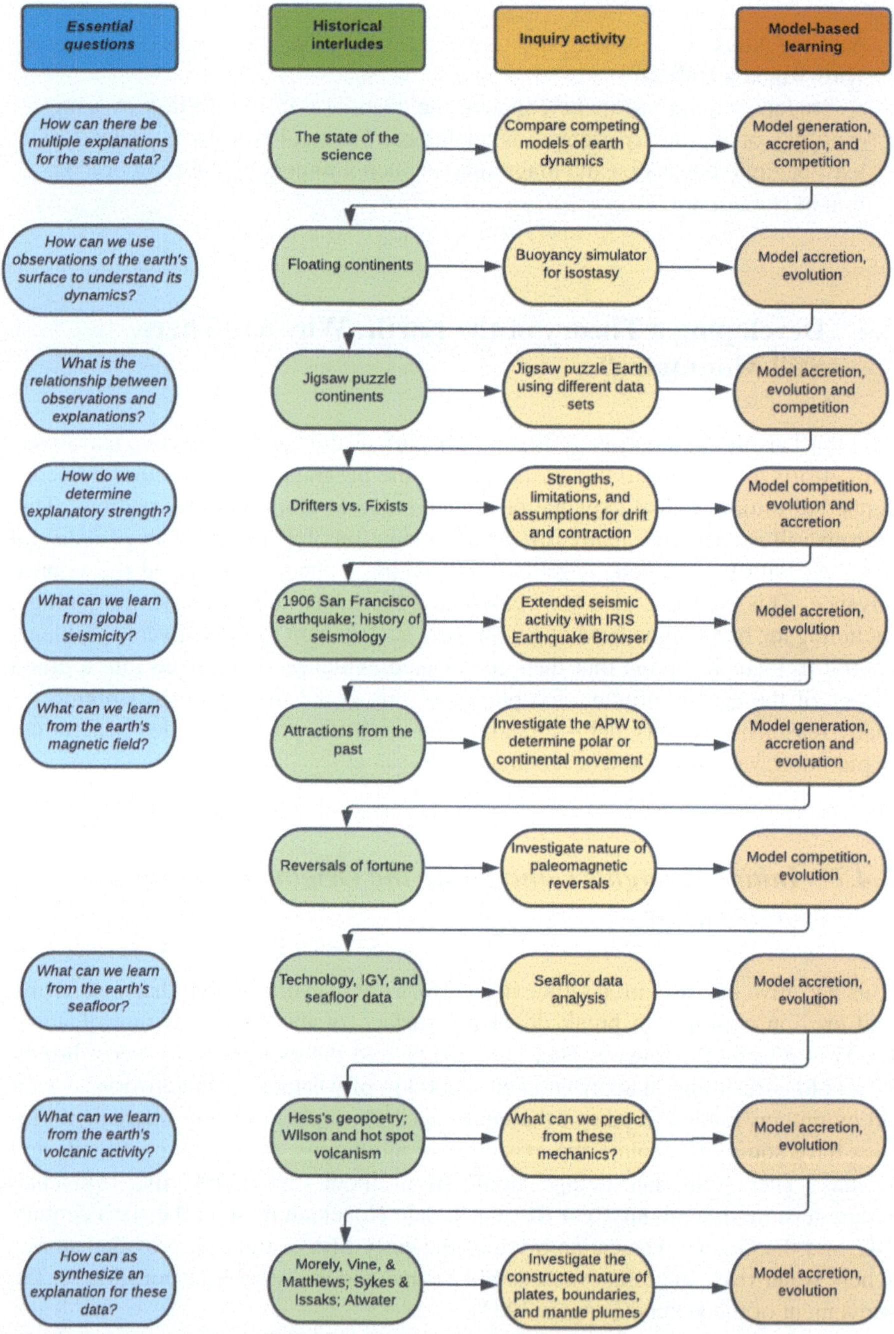

Fig. 3.3 Why are there still mountains?

understand the earth as being porous, with much open space underground to allow for the flow of fluids, whether it be lava, water, or air. Aristotle surmised from his observations of volcanic eruptions and winds blowing in and out of caves that the fluid was dynamic—ebbing and flowing and collecting within or flowing out of underground caverns and galleries—that would cause lowlands to rise into mountains, and highlands to collapse into oceans basins. Along the narrow zones between the rising and sinking lands deformation occurred from the stresses placed on the rocks as they moved. This resulted in narrow zones of earthquakes and volcanoes, mainly along coastlines (Şengör, 2003). Not only did this explain the location of earthquakes and volcanoes on the earth, but it also explained how signs of marine life, such as fossilized shells, could be found high on continental margins.

Aristotle's porous earth model was so powerful that it lasted into the nineteenth century and was the model for Jules Verne's (1874) famous novel, *Voyage au centre de la Terre* (*Journey to the Centre of the Earth*), where characters used passageways to travel to and explore the earth's interior. However, new ideas about the earth and its dynamics eventually emerged. James Hutton imagined the earth's interior as molten (Plutonism): its flux caused forces that deformed the rocks into mountains by causing motions in a vertical direction (part of his wasting, deposition, and deformation cycles), consistent with Aristotle's theory (Hutton, 1788).

This Plutonic view was related to the nebular hypothesis of Earth's formation—Emanuel Kant's proposal that swirling gases (nebulae) collapse due to gravity into planets and stars. According to this idea, as the earth coalesced from space debris, the gravitational collapse of material caused it to heat up and melt. By the late nineteenth century, two famous geological contemporaries used the earth's internal heat to explain the existence of continents and oceans. The first was ***Eduard Suess (1831–1914),*** an Austrian completing his fieldwork in the Alps, who imagined a molten Earth that began to cool, causing a homogenous and brittle crust to form on the earth. As the earth continued to cool, it contracted, causing a "wrinkling" of the brittle crust into uplands and basins. Over time and more thermal contraction, the uplands could eventually sink into ocean basins and the basins could be pushed up into mountains. The crumpling of the crust would also cause earthquakes and volcanoes (Suess et al., 1904).

The second geologist was ***James Dana (1813–1895),*** an American. He did much of his work in the Appalachian Mountains along the eastern seaboard of the United States. His observations of the fold and thrust belts of the Appalachians led him also to suppose a thermally contracting Earth beneath a rigid outer crust. The shrinking caused the crust to wrinkle, as in Suess's hypothesis. Suess's and Dana's perspectives differed mainly regarding the nature of the brittle outer crust. Suess assumed a homogeneous crust that would allow mountains and basins to form anywhere, as well as allow basins to become uplands and vice versa; in contrast, Dana supposed that as the crust began to solidify, the less dense rocks solidified first (becoming the continents) and more dense crust solidified later, becoming the ocean basins. As the earth continued to cool, the mountains would be pushed higher and the basins dropped deeper. The implication here is that continents will remain continents and ocean basins will remain ocean basins (Dana, 1847). This became the model of "permanentism."

At issue with Dana's vision, compared to Suess's, was the notion that if continents were permanent, marine fossils in high mountain regions were hard to explain. Both explanations had yet another problem. Those collecting fossils found evidence of the same species on continents separated by vast oceans—referred to as biological disjuncts. If continents were permanently separated, how could fossils of the same organism on different continents be explained? Based on Darwinian evolution, the mechanism of randomness would deem exact co-evolution impossible. To account for the occurrence of biologic disjuncts, geologists invoked land bridges (Schuchert, 1932) or isthmian links (Willis, 1932) between continents—areas of crust extended between continents allowing organisms to travel from one continent to the next. Subsequent submergence of the bridge or link would sever the connection, isolating the two continents.

Decades later, in 1911, a German meteorologist, ***Alfred Wegener (1880–1930)***, was paying a Christmas visit to his colleague, Emil Take (1879–1925), at the Marburg Physical Institute. Take invited him into his office to view a Christmas present, a large-format world atlas. While perusing it pages, Wegener noted the apparent "fit" of the east coast of South America with the west coast of Africa, especially when considering the edges at the continental margins. We know of this moment of observation because Wegener later wrote of this in a letter to his wife. This observation was notable enough that he was compelled to pursue its possible meaning (Greene, 2015). Within a year, Wegener presented a paradigm-shifting idea, which was published in a 50-page paper.

Within a couple years, he wrote an entire book on his innovative idea that the continents could break apart and move, drifting like icebergs, through ocean crust. Movement through the oceanic crust would cause deformation, including earthquakes and volcanism, along the leading edges of the continents. Wegener's idea was revolutionary because it relied on horizontal displacement as opposed to vertical motions of the crust, which had previously been the framework. Using vast reviews of the literature in geology, paleontology, and paleoclimatology, Wegener put forth many lines of evidence to support previous connections of the continents, similar to Schuchert's land bridges. However, instead of the bridge falling away into the earth, the continents split and spread apart (Wegener & Skerl, 1924).

In addition to the evidence of biological disjuncts, Wegener also identified geological disjuncts. These were places on parallel sides of separate continents where geologic features, formations, and structures, such as older mountain chains, would connect quite seamlessly if one imagined the continents being brought back together. Additionally, Wegener found paleoclimatic evidence. General acknowledgement of the most recent ice ages was still fresh, but evidence for ice ages in the distant past was also accumulating, located on multiple large continents (Africa, India, South America, and Australia). Bringing the continents together into one super continent, a continent Wegener referred to as *Pangaea* (whole earth), connected the glacial deposits together. Furthermore, some deposits that signified tropical climates, such as coal, had been located on Antarctica, situated at the same geologic time as these glacial deposits existed in many locations now deemed tropical.

By positing the movement of the continents around the face of the earth, Wegener was able to explain how such records of the past could make sense. Lastly, the matter of isostasy, crustal buoyancy on the mantle, further bolstered his radical proposition. Having a strong background in physics, Wegener was familiar with crustal buoyancy (Airy, 1855; Pratt, 1855), and found that the notion of rising and falling land bridges did not fit with the physics attributed to the earth's interior.

This segment has detailed four distinct models of earth dynamics–Aristotle's porous Earth, the two different contraction models by Suess and Dana, and horizontal displacement of the continents by Wegener. Each one used basically the same observations and data—earthquakes, volcanoes, other types of deformation, biological, geological and climatic disjuncts—yet each posed a different explanation for mountains and ocean basins.

Activity Alert
The Models of Earth Dynamics

Students compare the four models of earth dynamics, make predictions, propose tests, and judge the most reliable of the models. See Chap. 4, Sect. 4.5 for a full description. This activity gives students the experience of looking at differing explanations for the same data, identifying assumptions, and rationalizing which explanation they think is best.

3.4.2 Drifters versus Fixists

Wegener published multiple editions of his book in German, but his idea received only lukewarm reception. By 1920, however, he published it in English, which led to some initially harsh criticism. Despite the amount and strength of his evidence for horizontally displacing continents, many geologists felt Wegener was too theoretical in his analyses. Specifically, he conceived the idea of displacing continents and then went looking for evidence to support this phenomenon. This approach went directly against the methodology of multiple working hypotheses (Chamberlin, 1965), where the geologist might have several possible explanations but looked at the evidence without bias and decided on the best or most likely possibility. This was the same kind of argument used against William Smith when he tried to use fossils to correlate different strata across England.

Wegener's theory of continental drift is now a common story in almost all introductory geological textbooks. Though some details vary from book to book, the general story is quite consistent. Wegener was a lone genius who had an epiphany about continental drift upon seeing the map of African and South American coastlines, a classic case of "myth-conception" (Allchin, 2003). The story goes that he published his ideas and was immediately derided by the rest of the geologic community, mainly because he could not come up with a workable mechanism to drive continental motion. However, Wegener's main motivation for publishing his idea was not simply declaring that continental drift was occurring; rather, he saw this interpretation as a *starting point* for inquiry. He was interested in substantiating that drift could be happening before trying to figure out what was causing it (Greene, 2015).

The resistance of geologists to the notion of continental drift because no valid mechanism had yet been identified was, however, inconsistent with their acceptance of other phenomena—such as ice ages, gravity, and the earth's magnetic field (Oreskes, 1999)—which similarly lacked an identified causal mechanism. In addition, ***Arthur Holmes (1890–1965)*** had proposed a viable mechanism, substratum mantle convection (Holmes, 1928) that, according to the physicists he asked, could move continents. Still, those same physicists said mantle convection was not happening, perhaps because regionalism played a large role in who accepted and did not accept Wegener's hypothesis (Frankel, 2012d).

Many geologists were interested only in the regions where they conducted their research, and thus did not think globally. Of course, Wegener's idea *was* a global hypothesis. Interestingly, geologists from South Africa, South America, and India were far more sympathetic to continental drift, as the rocks in these areas had much stronger evidence for continental connections compared to North America and Europe. However, since North American and European geologists had most influence over the direction of geology, including what constituted important ideas, the research underway in regions south of the equator had little impact on most geologists' rejection of Wegener's hypothesis.

Returning to Arthur Holmes, he tried to map potential convection currents based on how they would manifest as mountains and troughs at the surface, thus building support for continental drift. Though he was highly regarded as a geologist, his idea in this case did not garner much traction. Nevertheless, a few other notable geologists did support continental drift and published work that fortified the idea of large-scale continental connections and subsequent separation. One of these individuals was ***Alexander du Toit (1878–1948)***, a South African with an enormous capacity for doing field research. He spent multiple years mapping along the east coast of South America to find that its geological aspect matched very well with those on the west coast of Africa that had already been mapped (Frankel, 2012d). Not only could he correlate strata, ore deposits, and depositional environments across the Atlantic Ocean, but he also traced mountains belts, fold axes, and fossil assemblages (du Toit, 1937).

Another South African, ***Edna Plumstead (1903–1989)***, produced groundbreaking evidence for the connection of the continents, as one prior super continent. A noted paleontologist in South Africa, she was asked by a tile company to investigate when they had quarried some rocks with fossil leaves in them. The fossil was of a tree called Glossopteris. Plumstead researched the species and found it to be quite common in a particular age of rock, one that showed up in Africa, India, Australia, Antarctica, and parts of eastern Europe and Russia. Having the same tree species all over the world, across broad oceans, did not make sense with then prevailing ideas about Earth's crust. Being familiar with Wegener's and du Toit's previous work, she argued a case for continental drift based on the Glossopteris occurrence (Plumstead, 1961)

3.4.3 The Role of Geophysics to Understand Earth's Dynamics

A subtle but repeated theme in these three narratives is the relationship between scientific and technological advancement. In the context of this last narrative, the advancement of geophysics as a science, as well as its associated technologies (seismologic, oceanographic, geomagnetic, and so forth) has generated vast amounts of data; this, in turn resulted in new understandings of the earth and its dynamics discussed below.

3.4.3.1 Seismology

In 1755, a large magnitude earthquake (8.5–9 Mw, or Moment magnitude) shook Lisbon, Portugal. Since candles and oil lamps were the only source of artificial lighting in those days, this shaking caused many fires in the city. The earthquake's epicentre was a few 100 km off the coast of Portugal, and thus triggered a massive tsunami that flooded Lisbon approximately 45 minutes after the quake. The earthquake, and subsequent tsunami and fires destroyed most of the city of Lisbon and killed some 75,000 people.

Philosopher ***Immanuel Kant (1724–1804)*** took an interest in the earthquake, especially with respect to the raw destruction that it caused, as well as its minor impacts observed elsewhere in Europe. His analysis of the earthquake and its damage led to a shift in his philosophy: he turned away from a belief in a God who controls all actions on Earth, to a mechanistic philosophy, where nature behaved according to natural laws that God had designed. Within this worldview, studying nature is what would help humans understand, predict, and be able to mitigate destructive forces. This orientation would become one of the biggest contributions to the modernization of Europe during the Enlightenment (Larsen, 2006). Kant's explanation for the earthquake was reminiscent of the Aristotelian porous earth model. Kant asserted that God did not shake the ground, destroy the city, and kill the people, but that hot fluids travelled through underground caverns, causing the ground above to shift and shake. Kant's use of empirical data to build and then support an explanation for earthquakes would be the birth of seismology.

About 100 years later (1857), another destructive earthquake occurred, this time in Naples, Italy. This earthquake was extremely destructive, estimated at 7.0 Mw, causing about 11,000 deaths. An Irish civil engineer, ***Robert Mallet (1810–1881)*** was already doing studies of seismicity when this earthquake happened and was in the process of creating a map from records in the literature of global earthquake occurrences (Mallet, 1858). He was also conducting studies using early seismometers and explosives detonated in the ground to measure the speed of seismic waves through the ground. He used the results of these investigations to develop the mathematics of wave travel through the ground. Given all these activities, many see Mallett as the founder of modern seismology.

With the Naples earthquake, Mallet turned his attentions to Naples and surrounding areas to understand the phenomenon. He used hundreds of photos, both monoscopic and stereoscopic, to document the damage, interpreting such data to tell the direction and strength of the seismic waves involved in the event. He eventually published a report of the event, *Great Neapolitan Earthquake of 1857: The First Principles of Observational Seismology* (Mallet, 1862). To estimate the location and depth of the earthquake epicentre, he used the alignment of cracks in houses affected by the earthquake and the alignment of fallen statues and chimneys, as well as his measurements of travel times of seismic waves (Ferrari & McConnell, 2005). Another massive earthquake, about 50 years later, was instrumental for seismology to better understand how the earth works.

That famous earthquake unfolded in the very early hours of 1906, when the ground shook violently for just under a minute along a 270-mile stretch of southern California, travelling through the city of San Francisco. The city and surrounding area had been growing tremendously in the years leading up to this event as the promise of gold brought people west in droves. At the time of the earthquake, up to 25% of the country's population was living west of the Rocky Mountains. San Francisco rivalled New York City in terms of trade and was known as the "Jewel of the West," and the "Paris of the Pacific." The Wells-Fargo bank began during this time with trade in gold. As well, German immigrant Levi Strauss invented blue jeans, seeing the need for durable clothing for gold prospectors (Dalessandro, 2006).

All came crashing down in less than a minute when the earthquake occurred, estimated at 7.8 Mw. The earth's crust broke, and buildings collapsed from the shaking. Fires started due to broken gas lines that suppled light to the city, complicated further as two of three water mains broke, leaving the fire companies without a way to battle the blazes. Fire departments and military personnel used explosives to try to stem the spread of the fires. In all, about 80% of the city was levelled and 3000 people died. The cost was estimated at $400 million USD (value in 1906), a figure equivalent to the then budget for the entire United States.

George Pardee (1857–1941), the governor of California in 1906, commissioned a group of scientists to document the damage and develop an explanation for the earthquake. The aim was to diminish the threat of such destruction in the future. One person on the commission, the mathematician and glaciologist ***Harry Fielding Reid (1855–1944)***, led part of the commission to attempt to develop a mechanism for the earthquake. He used geographic work performed in the 1850s and 1890s, and finally again in 1906 after the big quake, which surveyed the land in a straight line across the faulted area Fig. 3.4.

Reid was familiar with using such lines of survey-markers, as he often used them to track the displacement of glacial ice over time. He noticed that survey-markers were originally emplaced in a single straight line in 1860, running east to west across this area of deformation. When the markers were revisited in the 1890s, the

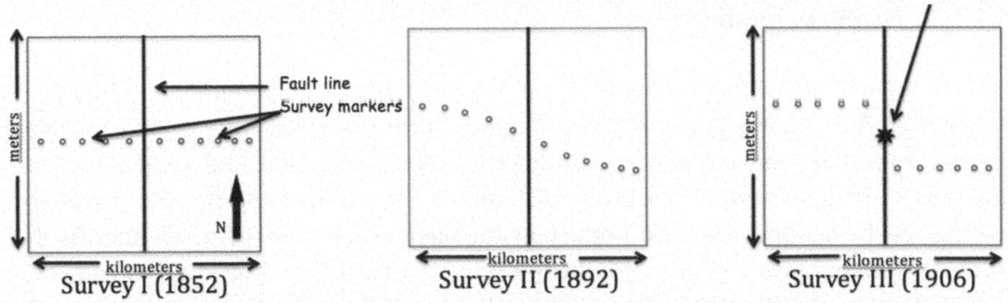

Fig. 3.4 Effects of crustal motion on a set of survey markers over time. Markers were placed in a straight line in 1852. By 1892, the survey markers indicate deformation in the crust. After the 1906 earthquake, the deformation was no longer visible, only the relative motion across the fault line

line of markers was no longer straight; instead, they showed deformation, as if the western side of the line was moving north and the eastern part of the line was moving south. The greatest amount of deformation was visible in the central portion of the line, where the impending break would eventually take place. After the earthquake, the survey-marker lines were again straight: they were broken across the fault line, with the western half of the line having moved about 27 feet north, relative to the eastern section of the line.

Reid took this to mean that the two sections of the crust (east and west of the fault line) had been in constant, but very slow motion and that the crust had deformed to accommodate the displacement. However, eventually the crust could not withstand the deformation and broke. The crust then returned to its original configuration *elastically* (straightening out the now displaced survey lines). To make sense of this data, Reid developed the explanation of elastic rebound as the mechanism causing earthquakes. The breakage, allowing for the release of stored up elastic potential energy, is what made the earthquake destructive (Reid, 1910). With the development of an accurate seismometer and the synchronization of seismic stations across the globe in conjunction with the International Geophysical Year project (1957–1958) and the Nuclear Test Ban Treaty (1963), seismologists gathered data they would use to develop understandings about the dynamics and the structure of the earth.

Activity Alert
Extended Seismic Activity

Using the IRIS (Incorporated Research Institutions for Seismology) Earthquake Browser, students develop a testable question and collect data from the browser to answer their question. This activity familiarizes student with a model of global seismicity and has them develop and test a question of their choice. See Chap. 4, Sect. 4.2 for a full description.

3.4.3.2 Seafloor Exploration

During the middle of the nineteenth century, transatlantic communication began to be popular. A profitable enterprise was manufacturing and installing submarine telegraph cables that connected Europe with Africa, North America, and South America. Unknown incidents sometimes broke the connection. Understanding what was happening on the seafloor became important for business, but also for science. At the time, scientists knew far more about the universe than they knew about our own planet. To remedy this gap in knowledge, ships travelled the oceans collecting data to map the seafloor. However, they had methodological challenges: the only way the ocean depth could be measured in those days was by dropping a weighted rope into the ocean, letting it sink until it hit the seafloor, and then measuring the length of rope as it was pulled back up.

This method had many limitations. First, it was extraordinarily time consuming, as a full day might be required to get one measurement. Second, knowing precisely where the boat was located when making the measurement was somewhat challenging. Finally, the underwater current could affect whether the weighted line dropped straight to the bottom; if it was pushed by a current, making an angle to the vertical, this would result in registering a deeper measurement (Höhler, 2003). Eventually, however, new technologies new technologies that greatly assisted in efforts to map the ocean floor came into use.

For near-shore explorations, ***Maurice "Doc" Ewing (1906–1974)*** interpreted seismic waves produced by explosives to describe the continental shelf along the eastern seaboard of North America. Ewing was also responsible for helping develop sound navigation and ranging (SONAR) technology to collect seafloor data, which was originally used in WWI and WWII to detect submarines. This technology meant that ships could finally transverse the Atlantic Ocean and take continuous depth measurements to the seafloor. Ewing eventually hired ***Marie Tharp (1920–2006)*** to synthesize the transects into a three-dimensional drawing of the Atlantic seafloor (Frankel, 2012b). She formalized an anatomy of the seafloor (continental shelf, continental slope, continental rise, and abyssal plain). From her data, Tharp drew a map of the seafloor of the North Atlantic Ocean. She also noted a long and narrow mountain range traversing the centre of the Atlantic Ocean, with a deep rift valley running along its centre.

This rift was a significant discovery, though immediately downplayed by Ewing because it was too suggestive of drift, and Ewing was an ardent fixist. Tharp eventually placed Atlantic Ocean seismicity data over her map of the seafloor and found that earthquake epicentres coincided with the rift perfectly. However, when she put forward these data as support for continental drift, her colleague at the Lamont-Doherty Geological Observatory, ***Bruce Heezen (1924–1977),*** dismissed them with the sexist characterization that this was just "girl talk" (Lawrence, 2002). Heezen then was also a fixist, though his understanding of earth dynamics would change with the accumulation of seafloor data collected through the 1950s and into the early 1960s.

Seafloor exploration began for commercial purposes, enhancing transatlantic communications, but it switched to defence purposes during WWI and WWII. Eventually, by the 1950s, understanding the seafloor also became an object of study for pure science. During this span of time, and especially with the International Geophysical Year (1957–1958), researchers focused on determining the nature of the seafloor primarily for the sake of understanding the earth better. The data they accumulated encompassed the geochronology of the seafloor, heat flow, sediment thickness, bathymetry, and even drill cores of rock and sediment.

These data often yielded unexpected results. For instance, radioactive age-dates for seafloor core samples, which were calculated routinely by the 1960s, showed that the earth's oceans ranged in age from modern day, to only 200 million years old. This was troubling because the earth's continents had samples of rock that were more than an order of magnitude older—that is, four billion years old. If the entire crust solidified at approximately at the same time, then why is the seafloor so much younger than the continents? In addition, the crustal age was youngest at the ridges and got symmetrically older away from the ridge.

Observations that sediment was thickest at the margins of the ocean basins and decreased to nothing towards the centre of the oceans and the mid-oceanic ridges was also puzzling. This is because the sediment thickness should be thickest along the coasts, since that is where the eroding continental sediment deposits from rivers. Researchers certainly did not expect to find that it decreases to zero thickness towards the centre of the ocean basins. If the seafloors formed all at the same time so long ago, why is sediment not found in the centre of the ocean basins? Other data showed that the crust towards the ridges is hotter compared to crust that is distant from the ridges; furthermore, some of the deepest parts of the oceans are not in the centre of the basin, but along the margins of some oceans. Observations also show an extensive network of extremely rugged mountain ridges, not just through the mid-Atlantic Ocean, but throughout the central portions of all the world's oceans.

Activity Alert
Seafloor Data Activity

Students get maps of various geophysical data sets and try to discern patterns within each map and then between maps. They try to apply one of the models of earth dynamics they are familiar with or develop their own model to explain the data. In this activity, students make descriptive claims based on the data provided by each map as well as making connections between maps. Students then make explanatory claims, a model to explain their descriptive claims. See Sect. 4.1.3 for a full description.

These observations raised many questions, which resulted in diverse but related answers. In 1960, ***Harry Hess (1906–1969)*** proposed a "verse in geopoetry" as he called it, because of its fanciful nature. In particular, he proposed convection currents within the mantle. Though he likely read about convection in Holmes's publications, he did not cite Holmes's work (Frankel, 2012a). Hess envisioned the mid-oceanic ridges as spreading centres, wherever the convection currents rose to the crust and spread apart. The crust would be very young, hot from just forming, and sitting up higher on the mantle due to its lower density. Constant spreading at the ridges would result in long, narrow zones of earthquakes and volcanoes at the rifts within these mid-oceanic mountain ranges. However, the new crust would, at some point, need to sink back into the mantle so that the earth did not continue to expand. Hess proposed that this happened where the convection currents sank into the mantle, which he identified as deep-sea trenches (Hess, 1962). This idea would eventually be coined "seafloor spreading" (Dietz, 1961).

Another possible explanation came from ***Samuel Warren Carey (1911–2002)***. He was one of the strongest proponents of a new hypothesis of earth dynamics: the expanding Earth (Carey, 1976). Originally an early adopter of Wegener's idea of horizontal displacement of continents, Carey saw this new data from the seafloor as indicating about 250 million years ago the earth was about two-thirds its current size and then began growing. Where the original crust split open would be the sites of new and relatively young ocean basins. Carey explained that the ocean basins were created by breaking at the ridges and splitting apart. New seafloor would form at the breakage and move in either direction, farther away from the ridge.

3.4.3.3 Geomagnetism

Contemporaneous with the later developments of seismology and seafloor exploration, another branch of geophysics was coming into its own: geomagnetism. In response to war tactics in WWII, allied forces used ships dragging magnetometers behind them to try to detect German submarines that were interrupting maritime shipping across the Atlantic Ocean. In some cases, these vessels produced data that were hard to interpret as the magnetometers picked up seemingly residual magnetism in the rocks making up the seafloor. What the magnetometers were actually reading were iron-bearing basalt rocks that displayed parallel to sub-parallel stripes (like zebra stripes) of alternating magnetic polarity.

Half of the rocks were polarized parallel to the current magnetic field of the earth and the other half seemed to express a polarity 180 degrees away from that current magnetic field—one that was exactly opposite. Physicist ***Patrick Maynard Stuart Blackett (1897–1974)*** at Manchester College, England, was already studying the phenomenon of the earth's magnetic field with his student, ***Keith Runcorn (1924–1995)***. Runcorn would graduate and move onto Cambridge University, where he continued these investigations with his colleague, ***Edward Bullard (1907–1980)***. These two research groups, along with a group at the Carnegie Institute in the US, would substantially develop the field of geomagnetism and paleomagnetism (Frankel, 2012c).

Aiming to better understand the nature of the earth's current magnetic field, Blackett, Runcorn, and Bullard began studies of the magnetic signatures in older rocks around England. They found that they could get paleomagnetic signals from many kinds of rocks, but the strongest and most consistent ones came from iron-bearing basalts and red—thus, iron-rich—sandstones. The process of extracting samples and measuring magnetic fields was tedious and required the development of more sensitive magnetometers than were in then existence. Eventually, they found that by placing thin slices of core samples within a very sensitive magnetometer they could get consistent results.

As they analyzed more and more samples, issues emerged that demanded resolution. Samples taken from folded strata had to be corrected for the folding. Also, they observed evidence of overprinting of multiple magnetic signals on the same sample. The challenge was discerning which was the original magnetic signal out of multiple such signals. Lastly, some samples yielded readings 180 degrees different ("reversed") from what was expected. Investigators were not sure if this was due to the extraction process or some other chemical phenomenon. Oftentimes, for the reversed signals, researchers just flipped them by 180 degrees, thinking the switch was an artifact of the sampling process (Oreskes, 1999). As researchers agreed that the signals were a paleomagnetic record, the Carnegie group began to use math to mitigate for the folded strata when discerning the paleomagnetic signal. Others worked on ways of "cleaning" samples of subsequent magnetic signals overprinted on the original (Frankel, 2012c).

These mitigation strategies were employed for multiple years, and the paleomagnetics community was very comfortable with the reliability of the data they were producing. However, for those not following the years of literature that explained this branch of geophysics—which was largely almost all other geologists—this was all problematic. Given their poor understanding of geophysics and geomagnetism, they gave little credence to these results.

Significantly, Blackett's and Bullard's teams found early on that rocks of different ages from around Britain showed signs that the north magnetic pole had shifted over time. They referred to it as the Apparent Polar Wander (APW) path. They did not know if this meant (a) the magnetic field had indeed shifted over time, or (b) the field stayed consistent, and it was the continents moving over top that changed with time. If the explanation was the latter, it would be an incredible piece of evidence in support of Wegener's hypothesis of horizontal displacement of the continents. To solve the riddle, they began collecting data from multiple places on Earth (North America, India, Japan, Australia, and Europe) to compare APW paths from one location to the next. If the paths all aligned on one map, this would signify that the continents were stable, and the field was in flux. If the APW paths diverged, then this would be evidence for the notion of moving continents.

The superimposed APW paths showed a divergence with age from modern times to the distant past. This discovery should have ended the debate of permanentism versus drift (Frankel, 2012c), but it did not. As noted earlier, many geologists were highly skeptical of the results because of the problems in the data—specifically, the multiple signals, deformation within sample sites, and reversed signals. They were not familiar with the decades-old geomagnetism data that had already addressed all these problems. More data from a part of the earth that scientists knew least about, the seafloor, would be required before most geologists were compelled to change their ideas.

As work on discerning the APW commenced, another group of geophysicists in California investigated the problem of the alternating magnetic signals from the rocks. ***Allan Cox (1926–1987)***, ***Richard Doell (1923–2008)***, and ***Brent Dalrymple (b. 1937)*** attempted to explain the observation of magnetic signals in continental basalts that were 180 degrees different from what was expected. Was it a local phenomenon? Did it have anything to do with the sampling process? Could it be a more global phenomenon? As the magnetometers became more sensitive and radiometric dating of rock samples became more precise, scientists discerned a pattern in the samples through time. When a particular age of rocks showed normal magnetization (aligned with the projected north magnetic pole), all rocks of that age, globally, were normally magnetized. When a particular sample of rocks showed a reversed magnetic signal (magnetic north in the sample points towards the projected magnetic south pole during that age), they were also reversed globally. It was not a local phenomenon, nor was it an artifact of sampling.

The only possible explanation for the observations was that the earth's magnetic field periodically reversed, and after some time then switched back. Because no one at the time had a good explanation for the cause of the magnetic field, an explanation for why it would reverse itself over time was even more remote. However, this did not prohibit researchers from developing a magnetic reversals timeline for the earth, based on an increasing number of samples and further advancements in radiometric age-dating (Glen, 1982, 2002). Cox and his group had documented at least 10 reversals in the past three million years, with stable periods from 10 s of thousands of years to 100 s of thousands of years long. However, the alternating magnetic signals were not just in continental basalts. Drill core samples of thick sediment packages along continental coastlines also maintained the same alternating signals with depth, with the reversals happening at the same time as those in the continental basalts.

Another part to this puzzle remained, however. During and after WWII, ships were dragging magnetometers across the ocean to identify possible threats from submarines. Readings coming from these magnetometers indicated that the rocks on the seafloor also showed reversals of magnetization. This data differed from what had been recorded within the continental basalts. Instead of the reversals taking place in horizontal layers—which was how these continental basalts had been laid down, one lava flow on top of the previous—these submarine readings appeared as stripes, referred to as zebra stripes, of alternating magnetic fields.

Furthermore, the pattern of reversals was symmetric on either side of the mid-oceanic ridges. ***Ron Mason (1916–2009)*** and ***Arthur Raff (no dates available)*** of Scripps Oceanographic Institute in San Diego, California, used magnetometers to explore the seafloor off the coast of the Pacific Northwest. They documented the same zebra stripes and symmetries across ridges in this area as had been described elsewhere (Mason & Raff, 1961; Raff & Mason, 1961). These studies also demonstrated that the pattern of stripes moving away from ridges matched the pattern of reversals created from continental basalts and coastal sediment cores (Frankel,

2012c). This is significant given that the age of these basalts was also increasing symmetrically away from the ridges. These observations seemed to be all connected. But how?

3.4.3.4 Putting the Pieces Together

Lawrence Morley (1920–2013), a Canadian geophysicist doing research in geomagnetism, took an interest in the reversal problem. He was sympathetic to the idea of continental drift, mainly because of his background in geomagnetism and the APW path implications. He noted that the zebra stripes along the seafloor matched the pattern of the magnetic reversals timeline. He was also familiar with Harry Hess's ideas of seafloor spreading. He assembled all the data that collectively indicated new rock formed as the seafloor spread from the rifts; since the newly forming basalt rock forming at the rift would record the earth's ambient magnetic field at that time, it would result in the stripes of reversals parallel to the rift, just as what had been observed. These findings would greatly support seafloor spreading as a mechanism for separating continents.

However, when Morley submitted a paper with his ideas to the journal *Nature*, it was rejected in April of 1963. The reviewers did not think he supported his argument with enough data. Undaunted, he rewrote the paper and resubmitted it to the *Journal of Geophysical Research*. That paper was also rejected because reviewers thought the idea was too far-reaching. Meanwhile, in the interim between these two submissions, other researchers at Cambridge University—Fred Vine (b. 1939), a doctoral student, and his supervisor, Drummond Matthews (1931–1997)—submitted their paper using the same data with the same independent interpretation to *Nature*. This time, the paper was accepted, in September 1963 (Vine & Matthews, 1963). This early version of what now is known as the theory of plate tectonics was referred to as the *Morley, Vine, and Matthews hypothesis* to acknowledge Morley's contributions, as well as those of Vine and Matthews.

Publishing this paper was highly significant, but only in hindsight. A majority of the North American and European geologists had yet to change their stance toward the drift theory. However, this paper gave a new framework for interpreting other anomalous data. ***J. Tuzo Wilson (1908–1993)***, a university of Toronto geologist and supervisor to Lawrence Morley, was a staunch fixist until he saw the paleomagnetic studies. After this, he started looking at geological problems in terms of drifting continents, which aligned with Wegener's original intent that continental drift be a starting point for investigation. Wilson identified volcanic island and seamount chains, like Hawaii and the Emperor Seamounts in the Pacific Ocean, that showed progressive aging in one direction along the chain; he hypothesized that they resulted from horizontal crustal displacement above a stationary plume of extra hot mantle material that burned a hole through the crust, releasing material to create the island chain (Wilson, 1963).

Wilson also used some geometry and the growing understanding of areas of crustal divergence (ridges) and convergence (trenches) to predict a third type of crustal dynamic boundary—a transform boundary. Wilson supposed that these linear entities (ridges and trenches) could not run indefinitely on a spherical planet. Instead, they must get connected by these transform boundaries. He predicted the likely boundary locations and the motion occurring along them, and then used seismic data to substantiate his predictions. Significantly, he was correct (Wilson, 1965). This was really the last piece of information needed to permanently reorient the geological community to a new paradigm, with a new paper outlining the new global tectonics (Isacks et al., 1968).

Up to the late 1960s, all data sources substantiating this new explanation were submarine, making them quite abstract for many geologists since they mostly work on land. The insights of ***Tanya Atwater (b. 1942)*** put the San Andreas fault into the context of the new tectonics (Atwater, 1970). She was investigating the San Andreas Fault, the site of the 1906 San Francisco earthquake, and asserted that it was a transform fault connecting two distant spreading centres—the East Pacific Rise to the south and the Juan de Fuca Ridge to the north. This became the new paradigm for research. Geologists who mostly cared about their small region of study could now put that region into a global context. In 1987, the United States Geological survey used GPS to measure crustal motion on earth, producing data that confirmed earlier predictions, ones founded only on indirect evidence (Frankel, 2012a).

Returning to the opening question of this third narrative, why are there still mountains? The current consensus view is that convection in the mantel drives crustal motions. In particular, a long narrow zone of deformation exists in between two large portions of crust, referred to as tectonic plates, that move relative to each other, whether these are divergent, convergent, or transform movements. Where plates converge, they may bring continents together, in which case subduction does not take place—that is, low density continental crust will not sink into higher density mantle—and the crust thickens and sinks isostatically into the mantle. Just as an iceberg has 90% of its ice under water, mountains have about 80% of their mass underground. The earth's dynamic interior continues to build mountains; once created, they take a long time to wear away. As the tops are removed due to weathering and erosion, the mantle isostatically pushes up the roots of that mountain exposing more rocks from within the earth.

This section has sketched out the diverse lines of inquiry that were crucial to the development of this scientific knowledge. This is to say, many people pursued various problems, but the information coalesced at the right time and in the right way to facilitate a huge shift in understanding how the earth works. Nevertheless, not all questions have been completely answered, since we still only have suppositions about how the mantle behaves and what its possible effects are on the lithosphere. Some geologists continue to question the role of mantle plumes (Anderson & Natland, 2005) on crustal dynamics (Anderson, 2006), suggesting that the plume was such an elegant explanation that it was accepted prematurely and thus propagated too soon (Glen, 2005). As always, the devil is in the details; research groups therefore continue to work out these details.

3.5 Summary

These three narratives highlight the development of the foundational geological knowledge leading to our current consensus view; they achieve this by describing the processes involved in creating the knowledge and explaining how and why that knowledge has changed over time. Each narrative incorporates some of the more important social, economic, and political forces that influenced the trajectory of each concept's development. The historical interludes lead to inquiry activities—six in all. Sometimes these are short-term activities the students write out and hand in. Other times, the activity may be questions that the students discuss in their small groups to bring back to the larger class discussions about the topic.

The critical aspect here is that I am not just presenting material that students must memorize and regurgitate on an exam; rather, I problematize the answers and the process, and present multiple possible answers to the same data at the same time. As a result, students must think about why the answers end up being what they are. They make an intellectual investment in getting the answer by considering a fuller picture of the circumstances surrounding the development of these foundational concepts. The trajectory also follows from direct observations to more theoretical approaches. Students can build their mental models in this same way: importantly, they start with the concrete and descriptive aspects of a concept and then move to more abstract and explanatory aspects (Dolphin & Benoit, 2016), which supports learning. This reflects that explanation is a far more complex way of thinking (Thagard, 2012).

Ideally, readers can follow the schematic I've presented in Figs. 3.1, 3.2, and 3.3 to summarize the narratives I presented or build their own versions of relevant biographies, tracking through essential questions, historical interludes, inquiry activity, and model-based learning. Depending on the subject matter being pursued, the instructional designer may need to identify and develop other narrative trajectories. This could mean different stories all together, or the same stories that have different emphases. The key is to ensure that whatever narratives are created, they yield the essential concepts in the field of study. The story should highlight the development of the target concept, including missteps and disagreements, giving evidentiary and contextual rationale for changes in the model.

The following chapter will detail the other strategies involved in teaching this course, beyond the interrupted case studies presented here. I developed GLGY 209 from scratch, my more than 10 years of previous teaching experience afforded me the opportunity to develop these strategies in small parts here and there. Immediately making all these changes is not necessary. Rather, draw from what is presented to do what is possible and manageable. Then, as part of a reflective approach, modify what needs to be changed, drop what is not working, and add more when and where feasible.

References

Airy, G. B. (1855). On the computation of the effect of the attraction of mountain-masses, as disturbing the apparent astronomical latitude of stations in geodetic surveys. *Philosophical Transactions of the Royal Society of London, 145*, 101–104.

Allchin, D. (2003). Scientific myth-conceptions. *Science Education, 87*(3), 329–351.

Allchin, D. (2017). Historical inquiry cases for nature of science learning. *Cadernos De História Da Ciência, 13*(2), 101–126.

Allchin, D. (2020). Historical inquiry cases for teaching the nature of science and analytical skills. In W. F. McComas (Ed.), *Nature of science in science instruction: Rationales and strategies* (pp. 595–606). Springer.

Anderson, D. (2006). Plate tectonics; the general theory: Complex earth is simpler than you think. In C. A. Manduca & D. W. Mogk (Eds.), *Earth and mind: How geologists think and learn about the earth* (*Special paper 413*) (Vol. 413, pp. 29–38). Geological Society of America.

Anderson, D., & Natland, J. (2005). A brief history of the plume hypothesis and its competitors: Concept and controversy. In G. Foulger, J. Natland, D. Presnall, & D. Anderson (Eds.), *Plates, plumes, and paradigms* (Special paper 388) (pp. 119–146). Geological Society of America.

Atwater, T. (1970). Implications for plate tectonics for the Cenozoic tectonic evolution of western North America. *Geological Society of America Bulletin, 81*(12), 3513–3536.

Carey, S. W. (1976). *The expanding earth*. Elsevier Scientific Pub. Co.

Chamberlin, T. C. (1965). The method of multiple working hypotheses (reprint of the 1890 version). *Science, 148*, 754–759.

Cutler, A. (2003). *The seashell on the mountaintop: A story of science, sainthood, and the humble genius who discovered a new history of the earth*. EP Dutton.

Dalessandro, J. (2006). *The damnedest, finest ruins – Episode 601* J. Dalessandro & R. S. Burton; KQED.org. https://www.youtube.com/watch?v=mvve8YiYGSo

Dana, J. D. (1847). Geological results of the earth's contraction in consequence of cooling. *American Journal of Science, 53*, 176–188.

Darwin, C., & Kebler, L. (1859). *On the origin of species by means of natural selection, or, The preservation of favoured races in the struggle for life*. John Murray.

De la Beche, H. (1830). *Duria Antiquior* [Watercolour].

Dietz, R. S. (1961). Continent and ocean basin evolution by spreading of the sea floor. *Nature (London), 190*, 854–857.

Dolphin, G., & Benoit, W. (2016). Students' mental model development during historically contextualized inquiry: How the 'tectonic plate' metaphor impeded the process. *International Journal of Science Education, 38*(2), 276–297.

du Toit, A. L. (1937). *Our wandering continents: An hypothesis of continental drifting*. Oliver and Boyd.

Ferrari, G., & McConnell, A. (2005). Robert Mallet and the 'Great Neapolitan Earthquake' of 1857. *Notes and Records of the Royal Society of London, 59*(1), 45–64.

Frankel, H. L. (2012a). *Evolution into plate tectonics* (Vol. IV). Cambridge University Press.

Frankel, H. L. (2012b). *Introduction of seafloor spreading* (Vol. III). Cambridge University Press.

Frankel, H. L. (2012c). *Paleomagnetism and confirmation of drift* (Vol. II). Cambridge University Press.

Frankel, H. L. (2012d). *Wegener and the early debat* (Vol. I). Cambridge University Press.

Geikie, J. (2012). *The great ice age and its relation to the antiquity of man*. Cambridge University Press. https://doi.org/10.1017/CBO9781139236560

Glen, W. (1982). *The road to Jaramillo: Critical years of the revolution in earth science*. Stanford University Press.

Glen, W. (2002). A triptych to serendip: Prematurity and resistance to discovery in the earth sciences. In E. B. Hook (Ed.), *Prematurity in scientific discovery*.

Glen, W. (2005). The origins and early trajectory of the mantle plume quasi paradigm. In G. Foulger, J. Natland, D. Presnall, & D. Anderson (Eds.), *Plates, plumes, and paradigms* (Special paper 388) (pp. 91–118). Geological Society of America.

Greene, M. T. (2015). *Alfred Wegener: Science, exploration and the theory of continental drift*. Johns Hopkins University Press.

Hess, H. (1962). History of ocean basins. In A. Engle, H. James, & B. Leonard (Eds.), *Petrologic studies: A volume to honor A. F. Buddington* (pp. 599–620). Geological Society of America.

Höhler, S. (2003). A sound survey: The technological perception of ocean depth, 1850–1930. In M. Hård, A. Lösch, & D. Verdicchio (Eds.), *Transforming spaces: The topological turn in technology studies* (pp. 1–17).

Holmes, A. (1928). Radioactivity and earth movements. *Transactions of the Geological Society of Glasgow, 18*, 559–606.

Hutton, J. (1788). Theory of the earth. *Transactions of the Royal Society of Edinburgh, 1*(2), 209–304.

Indurkhya, B. (1992). *Metaphor and cognition: An interactionist approach* (Vol. 13). Springer.

Isacks, B., Oliver, J., & Sykes, L. R. (1968). Seismology and the new global tectonics. *Journal of Geophysical Research, 73*(18), 5855–5899.

Jackson, P. W. (2006). *The chronologers' quest: Episodes in the search for the age of the earth*. Cambridge University Press.

Kuhn, T. S. (1993). Metaphor in science. In A. Ortony (Ed.), *Metaphor and thought* (2nd ed., pp. 533–543). Cambridge University Press.

Lakoff, G., & Johnson, M. (1980). *Metaphors we live by*. University of Chicago Press.

Larsen, S. E. (2006). The Lisbon earthquake and the scientific turn in Kant's philosophy. *European Review, 14*(3), 359–367. https://doi.org/10.1017/S1062798706000366

Latour, B. (1987). *Science in action: How to follow scientists and engineers through society*. Harvard University Press.

Lawrence, D. M. (2002). *Upheaval from the abyss: Ocean floor mapping and the earth science revolution*. Rutgers University Press.

Lederman, M. (1981). The early history of radiotherapy: 1895–1939. *International Journal of Radiation Oncology, Biology, Physics, 7*(5), 639–648.

Mallet, R. (1858). *Seismographic map of the world showing the distribution in space as discussed from the British association catalogue, plate 11 in fourth report on the facts of earthquake phenomena*. British Association for the Advancement of Science.

Mallet, R. (1862). Great Neapolitan earthquake of 1857. The first principles of observational seismology as developed in the report to the Royal Society of London of the expedition made by command of the society into the interior of the kingdom of Naples, to investigate the circumstance of the great earthquake of December 1857 (Vol. 1). Royal Society of London.

Mason, R. G., & Raff, A. D. (1961). Magnetic survey off the West Coast of North America, 32 degree N. Latitude to 42 degree N. Latitude. *Bulletin of the Geological Society of America, 72*(8), 1259–1265.

Mullner, R. (1999). *Deadly glow: The radium dial worker tragedy*. American Public Health Association.

Newton, I. (1846). *Newton's principia: The mathematical principles of natural philosophy* (pp. 1642–1727). Daniel Adee.

Oreskes, N. (1999). *The rejection of continental drift: Theory and method in American earth science*. Oxford University Press.

Plumstead, E. (1961). Ancient plants and drifting continents. *South African Journal of Science, 57*, 173–181.

Pratt, J. H. (1855). On the attraction of the Himalaya Mountains, and of the elevated regions beyond them, upon the plumb-line in India. *Philosophical Transactions of the Royal Society of London, 145*, 53–100.

Protenga, E. W., & Bierman, P. R. (2011). Understanding Earth's eroding surface with ^{10}Be. *GSA Today, 21*(8), 4–10.

Raff, A. D., & Mason, R. G. (1961). Magnetic survey of the west coast of North America, 40° N. latitude TO 52° N. latitude. *GSA Bulletin, 72*(8), 1267–1270. https://doi.org/10.1130/0016-7606(1961)72[1267:MSOTWC]2.0.CO;2

Reid, H. F. (1910). *The California earthquake of April 18, 1906: The mechanics of the earthquake* (Report of the state earthquake investigation commission in two volumes and atlas, Issue). N. Press.

Repcheck, J. (2003). *The man who found time: James Hutton and the discovery of the earth's antiquity*. Perseus.

Rudwick, M. J. S. (1985). *The great Devonian controversy: The shaping of scientific knowledge among gentlemanly specialists*. University of Chicago Press.

Rudwick, M. J. S. (2005). *Bursting the limits of time: The reconstruction of geohistory in the age of revolution*. University of Chicago Press.

Rudwick, M. J. S. (2014). *Earth's deep history: How it was discovered and why it matters*. Chicago University Press.

Schindler, S. (2007). Rehabilitating theory: Refusal of the 'bottom-up' construction of scientific phenomena. *Studies in History and Philosophy of Science, 38*(1), 160–184.

Schuchert, C. (1932). Gondwana land bridges. *Geological Society of America Bulletin, 43*(4), 875–915.

Şengör, A. M. C. (2001). *Is the present the key to the past or is the past the key to the present? :James Hutton and Adam Smith versus Abraham Gottlob Werner and Karl Marx in interpreting history* (Vol. 355). Geological Society of America.

Şengör, A. M. C. (2003). *The large-wavelength deformations of the lithosphere: Materials for a history of the evolution of thought from the earliest times to plate tectonics*. Geological Society of America.

Stinner, A., & Teichmann, J. (2003). Lord Kelvin and the age-of-the-earth debate: A dramatization. *Science Education, 12*(2), 213–228.

Suess, E., Sollas, W. J., & Sollas, H. B. C. (1904). *The face of the earth (Das antlitz der erde)*. Clarendon Press.

Taylor, C., & Dewsbury, B. M. (2018). On the promise and problem of metaphor use in science and science communication. *Journal of Microbiology and Biology Education, 19*(1), 1–5.

Thagard, P. (2012). *The cognitive science of science: Explanation, discovery, and conceptual change*. MIT Press.

Verne, J. (1874). *A journey to the Centre of the earth*. Henry L. Shepard & Co.

Vine, F. J., & Matthews, D. H. (1963). Magnetic anomalies over oceanic ridges. *Nature (London), 199*, 947–949.

Wegener, A., & Skerl, J. G. A. (1924). *The origin of continents and oceans*. Methuen & Co..

Wiggins, G. P., & McTighe, J. (2006). *Understanding by design* (Expanded 2nd ed.). Pearson Education, Inc.

Willis, B. (1932). Isthmian links. *Geological Society of America Bulletin, 43*(4), 917–952.

Wilson, J. T. (1963). A possible origin of the Hawaiian islands. *Canadian Journal of Physics, 41*, 863–870.

Wilson, J. T. (1965). A new class of faults and their bearing on continental drift. *Nature (London), 207*, 343–347.

Winchester, S. (2001). *The map that changed the world: William smith and the birth of modern geology*. Viking Press.

Chapter 4
Transformative Teaching Strategies

Immaturity is the incapacity to use one's intelligence without the guidance of another. Such immaturity is self-caused if it is not caused by lack of intelligence, but by lack of determination and courage to use one's intelligence without being guided by another. Sapere Aude! "Have the courage to use your own intelligence!" is therefore the motto of the enlightenment.

Immanuel Kant

Never doubt that a small group of thoughtful, committed people can change the world. Indeed. It is the only thing that ever has.

Marga ret Mead

4.1 Introduction: How Learning Theory Guides the Teaching Strategies

Due to the nature of large-lecture format courses, it is difficult to enact many effective pedagogic strategies large lecture halls provide students a feeling of anonymity, and this has been implicated in lower class attendance, reduced student mental engagement, and difficulties in promoting discussions among students and with the instructor. Moreover, authentic assessment strategies–like drawing, writing, and arguing all require more time and effort for evaluation in such large environments. Yet, research overwhelmingly makes clear that class attendance, mental engagement, and meaningful assessment are all associated with robust learning.

Because these activities and expectations are often new to students, they can feel anxious, causing them to pushback. Providing clear and compelling justifications, including research regarding how students learn; particularly how prior knowledge and meaning making are central to robust learning (Clement, 2008b; Franco & Colinvaux, 2000; Gilbert et al., 2000; Ibáñez & Masegosa, 2014; Núñez-Oviedo et al., 2008; Driscoll, 2005) can go a long way toward encouraging student buy-in.

G. Dolphin, *Teaching Geology Using the History and Philosophy of Science*, Science: Philosophy, History and Education,
https://doi.org/10.1007/978-3-031-68667-2_4

To encourage such buy-in, I reason with students how the pedagogical strategies I implement will help them leverage their own experiences–or that I will provide the necessary experiences to help them make more accurate meaning of the natural world (Clark, 2011; Niebert et al., 2012; Shapiro, 2011). I emphasize how analogical and metaphorical thinking (Amin, 2015; Beger & Jäkel, 2015; Indurkhya, 1992; Kuhn, 1993; Lakoff & Johnson, 1980, 1999; Nersessian, 2008; Niebert & Gropengiesser, 2015; Steen, 2014; Thagard, 2012) is crucial to understanding. I rationalize how the course's small-group structure incorporating a dialogic component (Redman, 2007) promotes necessary mental engagement, bringing forth students' thinking so I can better assist them in constructing reliable understandings of difficult science ideas (Kuorikoski & Ylikoski, 2015). The strategies below are employed throughout the course to achieve all this.

For convenience and clarity, I have described the following teaching and learning strategies separately, but in practice they extensively overlap (Sect. 9). These strategies need not be introduced all at once. Course and class context, personal comfort with each, and reactions from students all impact incorporation of the following strategies. I provide the background and descriptions based my context. The translation to other contexts might be relatively straight forward or require modification.

4.2 The Foundation of the Course: Team-Based Learning (TBL)

Team-based learning (TBL) consists of teams with five to seven students, grouped heterogeneously, to solve a relevant problem. Originally developed in the 1970s for business courses, the success of these teams led to other disciplines implementing them. Hyrnchak and Batty (2012) found that TBL was firmly grounded in how people learn. Sisk (2011) focused on TBL in her review of studies, finding that this teaching strategy showed positive results in terms of student satisfaction, student engagement, and student examination scores.

I organize students into small groups that remain constant throughout the entire semester. These smaller groups mitigate many of the impediments to effective teaching and learning. A small group structure allows the use of more authentic assessments and encourages students to discuss and argue during these assessments allowing me to hear their thinking (Nesari, 2015). Once in groups of five to seven, students are no longer anonymous individuals in a large class but have become members of a more tightly knit group, one that is working towards a common end. Many of my students expressed that they were familiar with group work and appreciated remaining in the same group for the entire term. In course evaluations, they noted that they learned a lot due to the constant communications among group members, and that it also reduced more trivial aspects of typical courses permitting them to focus more on important ideas.

As part of the TBL structure I also implement a team assessment called Individual-Team Performance Metrics (ITP Metrics),[1] a free online assessment platform developed by a psychology professor at the University of Calgary (O'Neill et al., 2017, 2019, 2020b). This team-building platform includes automated self- and team-evaluations. Students rate each other on five different criteria (commitment, communication, capabilities, standards, and focus), and the platform then calculates individual averages compared to the team average. This produces a multiplier that I used to adjust final marks, increasing the final average of team member who score higher than the team average by the multiplier). Students completed the team evaluation three times during the term (after each of two midterm exams and the final exam). This made students aware of their group effort and provided the opportunity and a motivation to enhance their team skills throughout the course (O'Neill et al., 2020a).

4.3 Case-Based (Problem-Based) Learning

Case-based teaching began as an experiment in Harvard Law School in 1870. Enormously successful, within a few decades this approach became the model for teaching in most law schools in the US and soon in schools of business. The method is successful because it situates learning in authentic cases that generalize to broader contexts and concepts, and it encourages student discussion and mental engagement.

Rejected by schools of education during the 1990s, including at Harvard, case-based learning, schools of education now often employ case-based teaching and learning, depending on their conceptual orientation (Merseth, 1991). This modality affords a means to explore the messiness of authentic situations where definitive theoretical answers might not exist, provide examples within different theoretical frameworks, and encourage self-reflection.

Although lecture still predominates in postsecondary science courses (Stains et al., 2018), case-based STEM teaching and learning is growing in. Many different case approaches exist (Herreid, 2012, 2018). Lecture-based cases have the least student involvement (Burko, 2016) as the story and data are presented with occasional pauses for student input. At such times, the students make hypotheses using the data and then check that hypothesis for coherence and amend it as necessary (Lerach & Yestness, 2019; Luster-Teasley et al., 2017). Highest student input occurs in small-group cases, or problem-based learning where students attempt to solve the posed problem through team collaboration (Herreid, 2017). Because students must, to at least some extent put forward their thinking, case studies are effective for addressing common misconceptions (Prud'homme-Généreux, 2017).

Conant (1947) argued that students should learn the "tactics and strategy" of science, and to accomplish this he developed and taught a course at Harvard using

[1] itpmetrics.com

historical case studies. Science instructors report that case studies help students become more engaged with course content (Faulkner et al., 2018; Terry et al., 2016), increases confidence (Holley, 2017) and enhances attitudes (Çam & Geban, 2017) regarding the case's content and skills. Increased critical thinking and problem-solving abilities have also been reported (Faulkner et al., 2018; Yadav & Beckerman, 2009). All this leads to better content understanding as evidenced in assessments (Blackburn, 2018; Carloye, 2017; Li & Yang, 2018; Terry et al., 2016) showing that students score higher on content assessments, and the gap between high and low performers narrows (Dori et al., 2003). The nature of cases can help overcome science gender bias (Çam & Geban, 2017; Şahin, 2019) and the marginalization of other underserved populations (Safapour et al., 2019). Instructors also benefit from teaching with cases. For instance, teachers have developed better attitudes about their own science teaching (Çam & Geban, 2017) and math teaching (Li & Yang, 2018) after creating and using cases for instruction, and case studies add to the repertoire instructors can use in class (Camill, 2006; Hodges, 2005).

In my course, historical case studies track the developmental trajectory of target scientific concepts (conceptual biographies) from very early to the modern consensus view (Dolphin et al., 2018). The three narratives in Chap. 3 outline the case studies as I present them. Following Allchin's (2013) approach[2] summarized in Table 4.1, I have supervised multiple graduate and undergraduate students to produce several historical case studies targeting specific science concepts (e.g., plate tectonics, development of geologic times scale, impact theory of dinosaur extinction).

The following summarizes how I teach the development of seismology.

Hook. I begin by having students view a YouTube video, *The Damnedest Finest Ruins* (Dalessandro, 2006)[3] about the 1906 earthquake, which affected a 200+ mile stretch of the western coast of California, and especially San Francisco. The epicentre was close to San Francisco and because of the city's great population density, the earthquake had a tremendous impact. The video is not so much about the geological aspects as it was about the human drama and possible hazards of earthquakes.

Personal narratives. Next, students read personal accounts of the earthquake written by two famous authors, Jack London (1906) and William James (1911). They wrote about the earthquake in very different manners, thus offering the students a broader view of the impact of this event. Students read the accounts and use a Venn diagram to compare them and explore why each writer has such unique perceptions.

Scientific investigation. The story of the disaster continues with Governor of California, George Pardee (1857–1941) commissioning a group of scientists, led by Andrew Lawson (1861–1952), then chair of University of California, Berkley's, geology department to investigate both the cause and the effects of such a disaster. Harry Reid, a member of the commission, used his observations (and past experiences with glacial dynamics) to develop the idea of elastic rebound theory.

---

[2] See https://ucalgary.ca/tamaratt-teaching-professorship/projects/tamaratt-case

[3] https://www.youtube.com/watch?v=mvve8YiYGSo

Table 4.1 Procedural description for researching and developing a historical case study

1. Choose a science concept or scientific controversy	Keep the concept focused and narrow the narrative to one scientists, or just a few (in the case of a controversy). This will help keep the case streamlined and doable, especially for those new at building cases
2. Locate and read high-quality historical sources	Look for authoritative sources, for instance the *Dictionary of Scientific Biography*, and the history of science, technology, and medicine search engine (HistSciTechMed). Be wary of popular history books, as their portrayal of science processes could be inaccurate (see Allchin, 2002, 2003)
3. Begin to outline a developmental trajectory and the nature of science (NOS) aspects	The goal of the case is to reveal "science-in-the-making" (Latour, 1987). This step focuses on highlighting a few salient NOS aspects for the main actors in the case, as well as within the surrounding context
4. Build a context around the developmental trajectory	Become familiar with the developmental stages of the target concept and then overlay with the social, political, economic, and other contexts to flesh out the development, showing why the question was important, and why things progressed as they did
5. Focus on one or two aspects of the context that are especially relevant to the trajectory	This might have to do with the history of the main figure involved in the discovery. What was their background, hobbies, or education that positioned them to make the discovery? Scientific advancement is more about "right place, right time" than it is about sheer intellect
6. Broaden the context, but keep it relevant to how different stakeholders accepted iterations of the concept	This could include those involved in any controversy and why they took the positions they did. How did their context set the stage for the controversy? What was perception/response of other scientists and the public to the discovery or development?
7. Develop NOS questions based on the main NOS features reflected in the case	Questions should guide students to explore the reliability of claims being made. They should try judging one side of the controversy against the other. Who makes the strongest case and why? Students should also explore how aspects of the context have played a role in the developmental trajectory of scientific knowledge claims
8. Develop questions and/or activities that help the students think like a scientist and develop knowledge	Activities within the case should model reality—In that the data are given and students engage in trying to make meaning of those data. If it is possible to model some of the examples where advancements happened, this would be ideal. The visible geology activity and the extended seismic activity are examples of this (see Sect. 4.4.1)
9. Consider how students will demonstrate what they have learned	What are the learning goals of the case (both disciplinary content and NOS), and how will students demonstrate they have met those goals? This could take many forms depending on the goals. The more authentic the assessment, the better

Ground abstractions in experience. The idea of rocks behaving like an elastic material is not intuitive to most students (Hubenthal, 2018). Thus, incorporating a concrete experience illustrating this abstract idea is important. I demonstrate elasticity using rock "tongs" using a 30-cm-long drill core of sandstone with a slit, sawn up the centre for about 20 cm. Pinching the sawn end together like tweezers, and the tongs open back up elastically. This is like the marble tongs described by Hubenthal (2018). I also have students observe their skin, pinch it together and shear it with their fingers, respond note its response to that stress and observe it return elastically to its original shape. Now I can refer to the crust of the earth metaphorically (Beger & Jäkel, 2015) as the *skin* of the earth (Dolphin & Benoit, 2016), using these embodied experiences (Niebert et al., 2012; Shapiro, 2011) to link to the elastic nature of rocks.

Finally, I have also used a molecular model with wooden balls as atoms, connected by springs that represent their bonds. This model demonstrates that the addition of a stress will result in a strain that reverses when the stress is released. This is a causal explanation for the release of energy when the strain exceeds rock strength and breaking happens, releasing that elastic potential in the form of seismic waves. The scaffolding of descriptive experiences to explanatory ones is important because thinking in terms of explanations is more complex and challenging for students (Thagard, 2012).

Try it out. To further emphasize the relationship between elastic rebound and earthquakes, students then engage with an earthquake machine (Hubenthal et al., 2008) consisting of a long piece of belt sandpaper fixed to a tabletop, abrasion side up. On top of the belt sandpaper lies a 2" × 4" × 6" piece of wood with sandpaper fixed to the bottom. Attached to the wood is a rubber band. Students pull the wood using the rubber band. The band stretches until the force pulling the block exceeds the friction from the sandpaper holding the block in place. The block lurches forward. The activity can run in many different configurations of inquiry—from very guided to very open—depending on available time and the intended learning outcomes.

I choose to have students analyze the model for all the forces involved in the system, identifying the forces they can and cannot control. They then develop their own question, manipulate variables in the earthquake machine, collect data, and develop an answer. Having accomplished their inquiry, students then analyze the model for strengths and limitations, primarily describing what the model highlights about elastic rebound theory and what it hides (Niebert & Gropengiesser, 2015). This analysis helps students develop a more nuanced understanding of the target concept (Boulter & Buckley, 2000; Buckley & Boulter, 2000).

I situate this case study at the beginning of the "history of seismology" section of the topic addressing the dynamics of the earth. The placement is significant as it sets up subsequent narratives about the development of the Worldwide Synchronized Seismic Network (WWSSN) in response to the nuclear test ban treaty. This new network allowed for unprecedented observations of earthquake signals, globally. It also made locating epicentres quicker and more precise, and enhanced understanding of the relative motion across faults. As these new data started to accumulate, a picture of earth dynamics came into sharper focus and global seismicity became a pillar in the data that supports our developing understanding of how the earth works.

4.4 Inquiry-Based Learning

Inquiry is a fundamental activity in science and therefore should be foundational when teaching science. Scientific inquiry is the process for developing an evidence-based explanation of some phenomenon observed in the natural world (Bybee, 2006). It includes using empirical evidence and logical reasoning, minimizing the role of bias, and presenting these explanations for review and critique by other scientists. Using inquiry as a science teaching strategy started with Conant (1947) and slowly became more popular in science classrooms since then. However, the definition of what counts as inquiry-based activity to science educators has become very broad, incorporating ideas like hands-on learning, experiential learning, discovery learning, and minds-on learning (Bybee, 2006).

Ironically, textbooks portray scientific inquiry very rigidly—scientific inquiry is a straightforward path from developing a question, researching background information, making a hypothesis, testing the hypothesis (usually through controlled experimentation), interpreting the data, and communicating results. To many, this outlines the monolithic scientific method (see Table 4.2). While this linear process may be reflective of how most science gets reported, it is not a reflection of how science gets done in real time (Medawar, 1963). This is especially true for geology. As already described in Chap. 1, Sect. 1.4, geology is not an experimental science like most of physics or chemistry. It is a historical and interpretive science (Frodeman, 1995), where geologists search for and interpret signs in the rocks, trying to discern causality (Cleland, 2013; Turner, 2013). This is not to say that experiments do not happen in the geosciences, but that the goal is to understand the earth's past; the only way to do that is by assigning meaning to artifacts left behind from past causes. There is no way to experimentally test how dinosaurs became extinct 65 million years ago, or if Pangaea existed and broke apart over 200 million years ago.

This stepwise portrayal of the process of scientific inquiry is revealed in how people describe doing of science. Many frame the process of scientific inquiry as largely this one method (see Chap. 6 Sect. 6.3.2.1). In addition, many school laboratory activities parallel the stepwise structure in their instructions; thus, what might be intended to be inquiry becomes an exercise in confirmation (National Research Council, 2006). For example, when students say they are baking—namely, following directions and mixing ingredients to get an expected outcome—this is equated

Table 4.2 The depiction of inquiry in different introductory geology textbooks

Inquiry steps	Marshak (2012)	Reynolds et al. (2013)	Fletcher et al. (2014)	Plummer et al. (2007)
1	Recognize a problem	Make observations	Make observations	Raise a question or problem
2	Collect data	State a problem	Collect data	Gather data
3	Propose a hypothesis	Ask a question	Create a hypothesis	Propose a hypothesis
4	Test a hypothesis	Create a hypothesis	Make a prediction based on hypothesis	Generate prediction based on hypothesis
5		Make predictions	Test predictions (data collection, organization, analysis)	Test prediction
6		Test predictions	Reject or accept a hypothesis	Hypothesis becomes theory

with doing science (see Chap. 6, Sect. 6.3.1.3). A more representative notion of scientific inquiry is increasingly acknowledged and is included in multiple national science teaching standards documents (see Table 4.3).

Grounding inquiry in the history of the concept through historical narratives or case studies provides students experiences to build their mental model, and experience science-in-the-making (Latour, 1987). Importantly, through this process they formulate knowledge, forge their way into the unknown just like scientists on the edge of discovery. Students construct associations and explanations of the data (Nersessian, 2008; Thagard, 2012), but are nevertheless also constrained by the history (Allchin, 2020). That is, as they interpret the data and develop an answer; they then continue with the historical narrative to see how the problem actually was resolved before moving to the next stage of development. When I finally present the consensus view, aligned with the learning goals, it makes sense to the students because they have been a part of the iterative process over the entire time of the narrative.

The following sections describe some of the activities in the course. With a range of student autonomy in scientific questioning, prioritizing evidence, explaining observations or phenomena, connecting to other scientific knowledge, and communicating or arguing an explanation with others.

The Visible Geology Activity (Sect. 4.4.1) and Extended Seismic Activity (Sect. 4.4.2) have students create questions they can explore and collect data to answer their questions. The seafloor data activity (Sect. 4.4.3) provides students data from which they must make evidence-based descriptive claims and explanations. The Age of the Earth Activity (Sect. 4.4.4) and Models of Earth Dynamics Activity (Sect. 4.4.5) have students judge the reliability of a given explanation based on the evidence provided. Table 4.4 shows how each activity addresses different inquiry activities, as outlined by the NRC (2000).

4.4.1 *Visible Geology Activity*

In this activity, students use a web-based application[4] to create a geologic section. The controls of the application allow students to do several things: add sedimentary rock layers; tilt, fold, and fault them; form domes or basins; and create igneous intrusions. The outcome of these actions is a three-dimensional rock section showing all the facets of its history. Students then exchange their geologic section for another group's geologic section, and both proceed with interpreting the sequence of events that led to the formation of that geologic section. They make observations of the sequence of layering, cross-cutting relationships, and other signs to interpret the order in which the formation of the section took place. They write down the sequence, with a rationale, invoking Nicholas Steno's four principles

[4] www.visiblegeology.com

Table 4.3 How inquiry is represented in multiple national science teaching documents

United States: *Next Generation Science Standards, Scientific practices* (NGSS Lead States, 2013).	Brazil: *Base Nacional Comum Curricular, Investigation processes and practices* (Brasil., 2017).	Australia: *Australian Curriculum Science inquiry skills (Earth and Environmental Science).* (ACARA, 2014);	South Korea *Korean Science Education Standards for the next generation, Sub-domain scientific literacy* (MOE, MSICT,, & KOFAC, 2019)
Asking questions	Identifying problems	Identifying, researching, and constructing questions for investigation	Identifying a problem
Developing and using models	Formulating questions	Designing investigations for collecting primary and/or secondary data; considering research ethics	Modelling capability (making and applying models)
Planning and carrying out investigations	Identifying relevant information of variables	Conducting investigations safely, competently, and methodically	Designing and conducting experiments
Analyzing and interpreting data	Proposing and testing hypotheses	Analyzing and representing data in meaningful and useful ways	Interpreting and transforming data
Using mathematics and computational thinking	Elaborating arguments and explanations	Interpreting and evaluating others' processes, claims, and conclusions, and data quality in constructing a scientific argument	Engineering design
Constructing explanations	Choosing and using measuring instruments	Selecting and constructing appropriate representations of data to solve problems and make predictions	Generating an explanation and making scientific argumentation
Engaging in argument from evidence	Planning and carrying out experimental activities and field research	Communicating appropriately to specific audiences	
Obtaining, evaluating, and communicating information	Reporting, evaluating, and communicating conclusions		
	Developing intervention actions based on the analysis of data		

Table 4.4 How activities in GLGY 209 align with the diverse essential features or variations of inquiry, as outlined in the national science education standards. (NRC, 2000)

Essential Feature	Variations			
1. Learner engages in scientifically oriented questions	Learner poses a question	Learner selects among questions, poses new questions	Learner sharpens or clarifies question provided by teacher, materials, or other source	Learner engages in question provided by teacher, materials, or other source
2. Learner gives priority to evidence in responding to questions	Learner determines what constitutes evidence and collects it	Learner directed to collect certain data	Learner given data and asked to analyze them	Learner given data and told how to analyze them
3. Learner develops explanations from evidence	Learner formulates explanation after summarizing evidence	Learner guided in process of formulating explanations from evidence	Learner given possible ways to use evidence to formulate explanation	Learner provided with evidence and how to use evidence to formulate explanation
4. Learner connects explanations to scientific knowledge	Learner independently examines other resources and forms the links to explanations	Learner directed toward areas and sources of scientific knowledge	Learner given possible connections	
5. Learner communicates and justifies explanations	Learner forms reasonable and logical argument to communicate explanations	Learner coached in development of communication	Learner provided with broad guidelines to use sharpen communication	Learner given steps and procedures for communication

Decreasing student autonomy →

Extended Seismic Activity | Seafloor Data Activity | Models of Earth Dynamics Activity | Visible Geology Activity | Age of the Earth Activity

(superposition, original horizontality, lateral continuity, and cross-cutting relationships) and why they chose to invoke each. Then they change back with their other group. They evaluate the other group's interpretation of their original section, noting especially the places where the other group might have misinterpreted the section, and describing why it was a misinterpretation.

4.4.2 *Extended Seismic Activity*

In this activity, students use the Incorporated Research Institutes for Seismology (IRIS) Earthquake Browser.[5] The browser displays earthquake epicentres on a world map and has controls for directing the data output. Students can choose the number

[5] https://ds.iris.edu/ieb/

of earthquakes displayed (up to 10 s of thousands), the time range of the data (from 1970 to present), the magnitude range (from 1 to 10), and the depth range. It provides the number of all the earthquakes that fit the specified parameters; further, a three-dimensional view helps student visualize the hypocentres as they appear below the surface. Epicentres appear on the map in different colours to represent depth and different sizes to represent magnitude. Finally, there is a "select new region" option that allows students to draw their own map to focus on a smaller location.

Students explore each of the variables for data output. Once they have become familiar with how the browser works, they develop a question they can explore using the browser. Possible questions include the following: What is the relationship between earthquakes and continents? Where do smaller earthquakes happen compared to bigger ones? Where do deep earthquakes happen in relation to shallow? What is the relationship between depth and magnitude of an earthquake? Students then manipulate the appropriate variables and collect data, either by counting the number of earthquakes for the set of parameters, or by making screenshots of the generated map. They then draw some conclusions about the questions, based on the data they collected.

4.4.3 Seafloor Data Activity

This activity is a variation of an activity called *Discovering Plate Boundaries* (Sawyer, 2002). Where Sawyer's activity allows student to rationalize plate boundaries based on the data, the approach here has students infer plate boundaries based on their observations. In this activity, students analyze six different world maps with mainly seafloor data of different kinds (seismicity, volcanicity, sediment thickness, geochronological, bathymetry, and heat flow). First, they make descriptive claims of relationships they see within each map. For instance, sediment seems to be thickest along the coasts of continents, and thinnest to non-existent, in the centres of oceans. The seafloor is youngest (brand new) in the centres of the basins and gets older (up to only 200 million years old), symmetrically toward the edges of the basins.

Having made these descriptive claims, they return to the maps and start to discern possible relationships from one map to another. Then, they make new descriptive claims reflecting these relationships. For instance, sediment is thickest along the coasts of continents, where the crust is also oldest and lowest. Once they have related the maps together using descriptive claims, they generate an explanation for the observations. This could be something they completely fabricated, or it could be related to some of the other historical ideas (contracting earth, porous earth, horizontal displacement, or expanding earth). Their explanation must be supported by the descriptive claims they made. The "right answer" is not what is important here. It is using the data to develop an answer and being able to support that answer with the data.

4.4.4 Age of the Earth Activity

I have five student volunteers perform for the class the dramatization, *Lord Kelvin and the Age of the Earth Debate* in class (Stinner & Teichmann, 2003). The dramatization is a fictitious enactment of the clash between some geologists/biologists and chemists/physicists—though it is based on the characters and actual controversy of the time—concerning how to calculate the age of the earth. After the play, students engage with the dramatization by recording the different assumptions made by each side of the debate: Charles Lyell and Thomas Huxley (geologists/biologists), and William Thomson and Hermann von Helmholtz (chemists/physicists). They analyze the assumptions for reliability and then make an argument for one side or the other of the debate, based on their analysis of assumptions and the observations that were current at the time of the debate.

4.4.5 Models of Earth Dynamics Activity

Students engage with the multiple arguments and evidence for the cause of continents and ocean basins, specifically Aristotle's *Porous Earth* model, Eduard Seuss's *Contracting Earth* model, James Dana's *Permanentism* model, and Alfred Wegener's *Horizontal Displacement* (continental drift) model. Students hear the observations about folded rocks, biological and geological disjuncts, paleoclimatic data, and how each one fits into (or does not fit into) the different propositions. They then predict from each model and devise a way to test each of those predictions. Finally, they select what they think is the most reliable explanation and explain why, using the data they have from the observations.

4.5 Model-Based Learning (MBL)

Mental modelling theory asserts that learning happens as a person reasons about an external reality via an internal mental model—a mental analogue of that external object or phenomenon (Rea-Ramirez et al., 2008). Clement (2008a) and Núñez-Oviedo and Clement (2008) identified multiple modes of activity within the MBL framework, including model generation, accretion, confirmation, disconfirmation, competition, and evolution (see Table 4.5). Dolphin and Benoit (2016) found that undergraduate general education students learning new content about earthquakes spent most of their time developing their mental models through model accretion (86%), with model competition (11%), model generation (2.5%) and model reduction (0.5%) happening much less often. This makes sense, because these students, as novices, would likely have heard of earthquakes but lacked formal education in the phenomenon. They would need to build that model during instruction.

Table 4.5 Modes of model-based learning

Modes of model-based learning	Description of activity (after Clement, 2008a and Núñez-Oviedo & Clement, 2008)
Generation	Given a target phenomenon, students develop an explanatory model
Accretion	Student generates one particular aspect of the target model at a time
Confirmation	Students provide hypothesis-supporting evidence
Disconfirmation	Students provide hypothesis-refuting evidence
Competition	Students compare aspects of one model against aspects of a rival model(s), with respect to available data. This mode may incorporate other modes, such as accretion and disconfirmation, for decision-making
Evolution	Students incorporate a series of modifications and revisions into their initial model

Dolphin and Benoit (2016) also found that students' descriptive claims from or about their mental models outnumbered explanatory claims by almost 5:1. This is an important finding that guided instruction to be initially more concrete and descriptive, but to progress to more abstract and explanatory content, giving students time to fortify their models and enable more sophisticated understanding. This diverges from the standard approach to teaching, especially in textbooks (see Sect. 4.7). Scientists typically use observations to generate explanations, but textbooks present explanations (ready-made science) and use the observations as justification. This distorts students' understanding of how science works. I used the MBL modes purposefully to develop the instruction so that students engage in them at the appropriate time, helping them to construct the intended learning about the target concepts.

Having established broad learning goals for the course (see course-scale learning goals in Fig. 2.3), I then considered finer-grained, unit-scale and class-scale learning goals. For students to develop a robust understanding of the integrated target concept—which is the highest order concept (Rae-Ramirez, 2008)—they must be taught lower-level elements that each support the integrated target concept. I created a hierarchy of elements, as follows: tertiary elements support secondary elements, which support primary elements, which ultimately support the integrated target concept. Figure 4.1 provides an example of developing one of the four course learning goals—*students will appreciate the dynamic nature of the earth*—into a hierarchy from the integrated target concept to the primary, secondary, and tertiary elements. The flow is especially focused on the theory of plate tectonics, based roughly on the level of conceptual concreteness.

In teaching for the integrated target concept, I start by presenting the most concrete elements and help students synthesize up to the next level of abstractness. This helps students build their mental model in a bootstrapping or bottom-up fashion (Carey, 2009), from more concrete (perhaps even embodied) experiences. This type of approach is very important for effectively developing conceptual understanding (Nersessian, 2008; Niebert & Gropengiesser, 2015; Shapiro, 2011). Similarly, this hierarchy leads from more descriptive knowledge (*what* is happening) to explanatory (*how* or *why* is it happening).

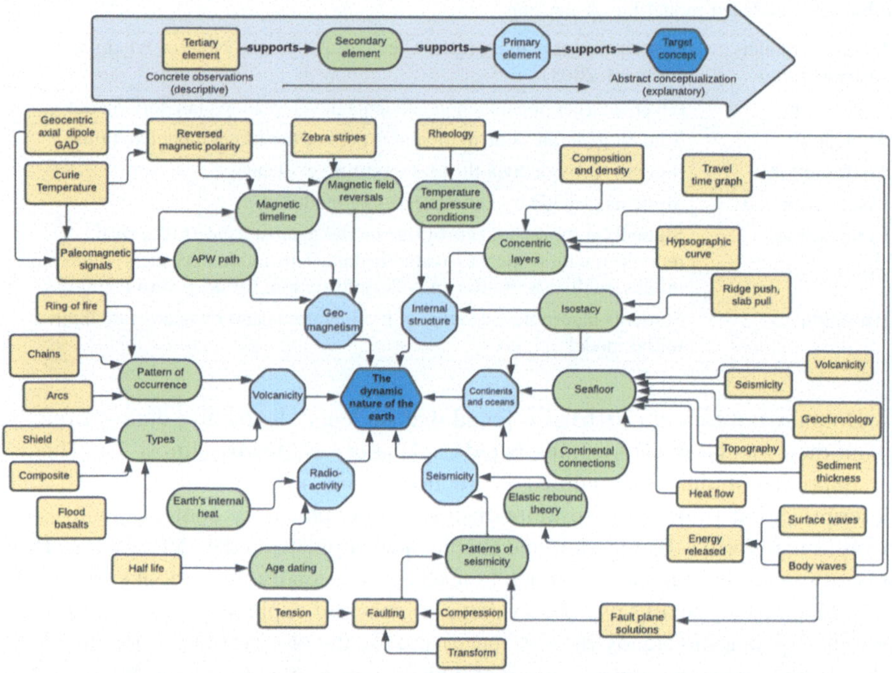

Fig. 4.1 Hierarchy of concepts constituting "The Theory of Earth Dynamics". (Note. This diagram shows the hierarchy of concepts that support the integrated target concept (dark blue). Tertiary elements (yellow) are the most concrete and support the secondary elements (green), which in turn support the primary elements (light blue). The diagram helps to map out what to teach and when, with respect to other elements)

After deciding on all the elements to be encompassed in the instruction, I took an MBL approach and mapped out how students would engage with them, both through the modes of MBL and the sequence (Núñez-Oviedo & Clement, 2008; Núñez-Oviedo et al., 2008). Clement and his group (2008) analyzed qualitative data about student/teacher dialogues to track student model development through progressive iterations—from M1 (initial model), to M2, M3, M4, and so on—and interpreted different modes of MBL (see Table 4.5). I used Clement et al.'s iterative model development, synthesized to plan both my activities and student activities at the scale of a regular class period (see Fig. 4.2).

Figure 4.2 presents two examples of the iterative structure of instruction—seismicity and geomagnetism—at the scale of a single class period, using the MBL framework. The flow of activity follows the arrows from left to right, projecting iterations of model evolution as M1, M2, M3, and the like, for two primary elements. What the instructor does or says appears along the bottom of each model evolution example. How students respond appears along the top of each example. At the right side of each example, the secondary and tertiary elements point towards the target learning outcome(s). In most cases, the generation mode and the accretion

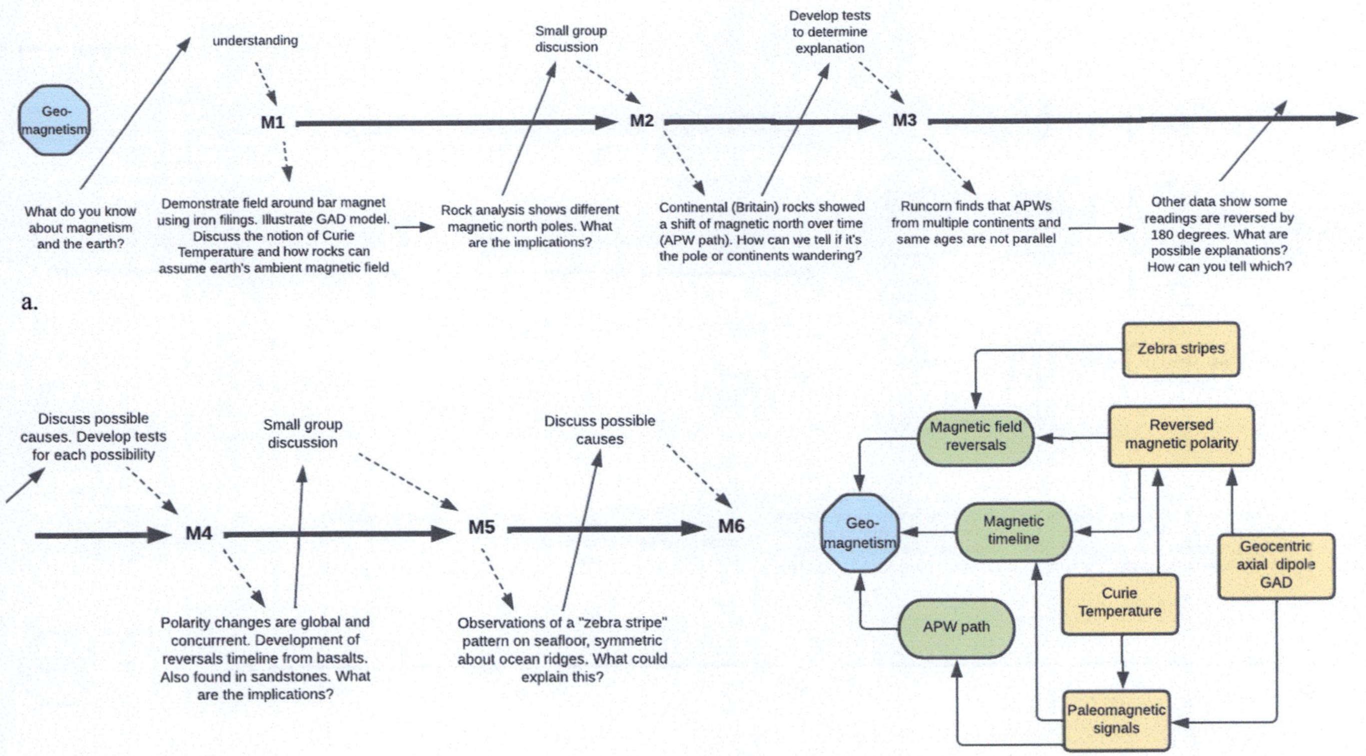

Fig. 4.2 Mapping instructor-student interactions facilitating mental model development. (Note. Two examples (**a and b**) of class period-scale instructional planning, using a model-based learning framework. The target concept is on the left (blue octagon). The instructor's actions are described in text along the bottom. Student responses are described in text along the top. Intended model evolution is given by M1, M2, M3, etc. Intended learning of tertiary and secondary elements is on the right side of each diagram)

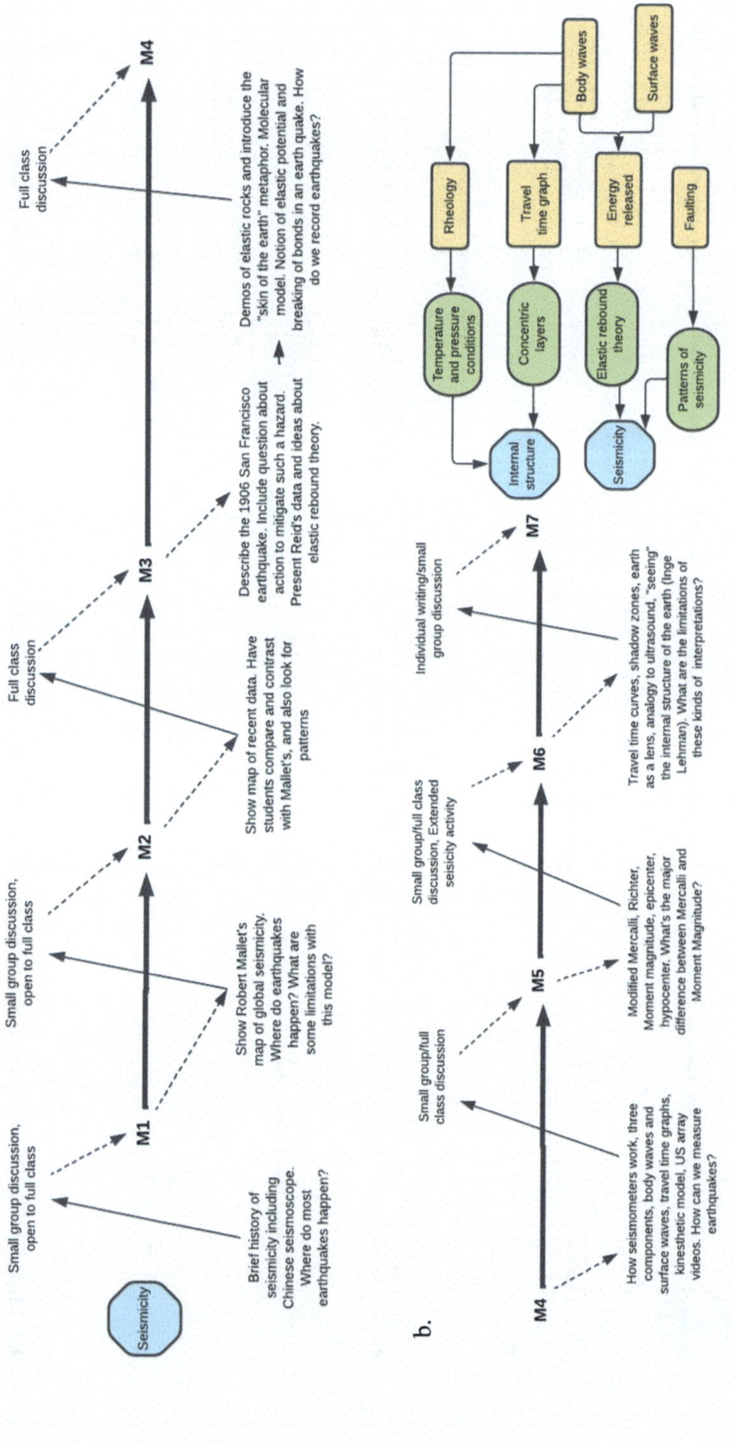

Fig. 4.2 (continued)

mode dominate the MBL activities, especially early on as students are learning about content with which they have little or no experience (Dolphin & Benoit, 2016). Also, models start out being mainly descriptive (*what* is going on?) before becoming explanatory (*how* or *why* is it going on?); this is because explanation is a more complex way of thinking than description (Thagard, 2012).

I combine historical case studies, inquiry activities, and model-based learning to metaphorically "braid" history, inquiry, and MBL (Dolphin et al., 2018), a notion I first presented in Chap. 2. In Chap. 3, I illustrated in Figs. 3.1, 3.2, and 3.3 this braiding structure for the three disciplinary content learning goals I set for the course (historical nature of the earth, the age of the earth, the dynamics of the earth, respectively). As those figures demonstrate, each iteration of the braid leads off with an essential question—one that is open-ended and frames the content being investigated (Wiggins & McTighe, 2006). A historic interlude (the narrative) follows the question and precedes the inquiry activity, which instigates the students' MBL. After this, the next iteration begins.

Readers interested in exploring this approach in their own teaching will find these exemplars a helpful demonstration of the finely articulated elements that work both sequentially and cumulatively to aid students' learning. Because this can be a demanding effort, it may be more manageable tackling small parts at a time and build with each time teaching the course.

4.6 Using Wikipedia to Highlight Women of Geology

Having started in 2001, Wikipedia rapidly became one of the most used peer-production communities on the internet: it has 6.4 million English articles, exists in over 200 different languages, and has over 42 million users (Wikipedia, n.d.). These types of platforms "have fundamentally changed the way that knowledge is produced, disseminated and consumed" (Zhu et al., 2020, p. 491). It is an important tool, as people get much of their science and health information from Wikipedia (Heilman & West, 2015; Maggio et al., 2020; Smith, 2020). Thompson and Hanley (2017) asserted that Wikipedia is one of the "most important sources of scientific review articles in the world" (p. 35), underscoring that the science appearing in Wikipedia shapes what subsequent science gets done and how scientists write about that science.

However, the dissemination of knowledge on Wikipedia is not equal in all domains (Warncke-Wang et al., 2015). For instance, women's biographical articles represent only about 15 to 17% of all the Wikipedia biographies (Rotramel et al., 2019). In 2010, the Wikimedia Foundation (which supports Wikipedia) implemented a public policy initiative to address these domain deficits. They created the Wiki Education Foundation, a group whose directive is to reach out to academics and encourage them to engage in developing Wikipedia. The Wiki Education Foundation offers support—specifically, information and training for academics and their students—to create, contribute to, and edit Wikipedia articles. In 2016,

Wiki Education implemented the Wikipedia Year of Science, which led to enlisting 285 higher education courses (about 6200 student editors) who added almost five million words to articles that have now been seen more than 260 million times.

In the fall term of 2016, I had my GLGY 209 students first participate in this effort. The two activities described below have two goals: Enhance the science (and NOS) content and increase the presence of women geologists on Wikipedia. Each group of students picks from a list of over 100 important women geologists and creates a page for one if it does not already exist or contributes to or edits one that already did exist. As of Fall 2021, I have run the Wikipedia activity five times, resulting in over 1100 students editing 294 articles, creating 78 articles, and adding almost 250,000 words, and 2300 references. Since 2016, their work has garnered over two million article views.

The Wiki Education Foundation has a wonderful staff that supports instructors and students through the process of editing. Participating courses get a dashboard that has all the information for students concerning the structure of the assignment and a schedule of module completion. The students sign up to be editors, and the instructor determines the timeline of events. The events include Wikipedia-produced modules for students to learn how to edit in Wikipedia. Students engage with modules and tutorials about editing and using the Wikipedia code, evaluating articles and sources, and writing biographies, as Wikipedia has a particular format for different kinds of articles. They learn about how to research a topic (only secondary research is allowed as a source), how to avoid plagiarism, and how to write from a neutral stance. Articles on Wikipedia are not supposed to be persuasive; rather, they are to be as objective a presentation of the information as possible. All changes that students make are saved in the history of the article, so the instructor can view and evaluate student progress. The instructor does not need to search for the edits by students. All the group's contributions are readily accessible from the course dashboard.

For GLGY 209, students have two opportunities to edit on or create content for Wikipedia:

- ***Short-term activity:*** All groups did this project. There is a short-term activity that lasts about 6 weeks, with six modules for students to complete. Students make some minor edits or other contributions to the page of their choosing. Groups are welcome to do as much as they want, and indeed, many new articles have been produced during this activity.
- ***Long-term activity:*** Groups could choose to do this as their end-of-term research project. This opportunity for students entails doing a more comprehensive Wikipedia project for the long-term writing assignment in the course (which is one choice of three different activities). If they choose this long-term activity, they complete four more modules, as well as a peer-review activity.

As Wikipedia projects have become more popular, so have investigations into the efficacy of the projects. Students report they are more engaged with Wikipedia projects than traditional writing projects (Vetter et al., 2019), especially because they have more autonomy (Azzam, 2017) and are writing for an authentic audience, as opposed to writing a paper seen only by the course instructor (Roth et al., 2013). Students found the modules and the editing activities helped them develop research and writing skills, especially writing from a neutral standpoint (Roth et al., 2013). They also found it contributed to some collaborative skills (Evenstein Sigalov & Nachmias, 2017; Vetter et al., 2019). The feedback I have received from students about these assignments has been mostly positive. I have even had students contact the person they were researching for an article, or the family of that person—engagements that have yielded very positive experiences for them. In the information age, such knowledge creation is important for students.

4.7 Textbook Aligned with Curriculum

The science textbook has been a mainstay in science classes for many decades. In fact, instructors often use the course textbook as the default curriculum (Ornstein, 1994). For instance, in a study of middle school math textbooks, researchers found a large majority of instructors use their course's textbook to develop objectives, create activities, select examples, and assign independent classwork and homework (Blazar et al., 2019). This can be problematic because "most science textbooks (particularly middle and secondary level books) are written in an impersonal, seemingly objective tone, which ignores the readers' needs. The style seldom has readers access and examine their prior knowledge about a topic" (Ulerick, n.d.). What follows describes in some detail the limitations of many standard introductory geology textbooks, especially if they are the main driver of the course curriculum. To be fair, textbooks are important as reference material, a location to find succinct information in a single place. However, due to this organization (density of information and economy of space), they are limited in their usefulness for understanding how science works, and their representation of science also has implicit messages about the nature of science process and products.

4.7.1 Limitations of Textbooks

- **Beginning content outside the mesocosm**

Table 4.6 lists several commercially popular geoscience textbooks and gives the first four chapter headings. In general, they have the same structure, though the information has been parsed into chapters slightly differently. An introductory geology textbook usually starts with a section like, "What is geology?" where it describes the science and how it is done. This is usually the only place where the book recounts NOS, invariably portraying it within the context of the monolithic scientific method (Table 4.3). Afterwards, the text will normally talk about the origins of the universe and then use this to describe the origin of the solar system and Earth. This

Table 4.6 First four chapter headings of commonly used introductory geology textbooks

	Chapter 1	Chapter 2	Chapter 3	Chapter 4
Grotzinger and Jordan (2014)	The earth system (includes NOS)	Plate tectonics: The unifying theory	Earth materials: Minerals and rocks	Igneous rocks: Solids from melts
Fletcher et al. (2014)	Introduction to geology (includes NOS)	Solar system	Plate tectonics	Minerals
Nance and Murphy (2016)	Foundations of modern geology (includes NOS)	Plate tectonics	Minerals	Origin and evolution of igneous rocks
Tarbuck and Lutgens (2017)	An introduction to geology (includes NOS and solar system)	Plate tectonics: A scientific revolution unfolds	Matter and minerals	Magma, igneous rocks, and intrusive activity
Marshak (2019)	The earth in context (includes NOS and solar system)	The way the earth works: Plate tectonics	Patterns of nature: Minerals	Up from the inferno: Magma and igneous rocks
Reynolds et al. (2013)	The nature of geology	Investigating geologic questions	Plate tectonics	Earth materials
Murck and Skinner (2012)	Earth as a planet (includes NOS, solar system, and geologic time)	Earth materials (includes minerals and rocks)	How old is old? The rock record and geologic time	Plate tectonics

problematically presents the magnitude of time and space involved in very abstract terms. It does not enable concrete experience from which students can build understanding. Instead, they must just accept these as facts given by the authoritative text.

Next, textbooks usually go into the foundational theory of plate tectonics—its history, perhaps, but mainly a description of the mechanics of earth dynamics. Again, the concepts here are completely abstract because of the scale of time and space of the phenomenon. From this, the texts address earth materials, starting with atoms, chemistry, and then turning to minerals. The scale of concepts has now gone from the macrocosm (too big for concrete experiences) to the microcosm (too small for concrete experiences). Finally, after the chemistry section, texts will discuss minerals, placing the content squarely in the mesocosm (at the register of human-scale concrete experiences), where it needs to be for effective learning to take place (Niebert & Gropengiesser, 2015).

- **Data used to rationalize theory**

As for the content in both the macro- and microcosm, students must accept this completed science so they can make sense of the rest of the book, which rationalizes theory by highlighting observations of its effects on Earth. Of course, this is in direct opposition to how the theory came about. What the text refers to as "effects" of some phenomenon (theory) were really the observations used to construct the theory in the first place.

Reynolds et al. (2013, p.56), asks this: "What do earthquakes and volcanic activity tell us about Earth's lithosphere?" (p. 56). There is a map of earthquake epicentres in yellow dots and active volcanoes in orange triangles. Of course, there are long, narrow regions of high activity along some coastlines and through the centres of oceans. Below this first map, is another one that highlights the tectonic plates of the world. The caption below that map states, "Compare the two maps, and note how the distribution of tectonic activity, especially earthquakes, outlines the shapes of the plates" (p. 59).

In other words, the text portrays the pattern of volcanoes and earthquakes as *confirmation* of the existence of plates—really, though, it was the identification of the narrow bands of tectonic activity that geologists used to construct the concept of tectonic plates and the location of plate boundaries in the first place.

- **Data don't speak for themselves**

Another problem with the previous example is that earthquake and volcanic activity do not *tell* us anything about the earth. Geoscientists, instead, create and assign meaning to their observations. They interpret the data and construct an explanation. If the data could actually "speak for itself," science would be a whole lot easier. But data are just data, lacking in meaning until the human mind attributes meaning to it. According to this rendering, the data tell the story, and scientists act more like technicians (which I discuss further in Chap. 6) needing only to confirm what should be obvious. They did not work to solve a problem but only showed up to make the discovery. The implicit messaging is that students need only accept these truths on faith. There is no need to critically evaluate them, no need to understand how they came about historically.

- **Rationally Reconstructed History**

In places where some history of a concept is given in a text, it usually appears as "rationally reconstructed" knowledge (Allchin, 2002). In other words, the author tells the story with the advantage of hindsight of the eventual consensus view. The path to knowledge appears inevitable, with the scientist predestined to make that discovery. A common example of this is the story of James Hutton's recognition of a large angular unconformity and his development of the idea of former worlds and deep time (Hutton, 1788). According to the account in Marshak (2019), Hutton

> used a boat to explore the seaside of Scotland, where shore cliffs provided great exposures of rock, stripped of soil and shrubbery....Here he saw a sequence of strata consisting of red sandstone and conglomerate that rested on a distinctly different sequence of gray sandstone and shale...the beds of gray sandstone and shale had a nearly vertical dip (slope), whereas the beds of red sandstone and conglomerate had a dip of less than 15°. (p. 347–348)

Marshak then described Hutton being perplexed, and as he "sat and stared at this odd geometric relationship, the tide came in and deposited a new layer of sand on top of the rocky shore. With the principles of uniformitarianism in mind, *Hutton suddenly realized the significance of the outcrop*" (p. 348, my emphasis). Marshak goes on to describe Hutton's revelation that the grey sandstones and shales had been deposited in horizontal layers, turned to stone, tipped up on edge and truncated by erosion, with the red sandstones and conglomerates deposited on top. The story mistakenly portrays Hutton's understanding as a sudden great insight; that the tides and sediments had spoken to him and revealed nature's truths. In reality, Hutton had spent years developing this idea and debating it using the exposure of other unconformities near Jedburgh and Glen Tilt (not Siccar Point). He even wrote about it a few years before searching the location at Siccar Point for further evidence to support his long-fought-for idea (Montgomery, 2003; Repcheck, 2003).

Another common story told in geology textbooks is Wegener's development of the hypothesis of continental drift. They start correctly, with Wegener having an insight after viewing a map of Africa and South America, but then most texts simplify the rest of the story so much that it obscures some real lessons in how science works. This is the consensus explanation in most books: Wegener could not devise a workable mechanism and therefore the idea was rejected. There was no way to "prove" his hypothesis as it was based only on circumstantial evidence. In this version of events, these texts portray science in a rigid manner where only after one gets all the proper supporting evidence, is it possible that the unbiased scientific community will automatically switch to the "right" answer. However, the lack of a mechanism to move the continents was not really a factor in rejecting the idea, as geologists at the time did accept the notion of global ice ages and Earth's magnetic field without having an accepted mechanism for either (Oreskes, 1999). Also, multiple geologists speculated (and published) about mantle convection and its implications to crustal stability, for example, the Reverend Osmund Fisher (1889), Reginald Daly (1923; Letsch, 2015) and Arthur Holmes (1928, 1929, 1944). Physicists agreed that convection in the mantle would be powerful enough to move continents. They just did not believe that it was happening in the mantle (Frankel, 2012b).

There were many reasons why geologists did not like the notion of horizontal displacement of the continents. Most important was that most geologists were only worried about the specific region they were doing their research in, and thermal contraction of the earth worked very well for them (Frankel, 2012b). Other reasons included the map projections that Wegener used were unfamiliar to many geologists, making them skeptical, and when Wegener's book was published in English (Wegener & Skerl, 1924), the translator used far more assertive language with Wegener's claims than Wegener had intended, turning many off the idea (Greene, 2015). R. T. Chamberlin (1928) gave 18 different reasons why he thought Wegener's idea should not be considered. While the question of the mechanism was among these reasons, other factors were true limitations for accepting Wegener's idea.

As part of the textbook story, the whole of the scientific community condemned the idea of continental drift. However, support did exist, though it was very regional (Frankel, 2012b)—coming from India, Africa, Australia (initially), and South America. These areas produced evidence for drift that was quite strong, but both North America and Europe continued to dismiss the notion, as the data were not so obvious in these regions. Moreover, North American and European geologists held much of the power over what got published and the ideas that became accepted.

Grotzinger and Jordan (2014) stated that "although Wegener was correct in asserting that the continents had drifted apart, his hypothesis about how fast they were moving and what forces were pushing them across the Earth's surface turned out to be wrong, as we will see, and those errors reduced the credibility [of this idea] among other scientists" (p. 28) also demonstrates the rational reconstruction of this history. Scientists could only know Wegener's estimates were wrong in hindsight from actual measurements once we had GPS—well after the acceptance of plate tectonics theory. Further, they assert that Wegener's reliance on purely circumstantial evidence produced skepticism. However, horizontal displacement of the continents (also known as "plate tectonics") was accepted before we had any direct measurements of continental movements. Indeed, the idea was based *only* on circumstantial evidence. In fact, it was after the publishing paleomagnetic evidence—circumstantial evidence published in the late 1950s—when geologists really should have accepted horizontal displacement almost a decade prior to Isacks et al.'s (1968) paper about the "new global tectonics" (Frankel, 2012a).

These stories, as recounted in most textbooks, reinforce a perspective of NOS and about what it takes to develop consensus around a new idea. It portrays scientists as completely rational and unbiased, only relying on objective and conclusive data. Scientific knowledge is obvious and straightforward, simplifying the role of the scientist in the process to do all of the intellectual heavy lifting.

4.7.2 Addressing the Limitations

To address these problems with geoscience textbooks, I wrote a textbook[6] (Dolphin, 2016) structured to parallel the lines of inquiry and historical contextualization necessary for promoting effective learning. The book has four sections, each representing one of the four learning goals of the course (Fig. 2.3)—the historical nature of the earth, the immense age of the earth, the dynamic nature of the earth and the process of science and how to judge reliability. The first three sections are all interrupted narratives (Allchin, 2017; Herreid, 2007) that supply important historical context while outlining the developmental trajectory of their respective overarching fundamental geology idea (Fig. 4.3).

In contrast to common geology texts, the historical narratives are the primary focus, and are periodically punctuated by sections entitled "Info Dumps" of content-heavy text, more akin to typical content presented in common textbooks. Info Dump segments have the following titles: *A modern categorization of rocks*; *A fossil primer*; *Let's talk about geologic maps;* and *Breaking down and transporting earth materials*.

[6] https://he.kendallhunt.com/product/stories-geology-what-we-know-and-how-we-figured-it-out

Preface
Section I: The Earth's Biography: Learning to Read the Rocks
  1 Simple Observations: Discerning the Signs of a History
  2 Clues of the Earth's Past: Giving Fossils Meaning
  3 The Human Pedigree: Icons, Extinctions, and Ice Ages
Section II: The Age of the Earth: The Discovery of Deep Time
  4 Breaking with Tradition: Empiricism Rises
  5 Deposition Rates: The Sands of Time
Section III: Developing a Theory of the Earth: Why Are There Still Mountains?
  6 The Mystery of Oceans and Continents: Initial Controversies
  7 Some History of Seismology: Putting a Finger on the Pulse of the Earth
  8 Attractions of the Past: The Importance of Paleomagnetism
Section IV: Putting It All Together: The "What" and the "How" of Science
  9 Epistemological Trajectories: Deconstructing Methods, Resolving Controversies
  10 Geology Within the Larger Scientific Terrain: Contributions and Questions
Bibliography
Glossary

Fig. 4.3 The table of contents for stories in geology: what we know and how we figured it out. (Dolphin, 2016)

In addition to the Info Dumps, the book incorporates other teaching strategies such as think questions, which ask students to reflect on aspects of the current problems presented in the text. As well, two science fiction short stories are included. One focuses on the extinction of the dinosaurs (Harrison, 1992), and the other looks at the scale of geologic time and how massive 4.6 billion years is (Preuss, 1992). The science in these short stories is solid and the stories make the content inviting and entertaining for all students to engage with. The book also links to (a) a number of historical narratives in an online collection named, *Stories Behind the Science* (Clough, 2020), (b) historical case studies (Allchin, 2011; Dolphin et al., 2018), and (c) a dramatization about a fictitious debate about the age of the earth (Stinner & Teichmann, 2003).

The final section of the book addresses the important aspects of the process of science in an explicit and reflective way. This part is crucial to help students develop a more robust understanding of NOS (Abd-El-Khalick & Lederman, 2000; Khishfe & Abd-El-Khalick, 2002; Lederman, 2007). This explicit and reflective instruction is also the final transformative teaching strategy that I will discuss in this chapter.

4.8 Explicit, Reflective NOS Instruction

Accurate Nature of Science (NOS) instruction is important for students to develop a more nuanced understanding for how science in general, and geology in particular, actually works. The aim is that, as a result, they will have a stronger capacity to determine the reliability of scientific claims they encounter in their daily lives. Literature shows that students will not readily develop a robust conceptualization of the NOS through purely implicit lessons, not to mention that the way science is portrayed in texts, labs, and lectures leaves much of the process and context of the scientific process out. Instead, students must draw their attention explicitly to and reflect upon the aspect of NOS that is part of the lesson (Clough, 2006; Khishfe & Abd-El-Khalick, 2002; Scharmann et al., 2005).

During the earlier portion of the course, I do make many overt references to NOS concepts, thus, I dedicate the last section of the course (about 2 weeks) to pulling those threads together and explicitly exploring aspects of NOS that were most prevalent in the three historical narratives that comprise the rest of the course. I placed this topic last because the previous three narratives problematized the more naïve views of NOS, and I wanted to leverage that cognitive dissonance so students might develop a more sophisticated understanding; being critical of the earlier narratives, picking out particular scenarios, and arguing how the context was important to the trajectory of the knowledge development. Doing it in this order gave the students the concrete experiences, the data, that they could then use to generalize descriptions of how science works. This section will describe some of the explicit and reflective activities incorporated within the course to help students develop a more nuanced understanding of the process of science.

4.8.1 Informing Students How They Learn

I spend the entire first-class period *not* doing geology, but instead address how learning requires mental engagement and meaning-making, even when information is presented. Among them are conceptual metaphor theory (Amin, 2015; Beger & Jäkel, 2015; Beger & Smith, 2020; Lakoff & Johnson, 1980, 1999; Niebert & Gropengiesser, 2015), embodied cognition (Clark, 2011; Hutchins & Renner, 2012; Shapiro, 2011), and cognitive linguistics (Indurkhya, 1992; Rohrer, 2005; Steen, 2014, 2007). I do not bring up these disciplines explicitly, but structure the class period around a story, *The Toolmaker's Paradigm* (Reddy, 1979), where I have students think critically about the way communication happens. They do not receive knowledge from the lecturer or from the textbook, but only information in the form of alternating compressions and expansions of air (from the lecturer's voice) and black signs and symbols on a white page (from the textbook).

I express to them that the information (sound or symbols) does not have any intrinsic meaning; rather, it is their mind that imposes meaning on the patterns of air vibrations or signs and symbols on the page. This constructed meaning is culturally dependent, with bias being an integral part of the meaning-making process. I contrast this with the way we talk about teaching and learning, which is very much in line with Reddy's (1979) "conduit" metaphor for communication. Reddy showed that how we talk about communication is highly metaphorical and the metaphors we use influence the way we understand communication and therefore the way we interact with each other. The conduit metaphor for communication includes the following metaphors (and accompanying examples):

- **IDEAS ARE THINGS:** I am giving you the information. These are the take-away points. I can't make it any more concrete.
- **WORDS ARE CONTAINERS FR IDEAS**: He had trouble putting his thoughts into words. I am having trouble finding the meaning in what you are saying.
- **COMMUNICATING IS SENDING:** Do not force your ideas down my throat. I am trying to figure out the best way to deliver this lecture.

This symbolic language matters because metaphors influence how we think about communication. The conduit metaphor, which emphasizes the physical transfer of ideas (like scientific concepts in a science classroom) reinforces the notion that as a teacher, *delivering* a lecture is the most efficient way to teach, and *taking* notes is the most effective way to learn. Once I was aware of this, I observed three specific metaphors present in discussions I have listened to or participated in about teaching and learning, shown in Table 4.7.

Making the use of these metaphors explicitly facilitates understanding about how they implicate and reinforce specific actions (Hauser & Schwarz, 2019; Kahneman, 2011; Lakoff & Johnson, 1999; Thibodeau & Boroditsky, 2013). For instance, one important insight is that students will know they cannot just sit in lecture and hope to *absorb* the knowledge or hope they *get it*. They need to realize that the work of meaning-making really lies with them. It is against this backdrop that I then talk about why I have implemented the strategies that I have. I also connect this idea to scientists and how the knowledge they develop does not come directly from the environment. In other words, data do not speak for themselves. Instead, information is mediated by the senses and imagination, based on past experiences; out of this process knowledge is constructed.

I emphasize the need for continual communication among group members as an effective way to help clarify and socially construct the disciplinary concepts; this interaction parallels what scientists do in their labs, at conferences and meetings, and through peer review. This kind of explicit and reflective reference to aspects of NOS occur regularly—such as when asking how an investigation might have unfolded, about the role of assumptions within scientific explanations, how multiple explanations for the same set of data can exist, how controversies resolve, how the social context might have influenced the developmental trajectory of scientific

Table 4.7 Examples of metaphors when talking about teaching and learning

Metaphor	Example sentences
Ideas are moving objects	Do you understand what I am trying to *convey*? Let the idea *sink in* for some time. What ideas *came out* of the reading? Try not to *mix up* these ideas. She only *regurgitates* information memorized from the notes
The head is a container	Try to keep an *open mind*. His mind was a *storehouse* of ideas. Try to *wrap your head* around these ideas. The lecture really *filled in a lot of gaps* in my mind
Possessing something means controlling it	Do you *get it*, what I am trying to say? She was not *receptive* to the idea. She was able to *soak up* the main points of the lecture. Did you *catch* what I am trying to say? My goal is to *gain* as much knowledge as possible

knowledge, or how the development of scientific knowledge might have influenced the social context. By the end of the course, students are ready to look at these aspects.

4.8.2 Geology as a Historical and Hermeneutic Science

This final section on explicit and reflective focus on NOS begins with a reminder from the beginning of the term that geology is about reading the signs in the rocks and giving meaning those signs as part of constructing the history of the earth. I remind students that reading a book is only observing signs and symbols with the encultured mind and creating meaning based on past experiences. Rocks contain signs that show the effects of past causes. Awareness of the signs appearing in rocks is necessary to assign appropriate meaning, building a history of the rock.

I also emphasize that more than one cause can be responsible for any sign (for example, a meteorite impact might create a crater, shocked quartz, an ash layer, glass spherules, and other things) and that some signs can result from more than one cause (for example, a clay with high iridium concentration could be caused by a meteorite impact, extreme volcanic activity, or a nearby super nova explosion). When we find multiple signs that could result from multiple possible causes, we need to determine which possible cause is the most likely—what Cleland (2013) refers to as a "smoking gun." Geologists do this by looking for effects that would be particular for one cause over the others. Figure 4.4 depicts this decision tree.

After reminding students of the multiple iterations of this approach within all of the narratives, I ask them how this compares to the stereotypical scientific method. In this discussion, I emphasize how determining the most likely explanation does not usually require a controlled experiment, nor would it be possible to conduct one. Often, it is seeking evidence appearing in rocks that supports cause 1 over cause 2. However, if the researcher is unable to find specific evidence for cause 1, cause 1 is not eliminated because the researcher could have missed the evidence, or it could have been destroyed by the many earth processes that work at breaking down and recycling earth materials. What helps us to decide one cause over another is the strength—that is, reliability—of the explanation, one that fits all relevant data better than any other explanation.

The implication of developing explanations from a set of signs leads to the description of geology as a primarily historical and interpretive science compared to most experimental sciences, such as physics and chemistry (Dodick & Argamon, 2006; Dodick et al., 2009; Frodeman, 1995). It means that what counts as evidence, testing hypotheses, and objects of study must diverge from experimental sciences (Dolphin & Dodick, 2014). In essence, the best tools depend on the research context. See Table 4.8 for details.

Accordingly, a geologist will observe some data, develop a hypothesis abductively, and decide what kind of data is needed to support that hypothesis or discriminate among multiple working hypotheses (Chamberlin, 1965). The new data would

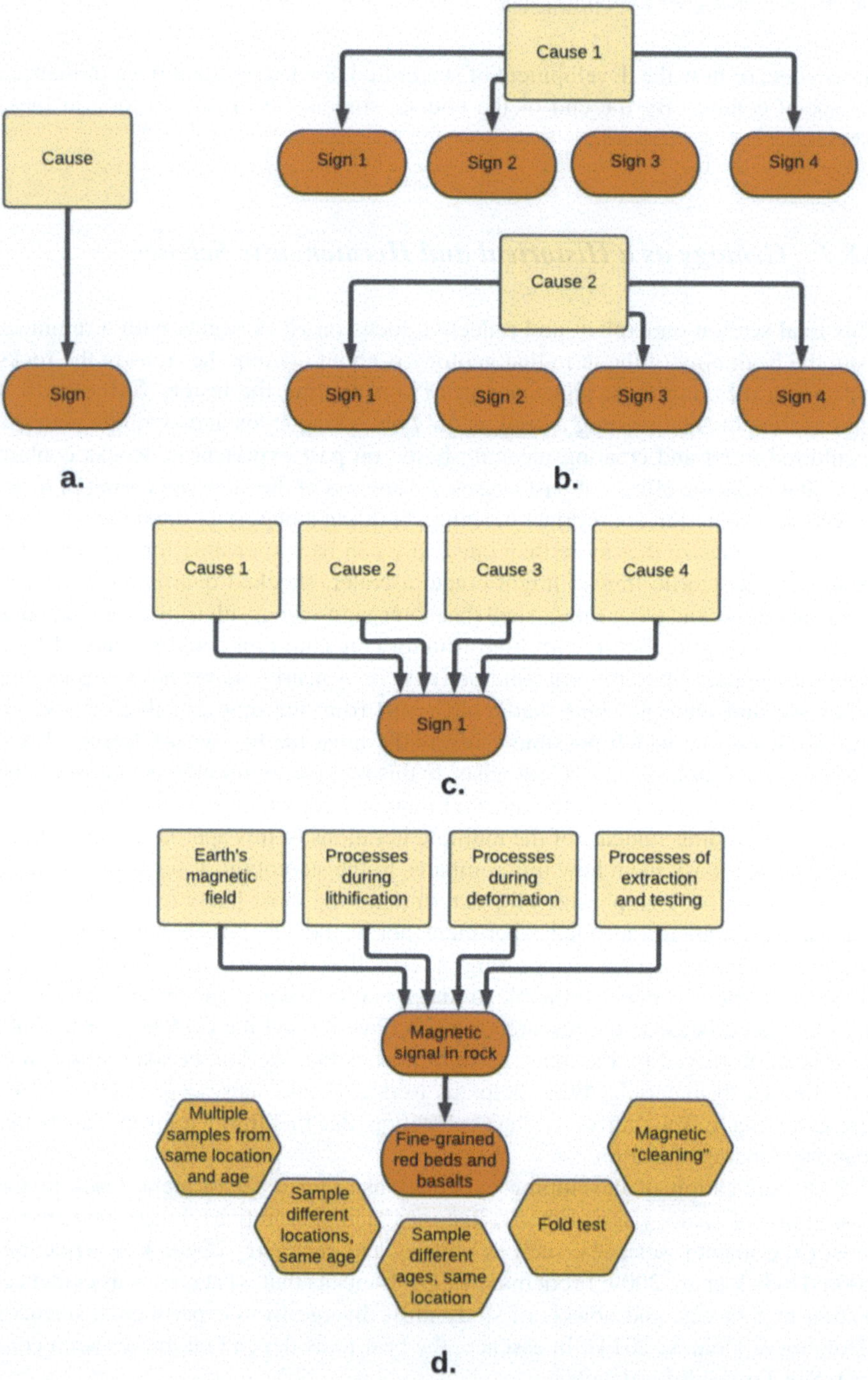

Fig. 4.4 The relationship between geologic causes and their signs. (Note. Graphic relationships between causes and signs: (**a**) any sign in a rock has a cause, (**b**) different causes can each have multiple signs, with some overlapping, (**c**) multiple causes could have one (or multiple) sign in common, and (**d**) multiple possible causes for geomagnetic signs in rocks. Different tests (gold hexagons) will help narrow down the most likely cause of the sign)

Table 4.8 Differences in dimensions between experimental and historical sciences. (After Dolphin & Dodick, 2014)

Dimension	Experimental science	Historical science
Research goal	Reduction to general (universal) laws and behaviours (inductive reasoning)	Explanations for ultimate and contingent causes (abductive reasoning)
How evidence is gathered	Controlled manipulation of nature (experiments)	Observing/analyzing pre-existing entities (signs)
How hypotheses are tested	Predictive accuracy	Explanatory accuracy (retrodictive accuracy)
Objects of study	Uniform and interchangeable entities	Complex and unique entities

narrow the choices of hypotheses or perhaps result in the researcher modifying their hypothesis that may necessitate new data collection, subsequently leading to further hypothesis modification. This iterative activity is very much parallel with how we have discussed effective learning, as individuals modify their mental model of a phenomenon of study. Their observations will either cause them to change their mental model, which is called "accommodation," or they will impose their ideas onto the observations to interpret them in a particular way, called "projection." This process is represented in Fig. 4.5. For example, Charles Lyell and George Cuvier demonstrated projection by interpreting mass extinctions through either uniformitarian or catastrophist lenses, respectively.

4.8.3 Resolving Controversy in Geology

The implication of this personal trajectory of mental model development is that multiple people can look at the same information and develop very different models. Each model will have a grade of reliability, how well it can predict effects of a phenomenon. We like to talk about "winners" and "losers" of controversies, but if we follow the trajectories of these separate ideas, we see that consensus models are often a hybrid of the models participating in the controversy. In this regard, I emphasize the controversy and how we came to our present knowledge rather than merely the "right" answer. For example, the brief sketches of controversies below illustrate what I address in class and the implications for science knowledge creation. Basically, *the scientific method* would have you think that a simple experiment would be the deciding factor for which side of a controversy is correct, but is that how it plays out?

James Hutton and Charles Lyell favored interpreting the signs in rocks as caused by small changes over a vastly long time, where Georges Cuvier recognized qualities in the rocks that signified mass extinctions and invoked the notion of big changes in short amounts of time, resulting from severe and quick catastrophes. Tensions persisted between the two views of how to bridge the present with the past.

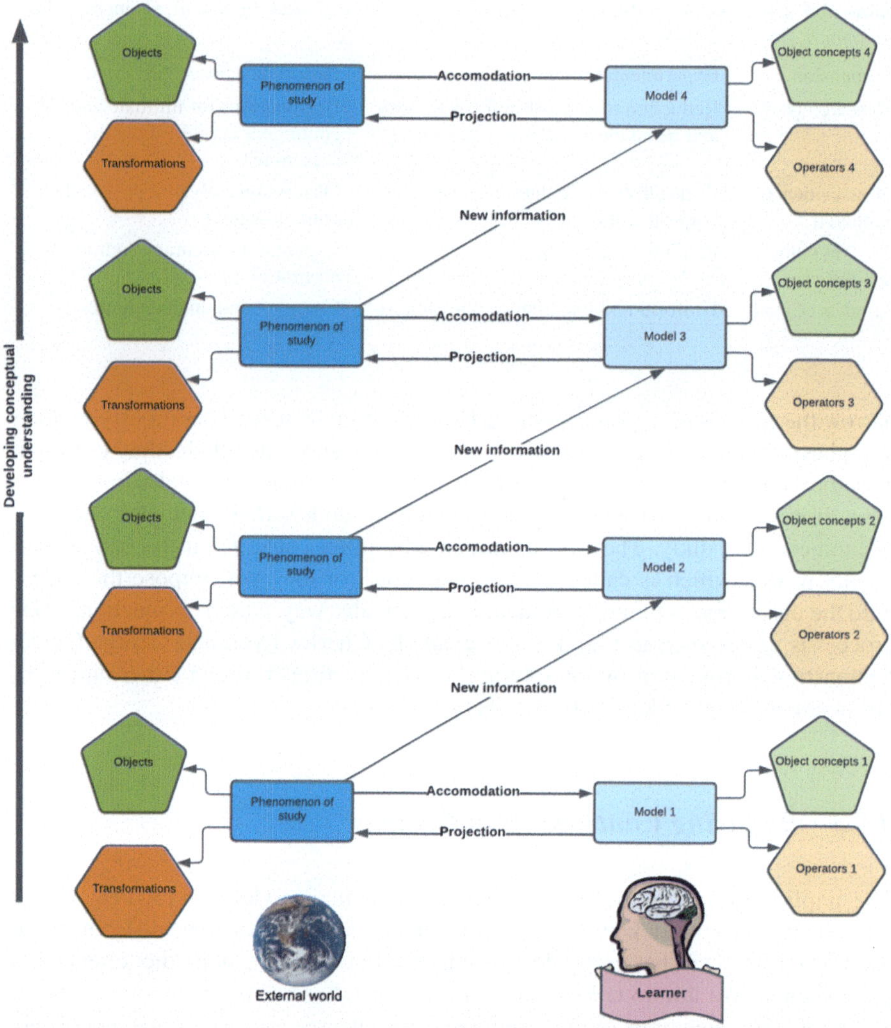

Fig. 4.5 The interface between the learner's mind and the external world in mental model building. (Note. Diagrammatic representation of mental model development over time. Learners (scientists included) study a phenomenon (dark blue rectangle), including objects (dark green pentagon) and transformations (dark orange hexagon), and develop a mental model (light blue rectangle), including object concepts (light green pentagon) and operators (light orange hexagon). "Accommodation" is how the phenomenon changes the mental model of the learner. "Projection" is how the learner projects their past experiences and biases onto the phenomenon of study, which will affect how they interpret that phenomenon. Introducing new information causes the learner to amend the initial model and inform successive models. (After Indurkhya, 1992))

Today we understand that much of the change we observe—weathering of rocks, deposition of mud, growth of mountains ranges—occur exceedingly slowly. However, the role of large, catastrophic events—super volcanoes, meteorite impacts, and dramatic climate fluctuations—are acknowledged as causes of some signs from the past. Both perspectives were correct about certain aspects, and the consensus view acknowledges portions of each wrapped into a singular concept, known as *actualism*.

Another example, beautifully illustrated by Rudwick (1985), occurred when ***Henry De la Beche (1796–1855)*** claimed, during his fieldwork in Devon, England, that coal measures existed in the Cambrian strata. This was surprising, as most geologists at the time considered all coal deposits to be Carboniferous in age, and thus much younger than Cambrian. De la Beche claimed this because the coal was found adjacent to the transitional rocks known as greywacke, and greywackes were considered the oldest sedimentary rocks. Meanwhile, ***Roderick Murchison (1792–1871)***, who had not seen the rocks personally, swore that the coal measures there must be much younger. He was using William Smith's method of correlation that depended on fossil occurrence.

Murchison asserted that an unconformity (missing rock due to erosion) must exist between the greywackes below and the coal measures above, which de la Beche obviously missed. This caused a great controversy in the geological community, mainly because it called into question the theoretical foundations of this type of relative dating of rocks by rock type or by the fossils they contained. After about 10 years of controversy and more field work, globally, De la Beche agreed that the coal measures were indeed younger than he had originally thought, and Murchison agreed that there was no unconformity; therefore, the coal measures were older than he originally thought. The problem was solved with the creation of a new age of rock, the Devonian Period, that allowed for both circumstances to be true. Despite the tremendous split it caused in the geologic community, no one really remembers it now because it was settled to the satisfaction of all involved (Rudwick, 1985). The message here is that reality does not determine the winner of the controversy: indeed, the winner of the controversy determines reality (Latour, 1987). This is not to say that we create the objective universe, but only our *conception* of the universe, which is always subject to change.

In sum, such examples (and there are many, many more) show how science knowledge is often *negotiated*, taking reliable aspects from both sides, and combining them into a new understanding that does a better job explaining the phenomenon than either model did separately. Establishing scientific ideas is not a case of voting for the better model, but rather which model gets most use because it is the most reliable; because it is the most pragmatic.

4.8.4 Cognitive Resources

As described earlier, Harry Fielding Reid developed elastic rebound theory as an explanation or mechanism for earthquakes. He noted the shifting of survey markers across the San Andreas fault over several decades and most likely combined this data with his past experiences using survey markers to track glacial movements.

This projection of past experiences onto a new situation is utilizing *cognitive resources* (Giere, 1988).

In like fashion, Wegener may have used his experiences in Greenland with drifting icebergs as a cognitive resource for projecting continents drifting through ocean crust (Greene, 2015). Cuvier may have used his experience living through the French Revolution to project large scale revolutions on Earth as being responsible for mass extinctions (Rudwick, 2014). Alexander du Toit and Edna Plumstead both learned about continental drift in school and lived in South Africa, where observations readily supported the idea of horizontal displacement of the continents. These cognitive resources helped them be open to and contribute to this idea (Frankel, 2012b).

Lawrence Morley was both sympathetic to drift, being a geophysicist with an interest in geomagnetism, but also could draw on his background in complex mathematics as a cognitive resource, which helped him formulate his ideas that would lead eventually to the notion of plate tectonics (Giere, 1988). All these examples highlight that it is not enough to just be smart or observant: having the right cognitive tools or a new kind of vision based on past experiences is key to making breakthroughs.

4.8.5 The Role of Serendipity

Serendipity is another aspect of the scientific process, though it does not get much attention, which is an oversight given the role it plays. Serendipity could be the chance experience or the novel combination of ideas that leads to some insight (Reiff-Cox, 2020; Thagard, 2012). Wegener was viewing a new world atlas when he saw the similarities between the eastern coastline of South American and the western coastline of Africa. This encouraged him to consider if they might have once been connected and set him out in search of data that would support such a notion.

Here are some other notable examples of serendipity:

- Mary Anning's chance discovery of the first Plesiosaurus, bringing great attention to her from other collectors.
- Henri Becquerel's placing pitchblende onto photo sensitive paper and having Marie and Pierre Curie explore the evidence of radiant energy.
- Walter Alvarez's discovery of the end Cretaceous clay layer and the iridium anomaly it contained.

Where textbooks portray science in a very linear, inevitable manner, these select examples demonstrate that scientific advancement is often dependent on being in the right place at the right time with the right cognitive tools.

4.8.6 Prematurity in Geologic Discovery

Serendipity can lead to novel insights, but sometimes the world is not ready for that insight. For instance, sometimes scientific claims get made, yet they are not taken up by the scientific community for years or even decades. This phenomenon is what Stent (2002) referred to as prematurity in scientific discovery. He showed that a new idea would not become part of the mainstream until an unequivocal and logical path from the new idea to canonical knowledge becomes evident.

Glen (2002) described a few examples of this in the earth sciences. He showed that though Svante Arrhenius had concluded by 1896[7] that adding CO_2 to the atmosphere would cause the atmosphere to warm, this idea only became part of the canonical knowledge in the 1950s and 1960s. In another example, Walter Alvarez and his group from the University of California, Berkley, discovered the iridium anomaly in the clay layer in Gubio, Italy in 1977. By 1980, they proposed the meteorite impact theory to explain dinosaur extinction. Though the idea did become mainstream in popular culture, it was not until 2010, after a lengthy review of the literature, that over 40 scientists declared the controversy over and stated definitively that the dinosaurs had gone extinct because of a meteorite impact (Schulte et al., 2010).

Another premature discovery in science was Wegener's idea of horizontal displacement of the continents. From its inception in 1912, it took nearly 60 years to develop into plate tectonics, as the consensus view. These ideas were most likely held up due geologists' adherence to the uniformitarian theoretical framework. In all cases, the notion of global warming, dinosaur extinction, and plate tectonics were not coherent with the rules governing uniformitarianism (Glen, 2002), namely small-scale change on a steady state earth.

4.8.7 Models and Modeling

The course concludes with a discussion of the nature of models and modelling within the process of science. We explore the notions that models are artificial, utilitarian, simplified, interpreted, and imperfect (Gilbert & Watt Ireton, 2003). Models highlight and hide aspects of what they represent—the target concept. Further, what gets hidden and highlighted depends on the kind of model in play (Boulter & Buckley, 2000). Models can be concrete and functional, meaning they behave in a way that is like the target concept, or they can be concrete and scale, meaning they are a proportionate in shape. However, scale models seldom are functional, because processes and materials do not often scale up or down appropriately.

[7] Note that recent research shows that Eunice Foote had made the same claim as Arrhenius, but in 1856, 40 years prior (Shapiro, 2021).

Visual models are another type—those that are two-dimensional representations, like maps and diagrams. Gestural models are when the body and its motions represent a target. Verbal models present in the form of similes, analogies, and metaphors, and mathematical models show relationships represented through equations. Using multiple modes to represent the same concept (Table 4.9) help modellers to develop a more robust understanding of the target concept.

The discussion then relates back to the beginning of the course and learning theory. We talk about how mental models, including their strengths and limitations, become different modes of expressed models, and their strengths and limitations as well, such as what they highlight and hide. Importantly, though experts think differently about their topics than novices (Clement, 2008b), many mental activities of experts are parallel to those of novices. This affirms that scientists are human and that any human, even a student, could be a scientist. The final assignment in the course is a long-term research assignment where students can review the NOS section of a typical introductory geology textbook (see Table 4.2) and write a letter to the author(s) of the book constructively critiquing their presentation of NOS content compared to what they have learned in the course.

All that has been described in this chapter demands that the beginning of the course be carefully structured so that all runs smoothly the remainder of the semester. The initial effort pays enormous dividends in promoting student attendance, mental engagement, and personal investment in the course. The evidence-based rationale for incorporating each pedagogical strategy and how they may be implemented in large lecture classrooms are to encourage and assist others in such efforts.

Table 4.9 Examples of multiple modes of representing aspects of earthquakes

Mode	Description	Specific target
Concrete-functional	Earthquake machine (Hubenthal et al., 2008)	The elastic and frictional aspects of earthquakes
Concrete-scale	Landscape model with hypocentres plotted	Earthquake locations
Visual	Map with epicentres plotted, or diagram with hypocentres plotted	Earthquake locations
Gestural	Hands sliding back and forth against each other	Nature of faulting and energy released because of an earthquake
Verbal	Simile-P-wave is like a sound wave Analogy-light is to air/water interface as seismic waves is to crust mantle interface Metaphor: The earth rings during a free oscillation	Seismic waves
Mathematical	The standard body: Wave magnitude formula is mb = log10(A/T) + Q(D,h) The standard surface: Wave formula is MS = log10 (A/T) + 1.66 log10 (D) + 3.30 Seismic moment: MO = μS‹d› Moment magnitude: MW = 2/3 log10(MO)—10.7	Energy released during an earthquake

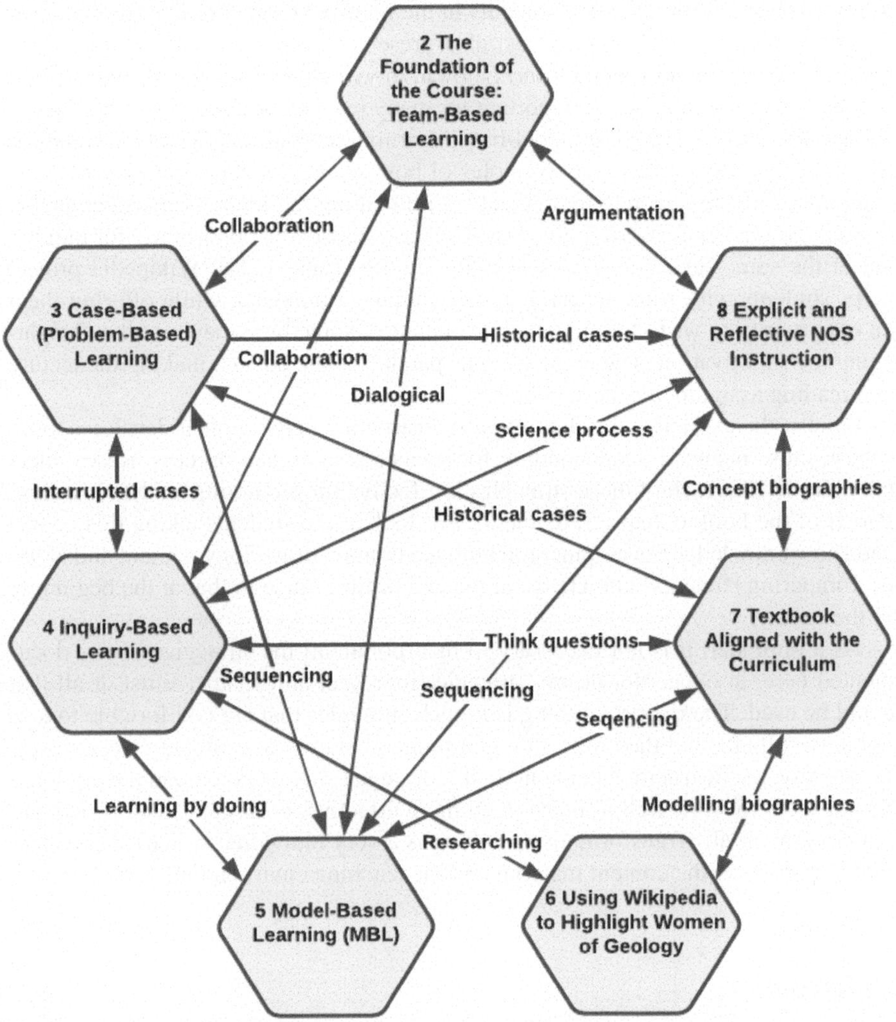

Fig. 4.6 The interconnectedness of the different teaching strategies. (Note. A diagram showing some of the major interconnections among the different teaching strategies described above. The numbers in the green hexagon reference the section containing the descripting of that strategy. Labeled arrows highlight the major relationships between any two strategies)

While addressed separately, they have many interconnections and interrelationships among them (Fig. 4.6).

For instance, team-based learning is the starting point for all other strategies because the students are in the same group for the entirety of the course. They experience the negotiation of knowledge as they argue and collaborate in their groups during exams, writing activities and in-class discussions. The textbook has the narrative and explicit/reflective NOS instruction as well as the inquiry activities and think questions, which involve students in the inquiry at that moment. Its sequencing, though conceptual biographies, also presents science content in order from more concrete and descriptive information to more explanatory and abstract.

The formation of teams is important for many reasons: smaller class feel, loss of anonymity, and encouraging model-based learning through dialogical communication. The historical cases give examples of how science is done and give students experience—from concrete to abstract—with making the kinds of discoveries that pushed the science forward in the past. The cases create the opportunity for inquiry, but at the same time constrain where the inquiry can go. The Wikipedia project helps students with some research and communication skills, while offering them an opportunity to write for an authentic audience, something they report as heightening their motivation. The textbook runs parallel to the course, making the lecture and reading assignments more coherent.

Finally, the explicit and reflective NOS instruction helps student develop a more robust, more nuanced understanding for science as a human process. Researchers have shown that each of these strategies is effective for promoting student learning. Part II of the book (Chaps. 5, 6, and 7) will look at the students taking this course and show knowledge development in both the domain of geology content and NOS, by comparing students' knowledge at the end of the course to that at the beginning of the course.

As a reminder, it is not necessary to incorporate all the strategies I have documented here, at once. Nor do my strategies represent an exhaustive list of all that could be used. The important thing is to pick strategies that are comfortable to use, implement them, and then reflect on that implementation to evaluate their efficacy. In this way, the instructor can maintain the strategies that work, amend the ones that do not seem to work well to make them more effective, or discard them if they are not working at all. Transformative teaching is a very individual activity, dependent on the instructor, the content material and the teaching environment.

References

Abd-El-Khalick, F., & Lederman, N. G. (2000). Improving science Teachers' conceptions of nature of science: A critical review of the literature. *International Journal of Science Education, 22*(7), 665–701.

Allchin, D. (2002). How not to teach history in science. *The Pantaneto Forum*, (7), 1–13. http://pantaneto.co.uk/how-not-to-teach-history-in-science-douglas-allchin/

Allchin, D. (2003). Scientific myth-conceptions. *Science Education, 87*(3), 329–351.

Allchin, D. (2011). The Minnesota case study collection: New historical inquiry case studies for nature of science education. *Science & Education*, Published on-line http://www.springerlink.com/content/v42561276210585q/. Accessed 1 Mar 2012, 1–19.

Allchin, D. (2013). *Teaching the nature of science: Perspectives & resources*. SHiPS Education Press.

Allchin, D. (2017). Historical inquiry cases for nature of science learning. *Cadernos De História Da Ciência, 13*(2), 101–126.

Allchin, D. (2020). Historical inquiry cases for teaching the nature of science and analytical skills. In W. F. McComas (Ed.), *Nature of science in science instruction: Rationales and strategies* (pp. 595–606). Springer.

Amin, T. G. (2015, April 15). Conceptual metaphor and the study of conceptual change: Research synthesis and future directions. *International Journal of Science Education*, Downloaded from internet: https://doi.org/10.1080/09500693.2015.1025313.

Australian Curriculum Assessment and Reporting Authority. (2014). *Earth and environmental science*. Education Services Australia Limited. Retrieved from https://www.australiancurriculum.edu.au/senior-secondary-curriculum/science/earth-and-environmental-science/?unit=Unit+1&unit=Unit+2&unit=Unit+3&unit=Unit+4

Azzam, A. (2017). Embracing Wikipedia as a teaching and learning tool benefits health professional schools and the populations they serve. *Innovations in Global Health Professions Education, 2017*(1), 1–9. https://doi.org/10.20421/ighpe2017.01

Beger, A., & Jäkel, O. (2015). The cognitive role of metaphor in teaching science: Examples from physics, chemistry, biology, psychology and philosophy. *Philosophical Inquiries, 3*(1), 89–112.

Beger, A., & Smith, T. H. (2020). Introduction. In A. Beger & T. H. Smith (Eds.), *How metaphors guide, teach, and popularize science* (pp. 1–40). John Benjamins Publishing Company.

Blackburn, R. A. R. (2018). Write my next lecture: Prelecture problem classes and in-lecture discussion to assist case-study teaching of synthesis. *Journal of Chemistry Education, 95*(1), 104–107. https://doi.org/10.1021/acs.jchemed.7b00528

Blazar, D., Heller, B., Kane, T., Polikoff, M., Staiger, D., Carrell, S., et al. (2019). *Learning by the book: Comparing math achievement growth by textbook in six common Core states.* C. f. E. P. Research. https://cepr.harvard.edu/files/cepr/files/cepr-curriculum-report_learning-by-the-book.pdf

Boulter, C., & Buckley, B. (2000). Constructing a typology of models for science education. In J. Gilbert & C. Boulter (Eds.), *Developing models in science education* (pp. 41–58). Kluwer Academic publishing.

Brasil. (2017). *Base Nacional Comum curricular.* Ministério da Educacão (MEC). Retrieved from http://basenacionalcomum.mec.gov.br/documento/BNCCAPRESENTACAO.pdf

Buckley, B., & Boulter, C. (2000). Investigating the role of representations and expressed models in building mental models. In J. Gilbert & B. Carolyn (Eds.), *Developing models in science education* (pp. 119–136). Kluwer Academic Publishers.

Burko, L. M. (2016). Using the case study method in teaching college physics. *The Physics Teacher, 54*(7), 413–415. https://doi.org/10.1119/1.4962777

Bybee, R. W. (2006). Scientific inquiry and science teaching. In L. B. Flick & N. G. Lederman (Eds.), *Scientific inquiry and nature of science: Implications for teaching, learning, and teacher education* (pp. 1–14). Springer.

Çam, A., & Geban, Ö. (2017). Effectiveness of case-based learning instruction on pre-service teachers' chemistry motivation and attitudes toward chemistry. *Research in Science & Technological Education, 35*(1), 74–87. https://doi.org/10.1080/02635143.2016.1248927

Camill, P. (2006). Case studies add value to a diverse teaching portfolio in science courses. *Journal of College Science Teaching, 36*(2), 31–37.

Carey, S. (2009). *The origin of concepts.* Oxford University Press.

Carloye, L. (2017). Mini-case studies: Small infusions of active learning for large-lecture courses. *Journal of College Science Teaching, 46*(6), 63–67.

Chamberlin, R. (1928). Some of the objections to Wegener's theory. In W. A. J. M. van Waterschoot van der Gracht (Ed.), *Theory of continental drift* (pp. 83–89). American Association of Petroleum Geologists.

Chamberlin, T. C. (1965). The method of multiple working hypotheses (reprint of the 1890 version). *Science, 148*, 754–759.

Clark, A. (2011). *Supersizing the mind:Embodiment, action, and cognitive extension.* Oxford University Press.

Cleland, C. E. (2013). Common cause explanation and the search for smoking gun. In V. R. Baker (Ed.), *Rethinking the fabric of geology: Geologic Society of America Special Paper 502* (pp. 1–9). Geologic Society of America.
Clement, J. (2008a). Student/teacher co-construction of visualizable models in large group discussion. In J. Clement & M. A. Rea-Ramirez (Eds.), *Model based learning and instruction in science* (Vol. 2, pp. 11–22). Springer.
Clement, J. (2008b). *Creative model construction in scientists and students: The role of imagery, analogy, and mental simulation*. Springer.
Clough, M. (2006). Learners' responses to the demands of conceptual change: Considerations for effective nature of science instruction. *Science Education, 15*(5), 463–494.
Clough, M. P. (2020). Using stories behind the science to improve understanding of nature of science, science content, and attitudes toward science. In W. F. McComas (Ed.), *Nature of science in science instruction: Rationales and strategies* (pp. 513–526). Springer.
Conant, J. B. (1947). *On understanding science; an historical approach*. Yale University Press.
Dalessandro, J. (2006). *The damnedest, finest ruins – Episode 601* J. Dalessandro & R. S. Burton; KQED.org. https://www.youtube.com/watch?v=mvve8YiYGSo
Daly, R. A. (1923). The Earth's crust and its stability. *American Journal of Science, 5*(29), 349–371.
Dodick, J., & Argamon, S. (2006). Rediscovering the historical methodology of the earth sciences by analyzing scientific communication styles. In C. A. Manduca & D. W. Mogk (Eds.), *Earth and mind: How geologists think and learn about the earth* (Geological Society of America special paper 413) (pp. 77–93). Geological Society of America.
Dodick, J., Argamon, S., & Chase, P. (2009). Understanding scientific methodology in the historical and experimental sciences via language analysis. *Science & Education, 18*(8), 985–1004.
Dolphin, G. (2016). *Stories in geology: What we know and how we figured it out*. Kendall-Hunt Publishing.
Dolphin, G., & Benoit, W. (2016). Students' mental model development during historically contextualized inquiry: How the 'tectonic plate' metaphor impeded the process. *International Journal of Science Education, 38*(2), 276–297.
Dolphin, G., & Dodick, J. (2014). Teaching controversies in earth science: The role of history and philosophy of science. In M. M. Matthews (Ed.), *International handbook of research in history, philosophy and science teaching* (Vol. 1, pp. 553–599). Springer.
Dolphin, G., Benoit, W., Burylo, J., Hurst, E., Petryshen, W., & Wiebe, S. (2018). Braiding history, inquiry, and model-based learning: A collection of open-source historical case studies for teaching both geology content and the nature of science. *Journal of Geoscience Education, 66*(3), 205–220. https://doi.org/10.1080/10899995.2018.1475821
Dori, Y. J., Tal, R. T., & Tsaushu, M. (2003). Teaching biotechnology through case studies: Can we improve higher order thinking skills of nonscience majors? *Science Education, 87*(6), 767–793.
Driscoll, M. (2005). *Psychology of learning for instruction* (Vol. 3rd). Allyn & Bacon.
Evenstein Sigalov, S., & Nachmias, R. (2017). Wikipedia as a platform for impactful learning: A new course model in higher education. *Education and Information Technologies, 22*, 2959–2979. https://doi.org/10.1007/s10639-016-9564-z
Faulkner, P. E., Meeshay, W.-W., & John, R. (2018). Case studies: A teaching strategy for promoting critical thinking in the 21st century agriculture student. *NACTA Journal, 62*(3), 286–287.
Fisher, O. (1889). In O. Fisher (Ed.), *Physics of the earth's crust, by the rev* (2d ed., altered and enlarged.). Macmillan and co..
Fletcher, C., Gibson, D., & Ansdell, K. (2014). *Introduction to physical geology: The science of earth* (Canada edition). John Wiley & Sons Canada, Ltd.
Franco, C., & Colinvaux, D. (2000). Grasping mental models. In J. Gilbert & C. Boulter (Eds.), *Developing models in science education* (pp. 93–118). Kluwer Academic Publishers.
Frankel, H. (2012a). *Paleomagnetism and confirmation of drift* (Vol. 2). Cambridge University Press.
Frankel, H. (2012b). *Wegener and the early debate* (Vol. 1). Cambridge University Press.
Frodeman, R. (1995). Geological reasoning: Geology as an interpretive and historical science. *Geological Society of America Bulletin, 107*, 960–968.

Giere, R. N. (1988). *Explaining science: A cognitive approach*. University of Chicago Press.
Gilbert, S., & Watt Ireton, S. (2003). *Understanding models in earth and space science*. NSTA Press.
Gilbert, J., Boulter, C., & Elmer, R. (2000). Positioning models in science education and in design and technology education. In J. Gilbert & C. Boulter (Eds.), *Developing models in science educuation* (pp. 3–18). Kluwer Academic Publishers.
Glen, W. (2002). A triptych to serendip: Prematurity and resistance to discovery in the earth sciences. In E. B. Hook (Ed.), *Prematurity in scientific discovery*.
Greene, M. T. (2015). *Alfred Wegener: Science, exploration and the theory of continental drift*. Johns Hopkins University Press.
Grotzinger, J. P., & Jordan, T. H. (2014). *Understanding earth* (Seventh ed.). W. H. Freeman and Company.
Harrison, H. (1992). Dawn of the endless night. In B. Preiss & R. Silverberg (Eds.), *The ultimate dinosaur* (pp. 264–273). Bantam Books.
Hauser, D. J., & Schwarz, N. (2019). The war on prevention II: Battle metaphors undermine cancer treatment and do not increase vigilance. *Health Communication, 35*(15), 1698–1704. https://doi.org/10.1080/10410236.2019.1663465
Heilman, J. M., & West, A. G. (2015). Wikipedia and medicine: Quantifying readership, editors, and the significance of natural language. *Journal of Medical Internet Research, 17*(3), e62. https://doi.org/10.2196/jmir.4069
Herreid, C. F. (2007). *Start with a story: The case study method of teaching college science*. NSTA Press.
Herreid, C. F. (2012). ConfChem conference on case-based studies in chemical education: The future of case study teaching in science. *Journal of Chemical Education, 90*, 256–257. https://doi.org/10.1021/ed2008125
Herreid, C. F. (2017). Puttin' on the ritz: How to put science into cases. *Journal of College Science Teaching, 47*(1), 54–56.
Herreid, C. F. (2018). Exercises in style: Is there a best way to Wwite a case study? *Journal of College Science Teaching, 48*(2), 34–38.
Hodges, L. C. (2005). From problem-based learning to interrupted lecture: Using case-based teaching in different class formats. *Biochemistry and Molecular Biology Education, 33*(2), 101–104.
Holley, E. A. (2017). Engaging engineering students in geoscience through case studies and active learning. *Journal of Geoscience Education, 65*(3), 240–249. https://doi.org/10.5408/16-145.1
Holmes, A. (1928). Radioactivity and continental drift. *Geological Magazine, 65*, 236–238.
Holmes, A. (1929). Radioactivity and earth movements. *Transactions of the Geological Society of Glasgow, 18*(Part 3), 559–606.
Holmes, A. (1944). *Principles of physical geology*. Thomas Nelson and Sons Ltd. 2nd edn 1965.
Hubenthal, M. (2018). Exploring undergraduates' conceptions of elasticity, within a plate tectonics context, before and after experience with rock's elastic behavior. *Journal of Geoscience Education, 66*(4), 261–277.
Hubenthal, M., Braile, L., & Taber, J. (2008). Redefining earthquakes and the earthquake machine. *The Science Teacher, 75*(1), 32–36.
Hutchins, E., & Renner, N. (2012). Situated and embodied learning in the field. In K. A. Kastens & C. A. Manduca (Eds.), *Earth and mind II: A synthesis on research on thinking and learning in the geosciences* (Geological Society of America special paper 486) (Vol. 2, pp. 181–182). Geological Society of America.
Hutton, J. (1788). Theory of the earth. *Transactions of the Royal Society of Edinburgh, 1*(2), 209–304.
Hyrnchak, P., & Batty, H. (2012). The educational theory basis of team-based learning. *Medical Teacher, 34*(10), 796–801.
Ibáñez, F. J. R. d. M., & Masegosa, A. G. (2014). *Cognitive modeling: A linguistic perspective* (Vol. 45). John Benjamins Publishing Company.
Indurkhya, B. (1992). *Metaphor and cognition: An interactionist approach* (Vol. 13). Springer-Science+Business Media.

Isacks, B., Oliver, J., & Sykes, L. R. (1968). Seismology and the new global tectonics. *Journal of Geophysical Research, 73*(18), 5855–5899.
James, W. (1911). On some mental effects of the earthquake. In H. James (Ed.), *Memories and studies* (pp. 207–226). Longmans, Green.
Kahneman, D. (2011). *Thinking, fast and slow* (1st ed.). Farrar, Straus and Giroux.
Khishfe, R., & Abd-El-Khalick, F. (2002). Influence of explicit and reflective versus implicit inquiry-oriented instruction on sixth graders' views of nature of science. *Journal of Research in Science Teaching, 39*(7), 551–578.
Kuhn, T. S. (1993). Metaphor in science. In A. Ortony (Ed.), *Metaphor and thought* (2nd ed., pp. 533–543). Cambridge University Press.
Kuorikoski, J., & Ylikoski, P. (2015). External representations and scientific understanding. *Synthese, 192*(12), 3817–3837. (Academia.edu).
Lakoff, G., & Johnson, M. (1980). *Metaphors we live by*. University of Chicago Press.
Lakoff, G., & Johnson, M. (1999). *Philosophy in the flesh: The embodied mind and its challenge to western thought*. Basic Books.
Latour, B. (1987). *Science in action: How to follow scientists and engineers through society*. Harvard University Press.
Lederman, N. (2007). Nature of science: Past, present and future. In S. Abell & N. Ledderman (Eds.), *Handbook of research in science education* (pp. 831–880). Lawrence Erlbaum Associates.
Lerach, D. G., & Yestness, N. R. (2019). Virtual storm chase: Bringing the atmospheric sciences to life through an interactive case study. *Journal of College Science Teaching, 48*(3), 30–36.
Li, Q., & Yang, X. (2018). Integrated case studies in teaching introductory mathematics courses at college. *European Journal of Educational Sciences, 5*(3), 1–13.
London, J. (1906). Story of an eyewitness: The San Francisco earthquake. *Collier's Weekly*. http://london.sonoma.edu/Writings/Journalism/
Luster-Teasley, S., Hargrove-Leak, S., Gibson, W., & Leak, R. (2017). Case studies in sustainability used in an introductory laboratory course to enhance laboratory instruction. *Journal of STEM education, 18*(2), 30–39.
Maggio, L. A., Steinberg, R. M., Piccardi, T., & Willinsky, J. M. (2020). Meta-research: Reader engagement with medical content on Wikipedia. *eLife*. https://elifesciences.org/articles/52426
Marshak, S. (2012). *Earth: Portrait of a planet* (4th ed.). W. W. Norton and Company.
Marshak, S. (2019). *Essentials of geology* (Sixth ed.). W. W. Norton & Company, Inc.
Medawar, P. B. (1963). Is the scientific paper a fraud? In P. B. Medawar (Ed.), *The threat and the glory: Reflections on Science and Scientists*. Harper Collins.
Merseth, K. K. (1991). The early history of case-based instruction: Insights for teacher education today. *Journal of Teacher Education, 42*(4), 243–249. https://doi.org/10.1177/002248719104200402
Ministry of Education (MOE), Ministry of Science and ICT (MSICT), & Korea Foundation for the Advancement of Science and Creativity (KOFAC). (2019). *Scientific literacy for all Koreans: Korean science education standards for the next generation*. KOFAC.
Montgomery, K. (2003). Siccar point and teaching the history of geology. *Journal of Geoscience Education, 51*(5), 500–505.
Murck, B. W., & Skinner, B. J. (2012). *Visualizing geology* (Third ed.). Wiley.
Nance, D., & Murphy, B. (2016). *Physical geology today*. Oxford Univeristy Press.
National Research Council. (2000). Inquiry and the National Science Education Standards: A guide for teaching and learning. *The National Academies Press*. https://doi.org/10.17226/9596
National Research Council. (2006). *America's lab report: Investigations is high school science*. T. N. A. Press.
Nersessian, N. J. (2008). *Creating scientific concepts*. MIT Press.
Nesari, A. J. (2015). Dialogism versus monologism: A Bakhtinian approach to teaching. *Social and Behavioral Science, 205*, 642–647.

NGSS Lead States. (2013). *Next generation science standards: For states, by states*. National Academies Press. Retrieved July 2 from https://www.nextgenscience.org/

Niebert, K., & Gropengiesser, H. (2015). Understanding starts in the mesocosm: Conceptual metaphor as a framework for external representations in science teaching. *International Journal of Science Education, 37*(5–6), 903–933. https://doi.org/10.1080/09500693.2015.1025310

Niebert, K., Marsch, S., & Treagust, D. F. (2012). Understanding needs embodiment: A theory-guided reanalysis of the role of metaphors and analogies in understanding. *Science Education, 96*(5), 849–877.

Núñez-Oviedo, M., & Clement, J. (2008). A competition strategy and other modes for developing mental models in large group discussion. In J. Clement & M. A. Rae-Ramirez (Eds.), *Model based learning and instruction in science* (Vol. 2, pp. 117–138). Springer.

Núñez-Oviedo, M. C., Clement, J., & Rae-Ramirez, M. A. (2008). Developing complex mental models in biology through model evolution. In J. Clement & M. A. Rae-Ramirez (Eds.), *Model based learning and instruction in science* (Vol. 2, pp. 173–194). Springer.

O'Neill, T. A., Deacon, A., Gibbard, K., Larson, N., Hoffart, G., Smith, J., & Donia, M. (2017). Team dynamics feedback for post-secondary student learning teams. *Assessment and Evaluation in Higher Education, 43*, 571–585.

O'Neill, T. A., Larson, N., Smith, J., Deng, C., Donia, M., Rosehart, W., & Brennan, R. (2019). Introducing a scalable peer feedback system for learning teams. *Assessment and Evaluation in Higher Education, 44*, 848–862.

O'Neill, T. A., Boyce, M. A., & McLarnon, M. J. W. (2020a). Team health is improved when peer evaluation scores affect grades on team projects. *Frontiers in Education, 5*, 1–10. https://doi.org/10.3389/feduc.2020.00049

O'Neill, T. A., Pezer, L., Solis, L., Larson, N., Maynard, N., Dolphin, G., et al. (2020b). Team dynamics feedback for post-secondary student learning teams: Introducing the "Bare CARE" assessment and report. *Assessment and Evaluation in Higher Education, 45*, 1121–1135. https://doi.org/10.1080/02602938.2020.1727412

Oreskes, N. (1999). *The rejection of continental drift: Theory and method in American earth science*. Oxford University Press.

Ornstein, A. C. (1994). The textbook-driven curriculum. *Peabody Journal of Education, 69*(3), 70–85.

Plummer, C., McGeary, D., Carlson, D., Eyles, N., & Eyles, C. (2007). *Physical geology and the environment*. McGraw-Hill Ryerson Higher Education.

Preuss, P. (1992). Rhea's time. In B. Preiss & R. Silverberg (Eds.), *The ultimate dinosaur* (pp. 88–99). Bantam Books.

Prud'homme-Généreux, A. (2017). Formulating questions that address student misconceptions in a case study. *Journal of College Science Teaching, 46*(4), 54–60.

Rae-Ramirez, M. A. (2008). Determining target models and effective learning pathways for developing understanding in biological topics. In J. Clement & M. A. Rae-Ramirez (Eds.), *Model based learning and instruction in science* (pp. 45–58). Springer Science.

Rea-Ramirez, M. A., Clement, J., & Núñez-Oviedo, M. C. (2008). An instructional model derived from model construction and criticism theory. In J. Clement & M. A. Rae-Ramirez (Eds.), *Model based learning and instruction in science* (Vol. 2, pp. 23–43). Springer.

Reddy, M. (1979). The conduit metaphor: A case of frame conflict in our language about language. In A. Ortony (Ed.), *Metaphor and thought* (pp. 284–324). Cambridge University Press.

Redman, C. (2007). *Towards a dialogical perspective on agency in student learning*. Australian Association for Research in Education.

Reiff-Cox, R. (2020). Exchanging the myth of a step-by-step scientific method for a more authentic description of inquiry in practice. In W. F. McComas (Ed.), *Nature of science in science instruction: Rationales and strategies* (pp. 127–139). Springer Nature Switzerland AG.

Repcheck, J. (2003). *The man who found time: James Hutton and the discovery of the earth's antiquity*. Perseus.

Reynolds, S. J., Johnson, J. K., Morin, P. J., & Carter, C. M. (2013). *Exploring geology* (3rd ed.). McGraw-Hill.

Rohrer, T. (2005). Image schemata in the brain. In B. Hampe & J. Grady (Eds.), *From perception to meaning: Image schemata's in cognitive linguistics* (pp. 165–196). Mouton de Gruyter.

Roth, A., Davis, R., & Carver, B. (2013). Assigning Wikipedia editing: Triangulation toward understanding university student engagement. *First Monday, 18*(6). https://doi.org/10.5210/fm.v18i6.4340

Rotramel, A., Parmer, R., & Oliveira, R. (2019). Engaging women's history through collaborative archival Wikipedia projects. *Journal of Interactive Technology & Pedagogy, 14*. https://jitp.commons.gc.cuny.edu/engaging-womens-history-through-collaborative-archival-wikipedia-projects/

Rudwick, M. J. S. (1985). *The great Devonian controversy: The shaping of scientific knowledge among gentlemanly specialists*. University of Chicago Press.

Rudwick, M. J. S. (2014). *Earth's deep history: How it was discovered and why it matters*. Chicago University Press.

Safapour, E., Kermanshachi, S., & Taneja, P. (2019). A review of nontraditional teaching methods: Flipped classroom, gamification, case study, self-learning, and social media. *Education sciences, 9*(4), 273. https://doi.org/10.3390/educsci9040273

Şahin, D. (2019). An evaluation of the intentions of preservice teachers to use the case study method in social studies classes. *Issues in educational research, 29*(2), 384–401.

Sawyer, D. (2002). *Discovering plate boundaries: A classroom exercise designed to allow students to discover the properties of tectonic plates and their boundaries*. Rice University. http://plateboundary.rice.edu/intro.html. Retrieved 17 December 2011 from http://plateboundary.rice.edu/intro.html

Scharmann, L. C., Smith, M. U., James, M. C., & Jensen, M. (2005). Explicit reflective nature of science instruction: Evolution, intelligent design, and umbrellaology. *Journal of Science Teacher Education, 16*(1), 27–41. https://doi.org/10.1007/s10972-005-6990-y

Schulte, P., Alegret, L., Arenillas, I., Arz, J. A., Barton, P. J., Bown, P. R., et al. (2010). The Chicxulub asteroid impact and mass extinction at the cretaceous-Paleogene boundary. *Science, 327*, 1214–1218. https://doi.org/10.1126/science.1177265

Shapiro, L. (2011). *Embodied cognition*. Routledge.

Shapiro, M. (2021). Eunice Newton Foote's nearly forgotten discovery. *Physics Today., 2021*. https://doi.org/10.1063/PT.6.4.20210823a

Sisk, R. J. (2011). Team-based learning: A systematic research review. *Journal of Nursing Education, 50*(12), 665–669.

Smith, D. A. (2020). Situating Wikipedia as a health information resource in various contexts: A scoping review. *PLoS One, 15*(2), e0228786. https://doi.org/10.1371/journal.pone.0228786

Stains, M., Harshman, J., Barker, M. K., Chasteen, S. V., Cole, R., DeChenne-Peters, S. E., et al. (2018). Anatomy of STEM teaching in north American universities. *Science, 359*(6383), 1468–1470.

Steen, G. J. (2007). *Finding metaphor in grammar and usage* (Vol. 10). John Benjamins Publishing Company.

Steen, G. (2014). The cognitive linguistic revolution in metaphor studies. In J. Littlemore & J. R. Taylor (Eds.), *The Bloomsbury companion to cognitive linguistics* (pp. 117–142). Bloomsbury.

Stent, G. (2002). Prematurity in scientific discovery. In E. Hook (Ed.), *Prematurity in scientific discovery: On resistance and neglect* (pp. 22–33). University of California Press.

Stinner, A., & Teichmann, J. (2003). Lord Kelvin and the age-of-the-earth debate: A dramatization. *Science Education, 12*(2), 213–228.

Tarbuck, E. J., & Lutgens, F. K. (2017). *Earth: An introduction to physical geology*. Pearson.

Terry, D. R., Lemons, P., Armstrong, N., Brickman, P., Ribbens, E., & Herreid, C. F. (2016). Eight is not enough: The level of questioning and its impact on learning in clicker cases. *Journal of College Science Teaching, 46*(2), 82–92.

Thagard, P. (2012). *The cognitive science of science: Explanation, discovery, and conceptual change*. MIT Press.

Thibodeau, P. H., & Boroditsky, L. (2013). Natural language metaphors covertly influence reasoning. *PLoS One, 8*(1), e52961.

Thompson, N., & Hanley, D. (2017). *Science is shaped by Wikipedia: Evidence from a randomized control trial*. MIT Sloan School of Management. Research Paper Series, No. 5238-17.

Turner, D. (2013). Hisotrical geology: Methodology and metaphysics. In V. R. Baker (Ed.), *Rethinking the fabric of geology* (Geological Society of America special paper 502) (pp. 11–18). Geological Society of America.

Ulerick, S. L. (n.d.). Using textbooks for meaningful learning in science. *Research matters–to the science Teacher*. https://narst.org/research-matters/using-textbooks-for-meaningful-learning

Vetter, M. A., McDowell, J. Z., & Stewart, M. (2019). From opportunities to outcomes: The Wikipedia-based writing assignment. *Computers and Composition, 52*, 53–64.

Warncke-Wang, M., Ayukaev, V. R., Hecht, B., & Terveen, L. G. (2015). The success and failure of quality improvement projects in peer production communities. 18th ACM conference on Computer Supported Cooperative Work & Social Computing (CSCW '15).

Wegener, A., & Skerl, J. G. A. (1924). *The origin of continents and oceans*. Methuen & Co..

Wiggins, G. P., & McTighe, J. (2006). *Understanding by design* (Expanded 2nd ed.). Pearson Education, Inc.

Wikipedia. (n.d.). *Statistics*. Retrieved 23 October from https://en.wikipedia.org/wiki/Special:Statistics

Yadav, A., & Beckerman, J. L. (2009). Implementing case studies in a plant pathology course: Impact on student learning and engagement. *Journal of Natural Resources and Life Sciences Education, 38*, 50–55.

Zhu, K., Walker, D., & Muchnik, L. (2020). Content growth and attention contagion on information networks: Addressing information poverty on WIkipedia. *Information Systems Research, 31*(2), 491–509.

Part II
Measuring Shifts in Students' Content Knowledge and Nature of Science Understandings

1.1 Introduction to Part II

In Part II, I analyze and interpret student performance data to measure the development of student understanding of geology content and the nature of science (NOS). This is accomplished via multiple strategies. For the content domain, students participated in a classroom response system (CRS), Top Hat, performed prior to the first class, and then again after the end of the final exam. The questions were "click-on" diagram questions. The students read each question and responded with an answer that is represented by a specific location on an accompanying diagram. Students employed their smart devices and clicked on what they thought is the appropriate location. The CRS then created a heat map showing density of student responses. More detail about this measurement method occurs in Chap. 5. They also responded to two different survey instruments, the Geoscience Concept Inventory (GCI) and the Views on Science, Technology, and Society (VOSTS), both before the start of the course and then once they had finished the course final exam.

Comparing pre-instruction responses to the CRS questions and the GCI to post-instruction responses showed that students made significant gains in understanding the target concepts–the earth is a historical body, the earth is very old, and the earth is dynamic. The results from the VOSTS survey showed that the students had a shift toward more nuanced understandings of science.

To further determine students' NOS understanding, I had students respond to two open-ended prompts prior to the first class and then again after having completed the final exam. Students posted responses to the prompts to an online platform called Padlet (www.padlet.com). Once student responses were completed, I transferred the data to Dedoose (www.dedoose.com), an online application for analyzing qualitative and mixed-methods data. Dedoose enabled the analysis of students' responses to reveal their scientific epistemologies—the nature of scientific knowledge and how it is generated—and allowed me to compare students' pre-instruction and post-instruction responses. This included a shift in students' responses from naïve realism (pre-instruction) to constructive realism (post-instruction). See Chap. 6 for this analysis.

Student responses also expressed multiple metaphors for science processes and products. In general, the metaphors did not change from pre-instruction responses to post-instruction. This analysis of students' use of metaphors also indicates a predominantly realist epistemology among students, comparable to the findings discussed in Chap. 5. Chapter 7 highlights the different metaphors students used when describing science knowledge and processes.

For the science education researcher, Part II documents the use of multiple instruments, both quantitative and qualitative used to operationalize student learning. The analyses also demonstrate the utility of using multiple theoretical frameworks–the lens through which one views and interprets the data. For the science educator, Part II also serves as a model of reflective practice. Some places in the analyses will get quite detailed, "in the weeds" so to speak. I am hoping this serves the purpose of demonstrating an approach (certainly not the only approach!) to critical analysis and theorizing about the results within the context of reflecting on the teaching involved and how that might change to improve the results, indicating more useful student learning.

Chapter 5
Measuring Student Learning

Glenn Dolphin and Nicole Ladue

Teachers support evaluations based on multiple measures: student growth, classroom observation and feedback from peers and parents.

Arne Duncan

Grades are a subjective rating masquerading as an objective evaluation.

Alfie Kohn

5.1 Introduction: Student Understandings of Content and NOS

This chapter outlines the three quantitative methods I used for measuring student learning of both geology content and NOS understandings: click-on diagrams (CODs) (LaDue & Shipley, 2018), the Geoscience Concept Inventory (GCI; Libarkin et al., 2011), and the Views on Science–Technology–Society (VOSTS; Aikenhead & Ryan, 1992). The first two instruments measured student learning within the domain of geology, and the VOSTS measured student understandings about science. I had two aims with this evaluation:

With Dr. Nicole LaDue

G. Dolphin (✉)
Department of Earth, Energy, and the Environment, University of Calgary, Calgary, AB, Canada
e-mail: glenn.dolphin@ucalgary.ca

N. Ladue
Northern Illinois University, DeKalb, IL, USA

G. Dolphin, *Teaching Geology Using the History and Philosophy of Science*, Science: Philosophy, History and Education,
https://doi.org/10.1007/978-3-031-68667-2_5

- assess the impact of the pedagogical approach noted in earlier chapters on students' understanding of geology content and the NOS.; and
- make data-informed decisions directed at improving the course structure and instruction.

This chapter will also describe the general student population in GLGY 209. The demographic profile of the students taking the course is remarkably consistent from 1 year the next. Though the student data will be an aggregate of the past five iterations of the course, we can assume the percentages are approximately the same from year to year. This is important because the survey questions were only administered in the most recent iteration of the course (n = ~100 students), while the qualitative data and analysis described in Chaps. 6 and 7 will be based on data collected from all five iterations (n = ~900–1200 students).

My intent with this chapter is to show that the transformative teaching strategies were successful in helping student achieve the intended learning goals. More than that, however, I want to demonstrate how I have approached the evaluation results on a finer grained level. In this way, I could be more directed in when tweaking the intervention. In some cases, it also helped me to analyze the evaluation questions themselves to see how they might be interpreted in different ways by the students. This is the most important part of reflective practice, as far as I am concerned. It has forced me to do my best to see through the eyes of the students. It also forces me to question my own understanding. If I can better own understanding of the content and come to understand how they are thinking, I can design my evaluations with those in mind. I can design my instruction with those in mind.

5.1.1 Student Participants and Learning Context

The current research took place at a top research-intensive university in Canada with a large Geoscience department. GLGY 209, a course for non-science majors, is capped at 250 students per section, with two sections offered per term.

I collected student data each time I taught the course—from 2016 to 2021 (Table 5.1). Students responded to the prompts at the beginning of the term and at the end. Because of the structure of my ethics protocol (REB17-1551), students

Table 5.1 Schedule of each GLGY 209 iteration and total responses at pre- and post-instruction

Term	Fall (Sept–Dec)	Winter (Jan–May)	Pre-instruction	Post-instruction	Mode of instruction
2016		X	138	232	Face to face
2017		X	153	197	Face to face
2019	X		290	398	Face to face
2020	X		180	224	Remote
2021		X	181	215	Remote
Total			942	1266	

decided at the end of the course about whether or not they would participate in the research. The COD survey was administered from 2019 on, and the GCI and VOSTS were administered only in Winter 2021. My intent was to investigate if these latter two instruments paralleled the COD questionnaire and the qualitative interpretation located in Chap. 6.

Tables 5.2, 5.3, 5.4 and 5.5 present the demographics of students for each course iteration during the study period. These data are broken down to show aggregate counts of students taking the course from 2016–2019 (three separate offerings, pre-COVID 19), and those taking the course during 2020 and 2021 (two offerings, both held remotely due to the COVID-19 pandemic).

The data remain fairly consistent from offering to offering. Females make up almost 60% of the student population, with most students (64%) between 17 and 20 years old. That increases to 95% by counting all students, 17–25 years old. Most students come from the Faculty of Arts (51%), followed by the Faculty of Science (15%), the School of Business (14%), the School of Education (6%), the School of Communication (3%) and Health Sciences (2%). At the time of taking the course, 14% of the students had not declared a major. Finally, most students taking the course (90%) claimed Canadian citizenship, followed by Chinese (1%), Nigerian (1%), and the US (1%). The remaining 7% came from various countries. Overall, students claimed citizenship of about 50 different countries.

These demographics are descriptions of all students enrolled in the course by the end of each given term. The participant population is a subset of this broader population, but I did not seek the demographics of the student participants.

Table 5.2 Students' gender distribution in GLGY 209 enrolments

Terms	Male (%)	Female
2016–2019	365 (43%)	482 (57%)
2020–2021	160 (42%)	225 (58%)
Total	525 (43%)	707 (57%)

Table 5.3 Students' age distribution in GLGY 209 enrolments

Terms	17–20	21–25	26–30	31–35	36–40	41–45	46–50	56–60
2016–2019	546	252	32	6	4	5	2	1
2020–2021	244	127	10	5	1	0	0	0
Total	790 (64%)	379 (31%)	42 (3%)	11 (1%)	5 (0.4%)	5 (0.4%)	2 (0.15%)	1 (0.05%)

Table 5.4 Students' faculty distribution in GLGY 209 enrolments

Term	Arts	Science	Business	Education	Communication	Health Science	Not Declared
2016–2019	418	115	159	54	27	11	112
2020–2021	209	68	19	23	12	9	61
Total (%)	627 (48.3%)	183 (14.2%)	178 (13.7%)	77 (5.9%)	39 (3.0%)	20 (1.5%)	173 (13.4%)

Table 5.5 Distribution of student nationalities in GLGY 209 enrolments (total of 50 countries)

Term	Canada	China	Korea	India	Nigeria	UK	Pakistan	Phillip-pines	US	All Others
2016–2019	759	12	4	5	11	3	7	4	7	36
2020–2021	350	4	0	2	4	2	1	2	5	17
Total (%)	1109 (89.8%)	16 (1.3%)	4 (0.3%)	7 (0.6%)	15 (1.2%)	5 (0.4%)	8 (0.6%)	6 (0.5%)	12 (1.0%)	53 (4.3%)

5.2 Geology Content Assessment—Click-On Diagrams[1]

In an attempt to garner more student engagement, especially in large-lecture courses in post-secondary schools, instructors have increasingly incorporated the use of student response systems (SRS), or "clickers," into instruction. These clickers allow instructors to ask multiple choice–type questions in class and have students respond by clicking their answer. Instructors can watch the results accumulate in real time and use them as a diagnostic tool, or as a formative or summative assessment. This helps both the instructor and the students by offering a quick and easy way to gauge the learning happening in the class (Caldwell, 2007).

Implementing clicker questions encourages more engagement from students (Caldwell, 2007; Powell et al., 2011). For students who feel anxious about answering questions in front of their student peers—a concern that may be more present in larger science classes—the clicker system provides anonymity. For the click-on diagrams, results appear in aggregate as a heat map, based on density of clicks in any one area. In this way, students can, without fear of embarrassment, respond to the questions, enhancing their engagement (Stowell et al., 2010).

Universities originally purchased SRS, which consisted of the classroom response system and a set of hand-held clickers for the students. Over the last several years, however, SRS have developed to include an online response system that students can access using their smart devices, like a smart phone, tablet, or PC. In addition, the structure of the questions instructors could ask has broadened, allowing for more open-ended answers from students.

The "click-on" question is one of these new types of questions (LaDue & Shipley, 2018) that became useful for this research. Click-on questions ask participants to identify an answer by locating it on an accompanying diagram. For instance, a question in the study featured a diagram of a section of outcropping layers of sedimentary rocks, with a query asking the student to identify which layer is the most resistant to weathering, based on its appearance in the diagram (see Fig. 5.1a). Students viewing the question and diagram on their device would then move the cursor to the location they think represents their answer to the question and click on the appropriate layer. The SRS then aggregates all of the responses into a heat map with warmer colours (oranges and reads) representing areas of higher densities of clicks and cooler colours (greens and blues) for lower densities of clicks (see Fig. 5.1b).

This question format grants students the freedom to pick their own location to answer the question, as opposed to forced multiple-choice questions that restrict student choice. With this system, students must consider all areas of the diagram. Furthermore, the heat map generated by the SRS allows for a quick and qualitative assessment of the understanding of the classroom population (LaDue et al., 2021).

---

[1] Clicker questions and diagrams were developed for a 2019 U.S. National Science Foundation grant (#1835930).

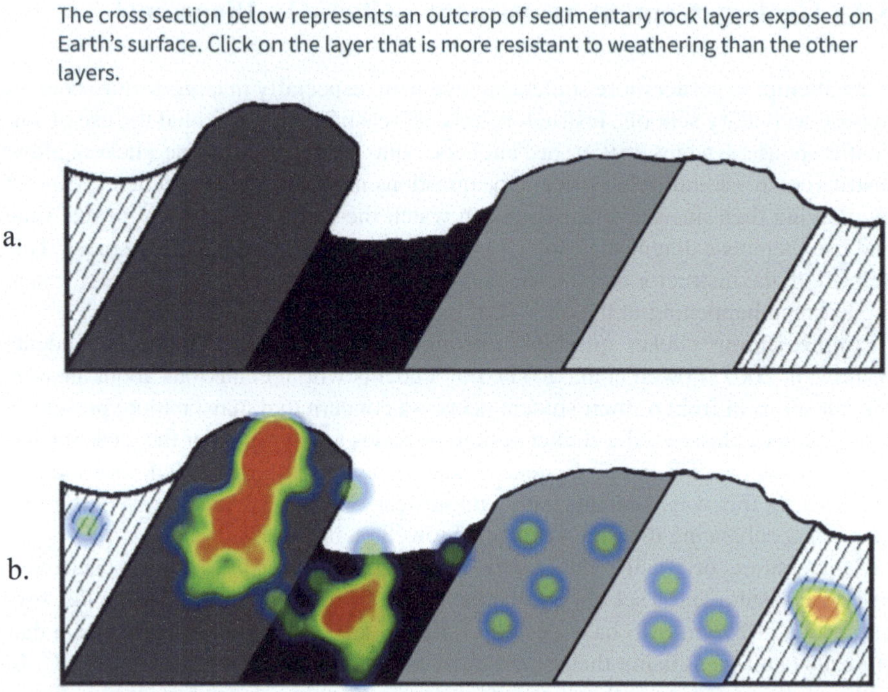

Fig. 5.1 Example of a question and accompanying click-on diagram. (Note. Image (**a**) is an example of a click-on diagram question, as students would see it on their devices. Image (**b**) shows the results in the form of a heat map after students have answered the question by clicking on the area of the diagram that best answers the question. Warmer colours represent a high density of student responses)

In the example above (see Fig. 5.1b), the SRS records the highest density of responses on the dark grey, protruding layer on the left side of the diagram, which is the correctly marked area for a layer that is more resistant to weathering. However, a sizable grouping of responses also shows in the adjacent layer to the right, which is depicted on the diagram as the least resistant to weathering. These collective responses may therefore indicate that students do not understand what "more resistant to weathering" means. The instructor can use these results to address areas in which student understanding does not match the learning goals. The instructor can also project the image showing the results to the class and use it as a way to help students understand what the intended answer was and why.

Since about 2017, I had been fortunate to collaborate with geoscience education researchers from Northern Illinois University (NIU) who have been investigating CODs, including how the diagrams themselves might play a role in student responses to questions. The large sample size of students in my class proved ideal as a population to test their hypotheses. Though the questions were designed with a different problem in mind, I could still use the results to inform my teaching by evaluating the results taken at the pre- and post-instruction stages (LaDue & Shipley, 2018).

5.2.1 Methods

LaDue and Shipley (2018) have described a process for analyzing SRS data from CODs aimed at operationalizing student learning. The diagrams present simplified images of material that is standard in introductory geology. My colleagues from NIU and I pilot-tested the simpler line drawings to make sure that students could understand their intended meaning, which we found was the case. Further, LaDue and Shipley also outlined a method for measuring the precision of students' clicks within the COD (to determine if the SRS recorded the click where the student actually made the click). They found that the system "demonstrated consistently high precision across devices" (p. 498).

I used the SRS known as Top Hat for this course. This system records the COD data as sets of x, y coordinates, starting with 0,0 at the upper left-hand corner of the diagram and 1,1 in the bottom right. Top Hat then plots the two-dimensional spatial data as a heat map. One can take the Top Hat data showing point locations and import them into a program like ArcGIS[2] (LaDue & Shipley, 2018). By plotting the data back onto the question image makes possible the drawing of polygons around areas to differentiate the intended correct response from the other regions (see Fig. 5.2), and thereby count "right" and "wrong" responses.

The questions aligned with the content learning goals of the course. A group of five questions focused on the earth's history, such as correlation, index fossils, and sequence of events. Two questions focused on the earth's age, mainly emphasizing the scale of that age. One question focused on weathering. Finally, three questions

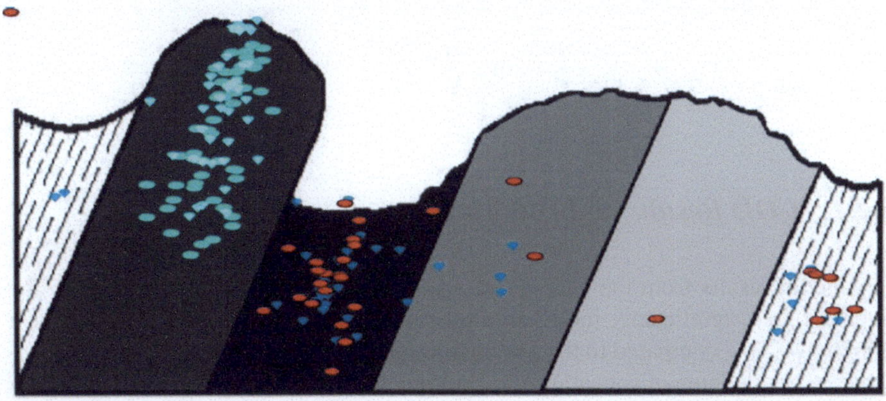

Fig. 5.2 Weathering COD, showing the location of individual student clicks. (Note. This is the same diagram seen in Fig. 5.1 but with individual click data—that is, each individual dot—shown. Each mark has a specific location coordinate assigned, making it easy to compare pre-instruction responses to post-instruction responses. Light blue is a "correct" click location. Ovals and triangles are pre- and post- instruction responses, respectively)

[2] ArcGIS is a cloud-based platform for analyzing and mapping data sets.

dealt specifically with the earth's dynamics (tectonics, isostasy, and paleomagnetism). I administered these 10 COD questions (Table 5.6) at the beginning of the course and then again at the very end, for participation credit only. Students had about 2 min to respond to each question. Because this particular COD administration happened during the COVID-19 pandemic (Winter 2021) when courses were being held online, the survey also happened online. Three of the questions were control questions. They were geology-related, but dealt with content that was not addressed in the course. They did not show any significant change from pre- to post-course. This builds the argument that the significant change recorded for the other 10 questions, most likely came from the course experiences.

5.2.1.1 Data Analysis

My colleagues at NIU collected and processed the data from both pre-instruction and post-instruction from the Winter 2021 administrations and performed a Chi-Square Goodness of Fit Test (X^2), using both pre-instruction and post-instruction data. This test measures the amount of difference between the number of students answering correctly.

$$X^2 = \sum \frac{(\text{observed} - \text{expected})^2}{\text{expected}}$$

To measure effect size, my colleagues then used the Chi-square value to calculate Phi (Φ).

$$\Phi = \sqrt{X^2 / n}$$

5.2.2 *COD Results and Discussion*

In general, results from the Chi-square analysis show that students clicked on the intended area significantly more often post-instruction —with medium effect size ($\Phi = 0.3$–0.5)— compared to the pre-instruction responses. This applies to all questions except the human appearance in geologic time and earthquakes in subduction zones questions (see Table 5.7). Below, I will discuss the results from three questions, one from each of the three foundational learning goals of the course: the historical nature of the earth (see Fig. 5.3), the immense age of the earth (see Fig. 5.5), and the dynamic nature of the earth (see Fig. 5.7).

Table 5.6 The 10 click-on diagrams administered in this investigation

Group	Subgroup	Image	Question
Earth history	Correlation strata	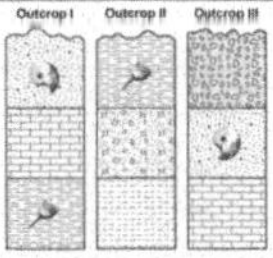	Fossils and rock types are evidence used to correlate rocks across great distances. By matching the outcrops below, click on the layer that is the oldest layer of all three outcrops.
	Correlation fossils	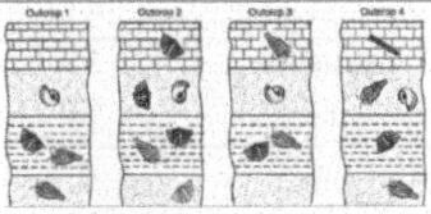	Based on the fossil distribution in the 4 outcrops below, click on the best index fossil.
	Differential weathering		The cross section below represents an outcrop of sedimentary rock layers exposed on Earth's surface. Click on the layer that is more resistant to weathering than the other layers.
	Geological events A	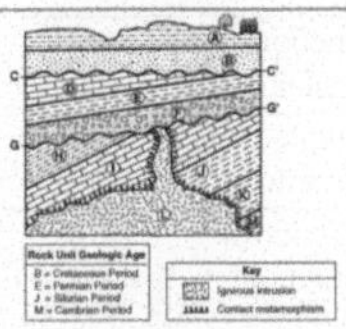	Click on the letter indicating the oldest rock in this outcrop.
	Geological events B	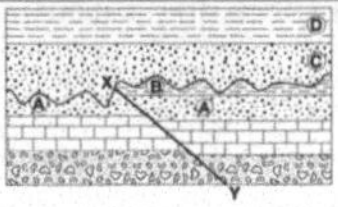	Click on the geologic feature that is older than C and younger than B.
Earth Age	Human appearance	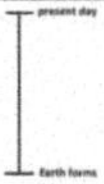	Click on the line where you expect humans appeared on Earth
	Dinosaur Appearance	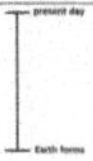	Click on the line where you expect dinosaurs appeared on Earth.
Earth Dynamics	Hop spot location	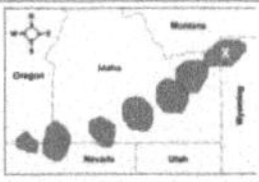	The North American tectonic plate moved southwest over the Yellowstone hot spot (location X). If the plate started moving north, click where the next volcanic caldera will form
	Magnetic reversals		Throughout geologic time, Earth's magnetic poles have reversed periodically generating a pattern in the sea floor rocks. Click where you expect to find rocks that are the same age as those labeled with the red dot.
	Subduction zone earthquakes	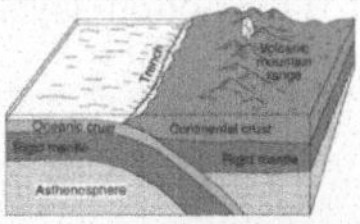	Click on the diagram where earthquakes will occur as a result of subduction.

Table 5.7 Analysis of responses to content focussed COD for GLGY 209 from winter 2021, comparing pre-instruction to post-instruction responses

			Responses			
	Topic		Incorrect	Correct	X^2	Φ
Earth History	Correlation of Strata	pre-instruction	75	13	30.73**	0.59
		post-instruction	40	48		
	Correlation of Fossils	pre-instruction	75	13	24.94**	0.53
		post-instruction	44	44		
	Differential Weathering	pre-instruction	51	37	6.58*	0.27
		post-instruction	34	54		
	Ordering Geologic Events A	pre-instruction	29	59	8.01**	0.3
		post-instruction	13	75		
	Ordering Geologic Events B	pre-instruction	64	24	13.63**	0.41
		post-instruction	39	49		
Earth Age	Human Appearance in Geologic Time	pre-instruction	66	22	1.001	n.s.
		post-instruction	60	28		
	Dinosaur Appearance in Geologic Time	pre-instruction	61	27	8.45*	0.3
		post-instruction	42	46		
Earth Dynamics	Hot Spot Location	pre-instruction	80	8	6.12*	0.26
		post-instruction	68	20		
	Magnetic Reversals	pre-instruction	28	60	10.87**	0.35
		post-instruction	10	78		
	Earthquakes in Subduction Zone	pre-instruction	51	37	3.84	n.s.
		post-instruction	38	50		

Note. ** = $p \leq 0.001$, * $p \leq 0.01$; Grey shading represents a non-significant difference

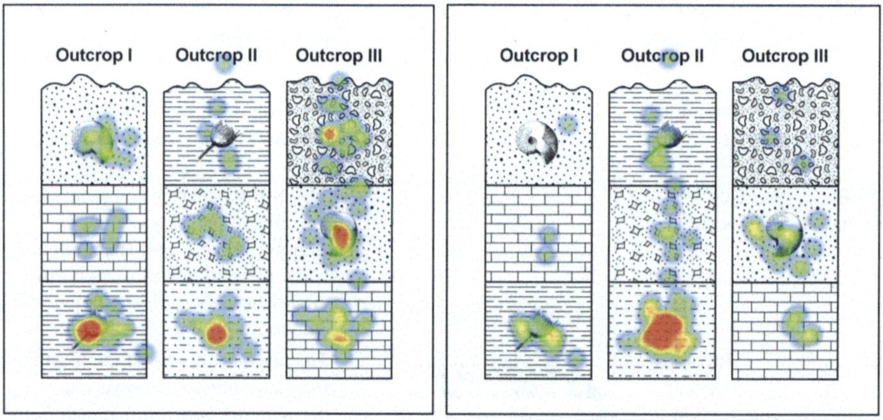

Fig. 5.3 Correlation of Strata, a COD Question. (Note. Showing pre-instruction (left) and post-instruction (right) responses to this question: *Fossils and rock types are evidence used to correlate rocks across great distances. By matching the outcrops below, click on the layer that is the oldest layer of all three outcrops.* The intended answer is the bottom layer of outcrop II)

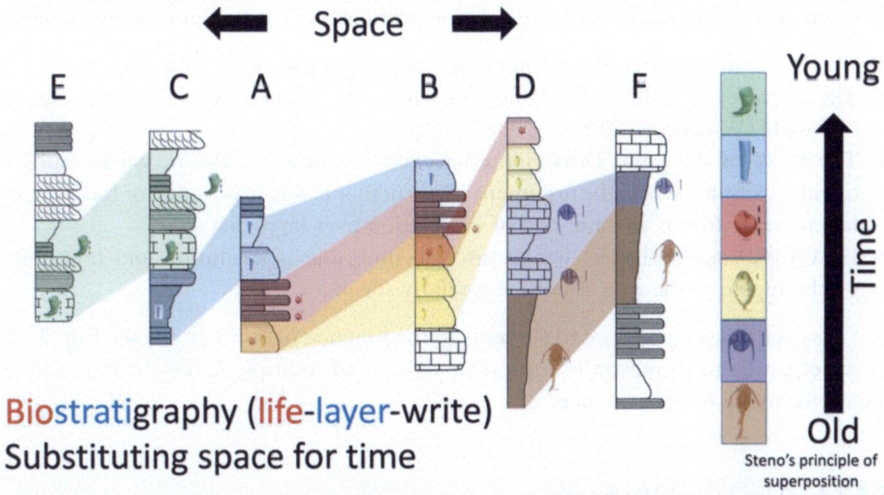

Fig. 5.4 Correlating outcrops using fossil evidence. (Note. Slide example, demonstrating how to use fossil evidence to correlate layers from one outcrop to the next (A–F), and eventually sequence the fossil organisms from oldest to youngest)

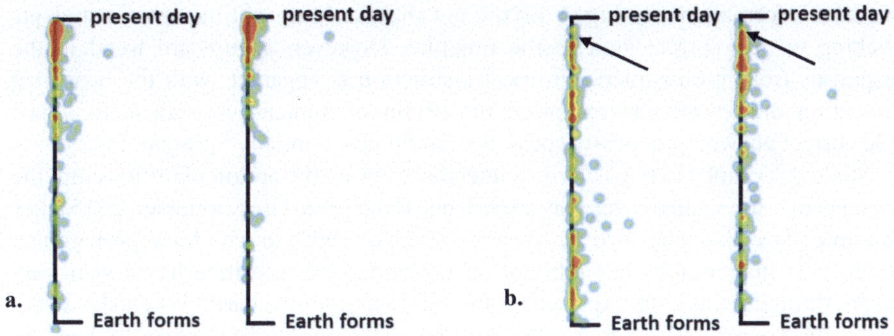

Fig. 5.5 COD Question About the Appearance in Geologic Time of Humans (image a) and Dinosaurs (image b). (Note. This figure shows students' responses to this question: *Click on the line where you expect humans* (**a**) *dinosaurs* (**b**) *appeared on Earth.* Pre-instruction responses are on the left side of each group, and post-instruction responses are on the right side. Correct answers are right at the corner for **a** and denoted with the arrows for **b**)

5.2.2.1 The Historical Nature of the Earth

Figure 5.3 is a correlation question, asking students to use rock type and fossil occurrence to determine the oldest layer, out of all three outcrops. The pre-instruction responses show most answers along the bottom of all three outcrops, and within all layers in outcrop III. This could indicate students' common-sense reasoning that the oldest layer is on the bottom, or to the right, as we read from left to right and sequence numbers from left to right.

Instruction intended to help explore this problem entailed several discussions:

- The principles of Nicholas Steno, especially the principle of superposition
- The activities of William Smith and his ability to correlate layers via the presence of fossils (Winchester, 2001)
- The story of the great Devonian controversy (Rudwick, 1985), where, subsequent to Smith's work, the argument was whether it was reasonable or too theory-laden to use fossils as a method of correlation over large distances
- How fossil assemblages change through time due to evolution and how they might appear in the rock record, especially for index fossils

Class activities also exposed students to the practice of correlation (see Fig. 5.4). This helped them to reason out the relative ages of multiple layers from multiple locations, using fossil evidence.

5.2.2.2 The Age of the Earth

The next question has to do with the age of the earth and more specifically the age of the origin of humans or dinosaurs relative to the age of the earth (see Fig. 5.5). The responses to the question of when humans appeared on the earth compared to the earth's formation (see Fig. 5.5a) did not show a significant increase in students clicking on the correct area of the timeline. However, an upward trend in the responses from pre-instruction to post-instruction is apparent, with the heat map indicating that most students place the origin of humans very close to today. The correct answer, practically speaking, would be essentially "present day."

Students completed two activities intended to bring the notion of billions into the mesocosm—the scale of everyday experience (Niebert & Gropengiesser, 2015). For example, they read a science fiction story (Preuss, 1992) in which the protagonist, Rhea, falls unconscious at midnight on December 30; she then has a year-long dream through which she experiences the full history of the earth. When she awakens in the hospital after a year, she describes how quickly life begins to develop, especially in the last few hours of year's history. The point was to show students the length of human existence: if Earth's history was 365 days long, humanity came along in the last 6 min or so. The second activity was the assignment *How much is a billion*? Here, students are charged with the task to bring the concept of a billion into concrete terms. For instance, a million seconds is approximately 13 days, whereas a billion seconds is approximately 31 years. I thought doing this math and making the comparison of abstract to concrete and experiential would be sufficient to help students develop a more accurate conceptualization of such magnitudes (Cheek et al., 2017), but this was not the case.

Research on this issue has shown that students are more likely to mark accurate locations on such a timeline if they know the actual dates of the events (Czajka & Mcconnell, 2018). Also, using multiple models that gradually bridge from familiar, concrete experiences of students to more expansive scales (millions then billions of years) may help students to map these different timings onto such a model, where space is substituted for time (Cheek et al., 2017). Students' range of answers to this query could be the result of a few teaching factors. Normally, I have used a long timeline (about 12 m long) stretched across the front of the lecture hall and hung signs on it for the different events happening on Earth at their scale age. With such a long timeline, the space between when humans appear is barely discernable from the part of the timeline representing present day.

In the Winter 2021 class, which was held remotely. Use of this concrete timeline was not possible. Instead, I created the timeline on a PowerPoint slide and used that to discuss the timing of specific events on the earth (see Fig. 5.6a). I surmise that students may have found the need to “zoom in” to the most recent times, so as to distinguish closely spaced events, confusing or ambiguous. In contrast to the large, in-class timeline, students would then perceive a greater degree of separation between specific events and the present day, as depicted on the expanded PowerPoint graphic. This experience may influence where they clicked on the much smaller timeline on the COD, attempting to leave some space between particular events and the present day.

As for the question about when dinosaurs first appeared, the post-instruction responses here showed a significant change to a more correct response. I noted a definite trend in the responses towards higher on the timeline and saw a much wider range of responses, indicating that students understood that dinosaurs are much older than humans. Still, they had difficulty in identifying dinosaurs’ appearance on Earth proportional to the age of the earth. Certainly, understanding such magnitudes is difficult for students (Cheek, 2012).

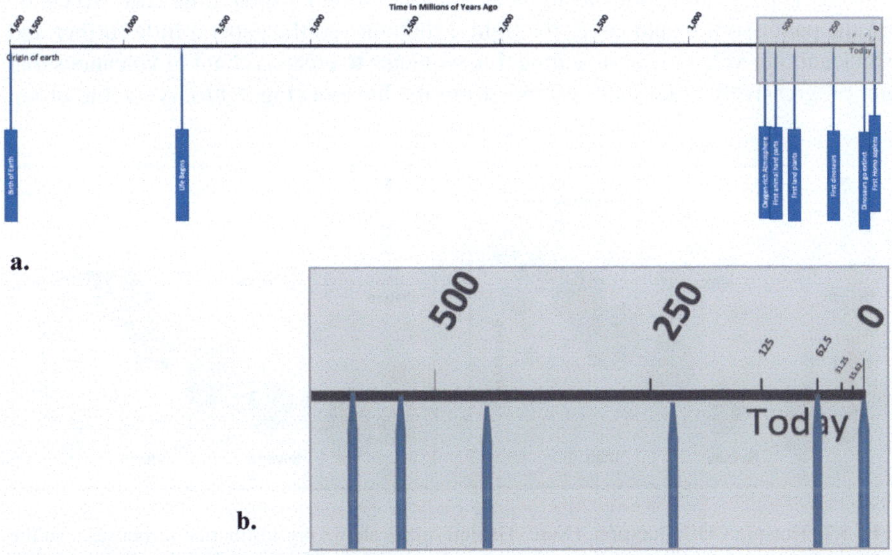

Fig. 5.6 Timing of major events during the earth’s history. (Note. The timeline across the top (**a**) shows the approximate locations where important events happened on the earth. The bottom of the diagram (**b**) is an expanded view of the most recent 500 million years of the timeline, showing that most important events, including the first dinosaurs (third arrow from the right) and the first humans (first arrow on the right) take place very close to this end of the timeline)

5.2.2.3 The Dynamics of the Earth

I used a hot spot question as one of the questions that explored students' understanding of the earth's dynamics (see Fig. 5.7). This question shows a chain of volcanoes resulting from a hotspot (located beneath point X). The item states that the crust has been moving to the southwest and will start moving northward, asking students to click on the location where they would expect the next volcano to be located. This is a challenging concept for students to visualize, and I had great difficulty getting students to predict the location.

I came to realize I was teaching the concept as I have seen in many textbooks—in a cross-sectional view, so students could see the structure of the crust and mantle and the effect of the plume on the crust. However, these diagrams are static, and it is difficult to explain the movement and subsequent volcano formation. This insight motivated me to try a new tack: I made a concrete functional model (Boulter & Buckley, 2000) comprising my arm, held horizontally as the crust, my iPhone with a video of a campfire burning as the mantle plume, and small paper volcanoes that I erected as my arm moved laterally across the iPhone flame (see Fig. 5.8a). Though students responded positively to the model, they still did not predict the new volcano very well. I then realized that I was teaching the concept in cross-section, but the question explored the concept in map view. Consequently, I also shifted to a map-view model.

My teaching assistant at the time, in Winter 2021, suggested that I take a sheet of paper as the crust and shine a small flashlight behind the paper as the hotspot. I moved the paper along the flashlight and paused to stick a modelling clay "volcano" on the paper so it would cover the light. I then moved the paper a little further and stuck another volcano on, repeating this sequence to create a chain of volcanoes that get progressively older with distance from the hotspot (Fig. 5.8b). A sizable group

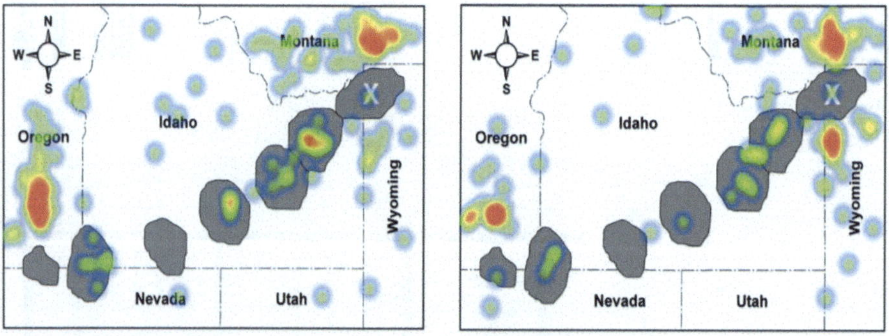

Fig. 5.7 Hotspot COD Question. (Note. The left image shows pre-instruction responses, and the right image shows post-instruction responses to this question: *The North American tectonic plate moved southwest over the Yellowstone hot spot (location X). If the plate started moving north, click where the next volcanic caldera will form.* The correct answer is the area immediately south of the X on the diagram)

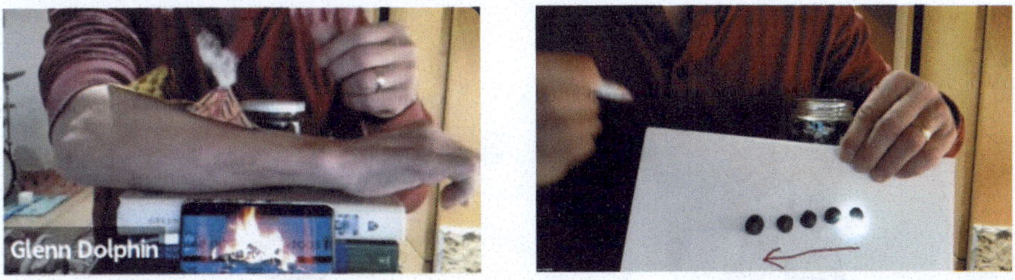

a. b.

Fig. 5.8 Teaching hotspot volcanism using two different concrete functional models. (Note. Two different orientations for teaching hotspot volcanism. Image **a.** shows cross-sectional orientation where my arm is the crust moving left over the fire (hotspot) creating a chain of volcanoes getting older, also to the left. Image **b.** is a map-view orientation with the paper (crust) moving to the left across the flashlight (hotspot), creating a chain of clay volcanoes that get older as they move to the left)

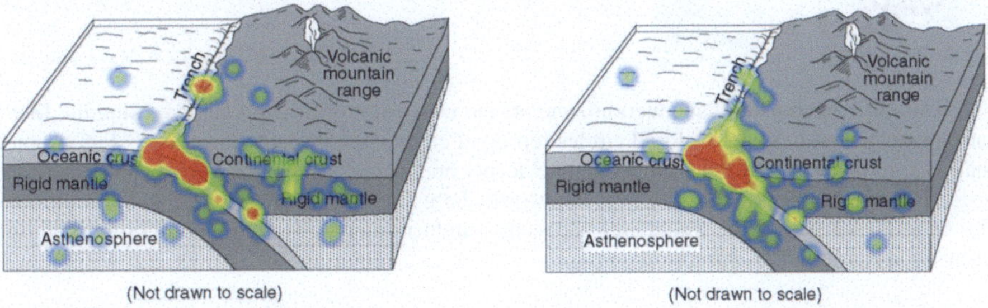

Fig. 5.9 Subduction COD question. (Note. The left image shows pre-instruction responses, and the right image shows post-instruction responses to this question: *Click on the diagram where earthquakes will occur because of subduction. The* correct answer is along the contact between the oceanic and continental crusts within the subduction zone (Original diagram from NYSED, 2013))

still clicked to the north of the X on the map and a large number of responses were to the north of the initial volcano. I could not discern why this would be a popular response, but it would be worth exploring further.

One question showed no significant change in response from pre-instruction to post-instruction; it asked the students to click on a cutaway portion of the earth's crust that showed a subduction zone (see Fig. 5.9) where they would expect to have earthquakes.

Prior to instruction, it is likely that students may reason that earthquakes would probably happen at a place where the two crusts meet, and the ocean crust looks like it is sinking under the continental crust. Though the data show that a majority of students did pick the correct area (from 38 to 50 correct clicks), there was still quite a spread of answers. I concluded from this that students might think almost any area of the diagram would be where subduction-related earthquakes occur. Though most earthquakes in an environment like this do happen along the contact between the

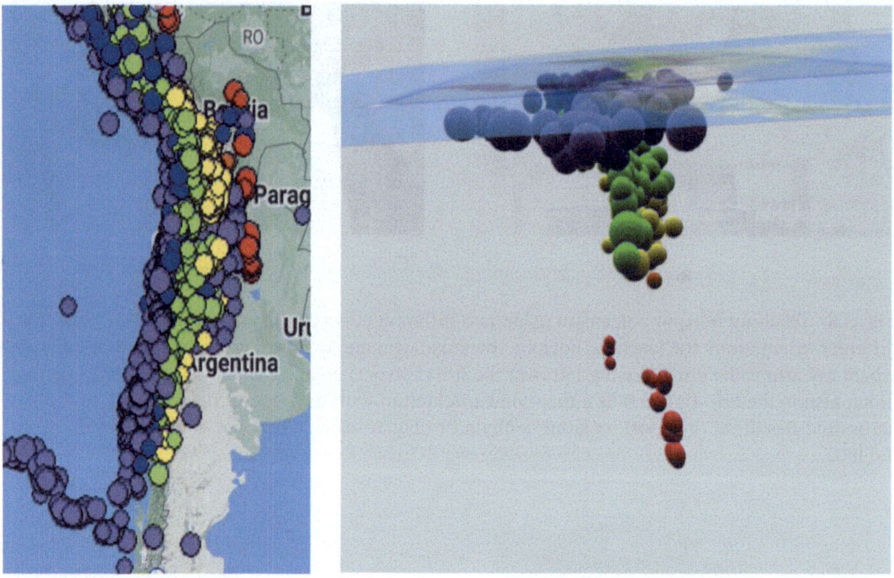

Fig. 5.10 Screen shots of a subduction zone off the west coast of South America, using the IRIS earthquake browser. (Note. Colored circles represent earthquake epicenters; cooler colors denote shallower hypocenters, warmer colors denote deeper ones. The left image is a map view; the right image is a three-dimensional view. It is noteworthy here that the shallow earthquakes (purple circles or balls) occur over the whole area across the subduction zone. Source: The IRIS Earthquake Browser: ds.iris.edu/ieb/index.html)

subducting and overriding crusts, some earthquakes happen in adjacent areas as well, well outside the target response area (see Fig. 5.10). Students' classroom experience with the extended seismic activity and the IRIS Earthquake Browser (Sumy et al., 2020) would have indicated to them the wider range of spatial possibilities for earthquakes in this geologic environment. In light of this, it would be wise to rework the question/diagram to limit the range of possible correct responses.

5.3 Geology Content Assessment—Geoscience Concept Inventory (GCI)[3]

The last couple of decades have seen an increase in the importance and development of disciplinary concept inventories—the sets of questions, usually multiple-choice, that address foundational concepts within separate science disciplines (Libarkin & Geraghty Ward, 2011).

[3] The GCI can be downloaded from this website: https://geocognitionresearchlaboratory.com

The GCI has been in development since the early 2000s (Libarkin & Anderson, 2005) and is based on *research about students' learning* of foundational geoscience concepts. It differs from traditional multiple-choice questions that a course instructor develops to evaluate student learning in their course, as those are based on what *that instructor* considers to be important. Moreover, those questions are in the instructor's words, grounded in the assumption that students interpret the questions and response options in the way the instructor intended.

Mitigating these biases is achieved by (a) getting data directly from students via interviews to inform question prototyping, which is necessary, (b) ensuring the validity of the items to actually measure the particular concept by having experts review and comment on them, and (c) checking that, to the extent possible, the test-taker adopts the intended interpretation of both the question and the available response options (Libarkin & Geraghty Ward, 2011). To develop GCI items, Libarkin and Geraghty Ward (2011) created open-ended questionnaires and interviews to get information about students' thoughts on different geoscience concepts; they then analyzed the qualitative data using grounded theory approaches (Bogdan & Biklen, 2007). This enabled them to identify themes in the data and subsequently develop question stems and response alternatives. They proceeded through several steps to achieve these various aims, as follows:

1. Generating the test questions—using question stems composed of appropriate and unambiguous language for the population being tested, with three to five plausible choice options of about the same word length that avoid absolute and/or complex answers.
2. Pilot-testing the questions—by holding "think aloud protocol" interviews (Cowan, 2017; Leighton, 2017; van Someren et al., 1994) with a small target student group, in which students shared their thoughts and perceptions about the pilot questions. Analyzing these qualitative data, along with quantitative analysis of responses contributed to question refinement.
3. Conducting iterative pilot-testing, as necessary, to ensure stable, satisfactory responses.

From its inception in the early 2000s until 2011, the GCI items have been tested on over 5000 students at over 40 universities. In 2011, there was a call to enhance the questions by adding diverse structures: for example, multiple-choice with multiple correct answers, instead of only one; and free response portions where student write rationales for the answers they choose (Libarkin et al., 2011).

5.3.1 Methods

I chose eight GCI items to use as markers for conceptual development that aligned with the course learning goals: three for historical nature of the earth, one for the age of the earth, and four for dynamics of the earth. These eight items were part of a larger survey that included questions from the VOSTS survey (see Sect. 4 for a detailed explanation of this instrument). I administered the survey through the university's online survey platform, Qualtrics, during the first week of classes and again after the students took the final exam. Students participated in the survey on their own, outside of class, having been told that the survey was for my information and that I was not grading for correct answers, but only for participation. I promised them an extra point to be added to an exam score for each survey they participated in, including this one.

After the date for challenging any course grades had passed—this was the date designated in the informed consent letter as the last day students could withdraw from the research—I downloaded the results of the survey into an excel spreadsheet and created histograms of the results. I have analyzed the results in terms of percentages for each answer, to directly compare the pre-instruction responses to the post-instruction responses.

5.3.1.1 Data Analysis

After the deadline for students to challenge course grades had passed, I accessed the data from Qualtrics. For the GCI data, I downloaded the students' individual responses, as well as results aggregated by question. For the students' individual responses, I first matched the pre-and post-instruction survey results of those students who had participated in both these surveys. In other words, I eliminated from the data the results of anyone who participated in the pre-instruction administration but then left the course before the post-instruction administration; similarly, results from anyone who entered the course after the pre-instruction implementation were also removed. This left 77 participants (n = 77).

I calculated a mean out of eight questions for pre-instruction responses and post-instruction responses. I also calculated the standard deviation in the data (Furey, 2022), where the standard deviation is the square root of the variance of the data set.

$$s = \sqrt{\frac{\sum_{i=1}^{n}(x_i - \overline{x})^2}{n-1}}$$

I then used the mean and the standard deviation to calculate a comparison of means (MedCalc Software Ltd., 2022). First, the pooled standard deviation (s) is calculated. The individual standard deviations from the pre-instruction results and the post-instruction results are s_1 and s_2, respectively, and the sample size of each corresponding group is n_1 and n_2.

$$s = \sqrt{\frac{(n_1 - 1)s_1^2 + (n_2 - 1)s_2^2}{n_1 + n_2 - 2}}$$

The standard error (*se*) is found using the following equation:

$$se\left(\bar{x}_1 - \bar{x}_2\right) = s \times \sqrt{\frac{1}{n_1} + \frac{1}{n_2}}$$

The standard error is part of conducting a *t*-test:

$$t = \frac{\bar{x}_1 - \bar{x}_2}{se\left(\bar{x}_1 - \bar{x}_2\right)}$$

The result of the *t*-test distribution is used to calculate the P-value, or significance level.

5.3.2 *GCI Results and Discussion*

Doing a *t*-test to compare means shows a significant difference from pre-instruction responses to post-instruction responses ($p < 0.005$) (see Table 5.8).

In addition to the statistical analysis, I will give some examples of the GCI questions, with histograms of pre-instruction and post-instruction responses, to demonstrate the shift in understanding in response to instruction.

5.3.2.1 The Age of the Earth

Parallel to the geologic timeline question regarding designating when dinosaurs originated and when humans originated, relative to the entire age of the earth (see Fig. 5.11), the post-instruction responses demonstrated a more accurate conceptualization of when these events occurred.

The correct response to the question is diagram D. The responses, shown in Fig. 5.12, reveal that students at the post-instruction stage demonstrated a major shift in understanding when dinosaurs originated, whereas in the COD above (see Fig. 5.5b), they placed dinosaur origins much earlier in Earth history than the consensus view—reflecting a commonly held belief (Francek, 2013). Pre-instruction, just over 40% of students chose diagram C, which diminished to 26% post-instruction; for diagram D, responses rose from just under 40% to just under 60%. Of note, diagram E—in which human existence happens very early in the history of the earth, coincident with the origin of all other life—also had fewer responses.

Table 5.8 GCI t-test results

	n	mean	Standard deviation (*s*)	Standard error (*se*)	t	P
Pre-instruction	77	3.7	1.81	0.294	3.060	0.0026
Post-instruction	77	4.6	1.84			

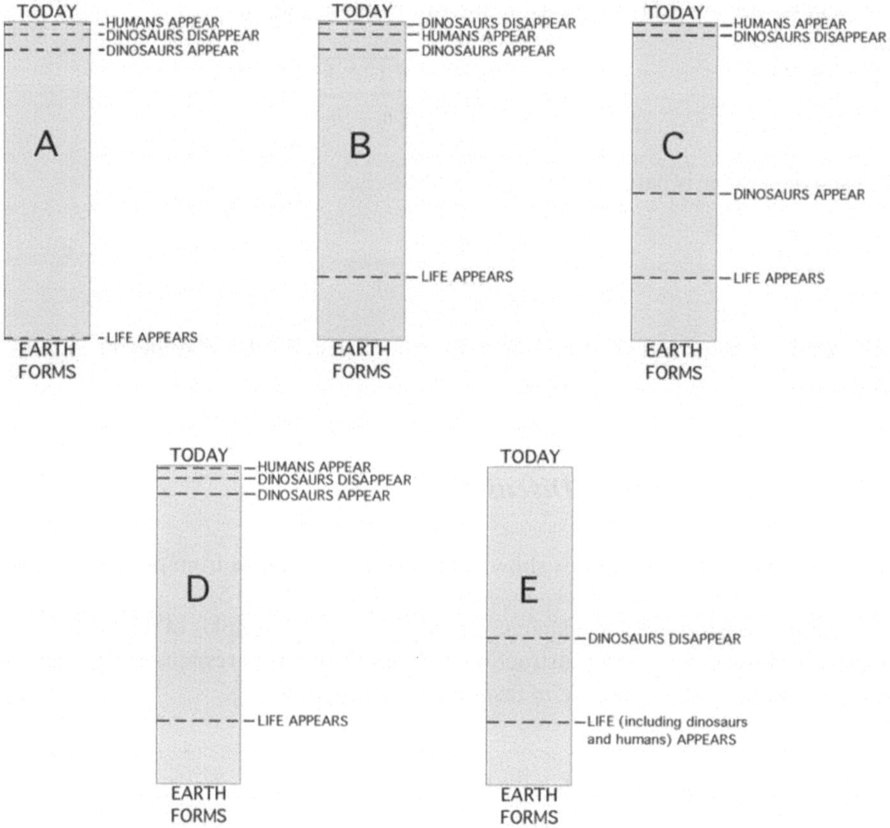

Fig. 5.11 The GCI Question about Geologic Time. (Note. The image shows the possible responses to the following CGI item: *Click on the diagram below that you think most closely represents changes in life on Earth over time*)

In addition to better understanding the sequence and scale of time over the earth's history, students also showed better understanding for how scientists discern these ages. Another GCI question probed this understanding (see Fig. 5.13): *Which technique for determining when the earth first formed as a planet is most accurate?* Possible responses were as follows:

A. Comparison of fossils found in rocks
B. Comparison of layers found in rocks
C. Analysis of uranium found in rocks
D. Analysis of carbon found in rocks
E. Scientists cannot calculate the age of the earth

The correct answer is choice C, of course: analysis of uranium. Figure 5.13 shows a shift in favour of this choice, from pre-instruction (16%) to post-instruction (26%),

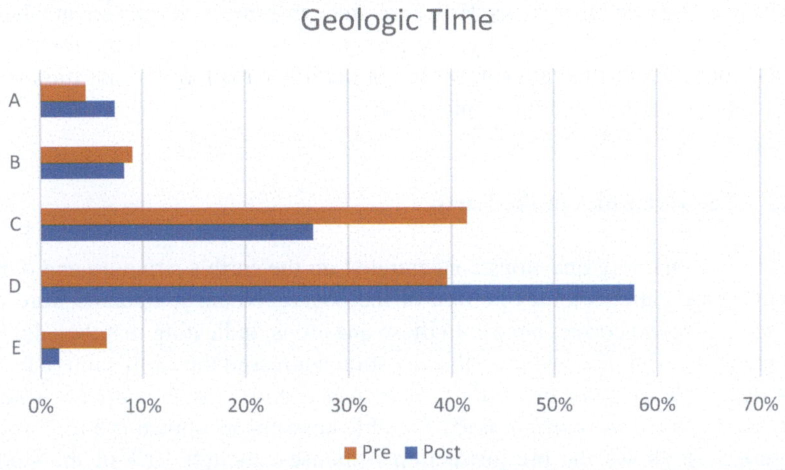

Fig. 5.12 Distribution of responses, by percentage, to GCI question about geologic time. (Note. This histogram shows both pre- (orange) and post-instruction (blue) student responses to this question: *Click on the diagram below that you think most closely represents changes in life on Earth over time.* "D" is the correct answer)

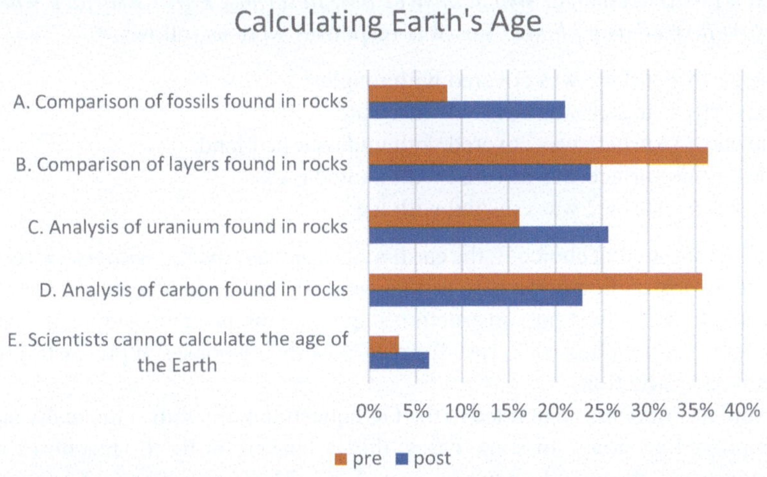

Fig. 5.13 Distribution of responses, by percentage, to GCI question about calculating when the earth formed. (Note. This histogram shows both pre- (orange) and post-instruction (blue) student responses to this question: *Which technique for determining when the earth first formed as a planet is most accurate?* "C" is the correct answer)

and away from choice D, using carbon, which is a common misconception (Francek, 2013). Another shift—in a less favourable direction—was towards answer A, the use of fossils. We did talk in the classroom about fossils and how they can assist in correlating time indicators. We even discussed dating the rocks with the index fossils they contain, and then using their occurrence in other rocks to understand an absolute age of those other rocks. However, since this question asks about when the earth first formed, no life would yet be present to eventually become fossils. This response may reflect that students were not carefully reading the question or that they did not yet understand the implications.

5.3.2.2 The Dynamics of the Earth

Most of the remaining questions concentrated on the earth's structure and dynamics, part of the plate tectonics portion of the course. Notably, students were more likely to choose the correct answer to these questions, indicating that they did learn the content material. For instance, this question addressed the earth's internal structure: *Click on the diagram below that you believe is most closely related to what you might see if you cut the earth in half?* Possible answers are shown in Fig. 5.14.

Figure 5.15 shows the pre-instruction responses: though 40% of the students picked the correct model, E, about 30% picked two models that do not represent the consensus view—one with the earth having a liquid core (B) and one with the earth having a gaseous core (D). In the post-instruction responses, over 60% of students picked the correct model.

The following question, concerning the surface of the earth at its formation, also showed a positive change: *Which answer best describes Earth's surface when the Earth first formed as a planet?* Possible responses were as follows:

A. The earth's surface was covered with jungles
B. The earth's surface was covered with water
C. The earth's surface was covered with undeveloped land
D. The earth's surface was covered with melted rock
E. The earth's surface was covered with ice

Figure 5.16 shows that choice B, the earth's surface was covered with water, received the most pre-instruction responses, at 42%—a call-back to the Neptunist view—but this shifted to 28% at the post-instruction stage. The intended answer, D, the surface was covered with melted rock, rose from 38% of the responses at pre-instruction to 56% at post-instruction.

To sum up, students responded to the GCI questionnaire with statistically significant improved accuracy. In some cases, this is backed up by the responses to the COD questions, discussed above in Sect. 5.2.2. These measures are not perfect; however, the trends largely indicate that students did learn foundational geologic concepts regarding the historical nature, age, and dynamics of the earth. As I did not set out to show that the transformative teaching practices were better at teaching science content than more traditional pedagogy, I did not have a control group to measure any such difference. Nevertheless, I did intend to demonstrate that structuring the learning environment as I did would not sacrifice student learning. Using the results from the VOSTS questionnaire (below), I will show that in addition to the content learning students also developed a more nuanced understanding of the

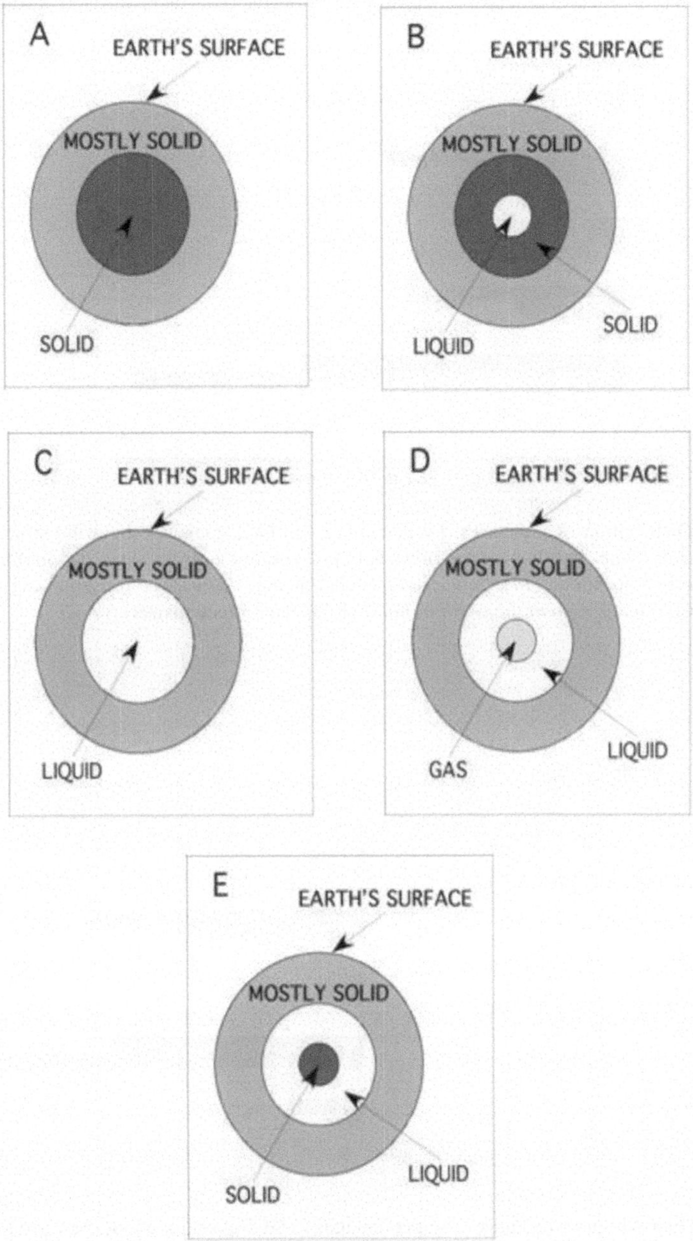

Fig. 5.14 One GCI question about the dynamics of the earth. (Note. The image shows the possible responses to the following CGI item: *Click on the diagram below that you believe is most closely related to what you might see if you cut the Earth in half*)

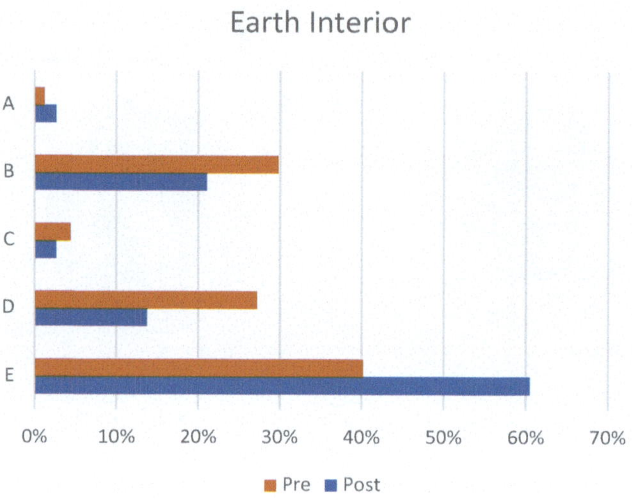

Fig. 5.15 Distribution of responses, by percentage, to GCI question about the structure of the earth's interior. (Note. This histogram shows both pre- (orange) and post-instruction (blue) student responses to this question: *Click on the diagram below that you believe is most closely related to what you might see if you cut the earth in half?* "E" is the correct answer)

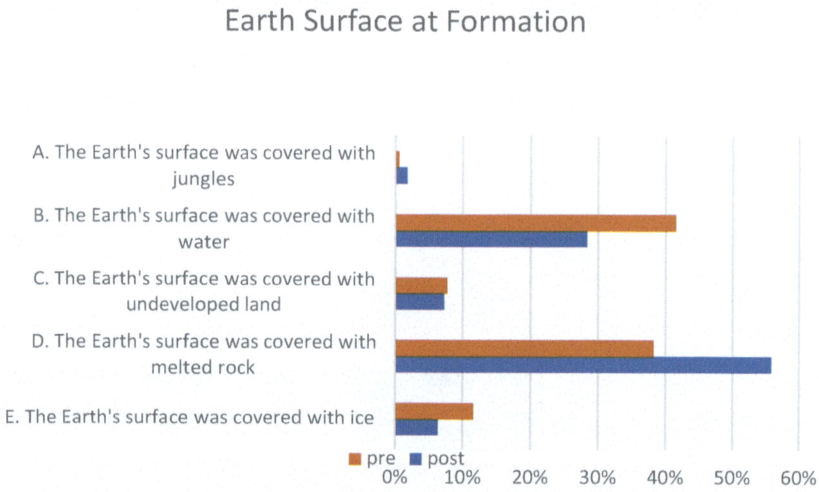

Fig. 5.16 Distribution of responses, by percentage, to GCI question about the earth's surface at formation. (Note. This histogram shows both pre- (orange) and post-instruction (blue) student responses to this question: *Which answer best describes Earth's surface when the Earth first formed as a planet?* "D" is the correct answer)

process of science and the nature of scientific knowledge. Thus, this supports my results from Chap. 6.

5.4 NOS Assessment—Views on Science–Technology–Society (VOSTS)

Measuring students' understandings of NOS has been a goal of science educators for several decades. I have already addressed this in Chap. 1 (Sect. 1.3), and delved deeper into this aspect in Chap. 6 (Sect. 6.1). As with the development of the GCI described above, instruments developed to measure students' NOS understandings suffered from the same deficiencies: limited choices, and variable interpretation of the questions and possible answers by respondents (Aikenhead & Ryan, 1992; McComas et al., 2020). To increase the reliability of the results, Aikenhead et al. (1989) developed an instrument called VOSTS that differed from other instruments because it relied on interviews and open-ended questionnaires with students to develop the questions and the answer options. Aikenhead et al. (1989) described the process they used to formulate the items of the VOSTS as follows:

1. Developing reasoned statements addressing aspects of science, technology, and their interaction with society, which are appropriate for Canadian students in grades 11 and 12. In each case, one contrary statement is also created. Through a Likert scale, students indicate agreement/disagreement with the statement and write an argumentative paragraph justifying their choice.
2. Analyzing 50 to 70 student-written arguments for each pair of statements to identify common arguments and justifications; then, they discuss interpretive differences to reach consensus. Final interpretations represent students' common viewpoints and beliefs, while retaining their familiar wording. The shared student responses become the multiple-choice answers to the original statement, with 5 to 13 responses reflecting reasoning that ranges from one extreme to the opposite extreme. They also added three additional choices to account for possible student responses that would not match the common answers. These were (1) I don't understand, (2) I don't know enough about this subject to make a choice, and (3) None of these choices fits my basic viewpoint.
3. Pilot-testing the questions with about 10 students per question. Students read the original problem statement and respond with an argumentative paragraph, followed by selecting multiple-choice options derived from the earlier round of student paragraphs. These students' paragraph responses are then compared with the multiple-choice responses to establish coherence.
4. Conducting interviews with student respondents to identify any ambiguities in the questions and responses, followed by amending the question wording, as necessary.

5. Pilot-testing a second time with about 10 more students per item, who verbally respond to the statements. This allows the researcher to confirm that the students interpret both the statement and the responses as intended. Based on this, items are reworded one last time, as necessary, to enhance clarity.
6. Finally, deleting those response options that were seldom chosen by respondents, to shorten each item.

5.4.1 *Methods*

The present survey consisted of 19 of the 114 items in the VOSTS. I picked questions most related to NOS aspects that were salient to the course. I used the VOSTS only for the Winter 2021 course, as with the GCI. I selected these instruments to supplement the instruments I was already using, as a way to corroborate my findings from the qualitative data (see Chap. 6). Table 5.9 lists the topics addressed by the VOSTS items, how many items invoked each topic, and the item numbers as found in Aikenhead et al. (1989).

I added the VOSTS items to the same Qualtrics survey, following the GCI items and students completed the survey in the first week of the course, and then again after they took the final exam. As I've already noted above, I offered students an extra credit point for each instrument they completed at both pre-instruction and post-instruction.

5.4.1.1 Data Analysis

As with the GCI questions discussed above in Sect. 3.2, I downloaded the results of the VOSTS survey after the deadline for students to challenge their grade for the course. This was the last day they could withdraw their participation from the research. I assessed the data by the percentage of respondents choosing each

Table 5.9 Number of VOSTS questions asked, the topics they addressed, and the identification number from VOSTS CDN.mc.5 (Aikenhead et al., 1989)

Number of items	Topic addressed	Item number in CDN.mc.5
1	Science and technology	10112
2	Influence of society on science/technology	20141, 20411
1	Influence of science/technology on society	40421
2	Characteristics of scientists	60211, 60511
2	Social construction of knowledge	70221, 70231
11	Nature of scientific knowledge	90111, 90211, 90311, 90411, 90511, 90521, 90611, 90631, 90721, 91011, 91021

possible answer to each question. These questions are different from the GCI questions because they have no intended, or "right" answer. The responses are opinion-based, derived from the qualitative responses of many Canadian students during the development of the survey. In other words, there is no score for this survey, as I calculated for the GCI. Instead, the following section will describe, in a qualitative fashion, general trends in the data for the class as an aggregate whole. These questions are useful because they afford qualitatively derived responses and therefore can be instituted for gathering data in much larger populations.

5.4.2 VOSTS Results and Discussion

When considering science in general, students entered the course conceiving of it as largely a body of facts or knowledge that help explain phenomena. However, upon completing the course, students were more likely to understand science as processes—as opposed to just products (see Fig. 5.17). Notably, students in pre-instruction responses view science as a body or a thing; afterwards, at the post-instruction stage, they see science more in terms of an activity, studying, exploring, or carrying out experiments. This reflects the structure of the course: Instead of giving students continuous lists and endless vocabulary to memorize, the narratives and assignments emphasized the active part of science, which entails exploring and solving problems. The main characters in the historical narratives *were* the main characters *because* they had a problem to solve, they actively engaged in inquiry, they argued with colleagues that their explanation was the better one, and so on. Indeed, even the assignments were more about process than product. Students had to pose questions, solve problems, and argue assertions. That activity is reflected in the answers here.

(10112) Defining science is difficult because science is complex and does many things. But MAINLY science is: Your position, basically: (Please read from A to J, and then choose one.)

A. A study of fields such as biology, chemistry, and physics.
B. A body of knowledge, such as principles, laws, and theories, which explain the world around us (matter, energy, and life).
C. Exploring the unknown and discovering new things about our world and universe and how they work.
D. Carrying out experiments to solve problems of interest about the world around us.
E. Inventing or designing things (for example, artificial hearts, computers, space vehicles).
F. Finding and using knowledge to make this world a better place to live in (for example, curing diseases, solving pollution, and improving agriculture).
G. An organization of people (called scientists) who have ideas and techniques for discovering new knowledge.
H. I don't understand.
I. I don't know enough about this subject to make a choice.
J. None of these choices fits my basic viewpoint.

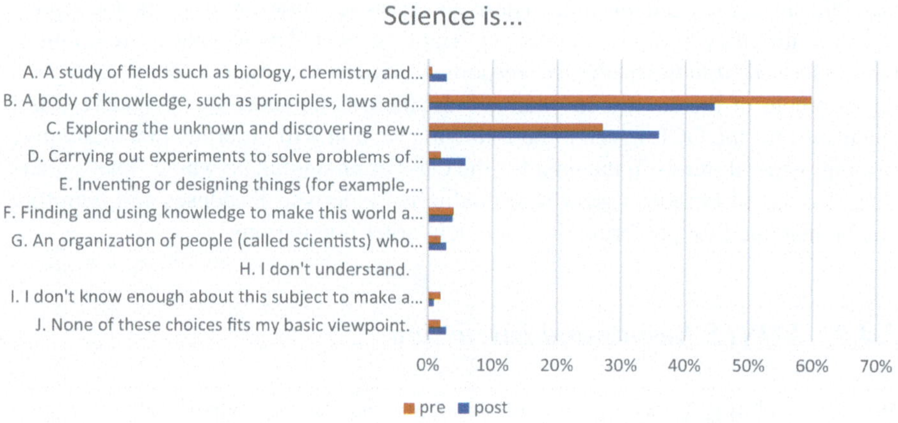

Fig. 5.17 Distribution of Student Responses, by Percentage, to the Question About What Science Is. (Note. This histogram shows both pre- (orange) and post-instruction (blue) student responses to the question)

Of course, when we do talk about science, we can mean both science as process and science as products. Students generally came to view the process of science as one of problem-solving, and not just for abstract problems unconnected to their lives, but as a relevant process they could incorporate into their lives. For instance, this question asks:

> (40421) In your everyday life, knowledge of science and technology helps you personally solve practical problems (for example, getting a car out of a snowdrift, cooking, or caring for a pet). Your position, basically: (Please read from A to J, and then choose one.) The systematic reasoning taught in science classes (for example, hypothesizing, gathering data, being logical):

A. Helps me solve some problems in my daily life. Everyday problems are more easily and logically solved if treated like science problems.
B. Gives me greater knowledge and understanding of everyday problems. However, the problem-solving techniques we learn are not directly useful in my daily life.
C. Ideas and facts I learn from science classes sometimes help me solve problems or make decisions about such things as cooking, keeping healthy, or explaining a wide variety of physical events.
D. The systematic reasoning and the ideas and facts I learn from science classes help me a lot. They help me solve certain problems and understand a wide variety of physical events (for example, thunder or quasars).
E. What I learn from science class generally does not help me solve practical problems; but it does help me notice, relate to, and understand, the world around me.
F. What I learn from science class does not relate to my everyday life: biology, chemistry and physics are not practical for me. They emphasize theoretical and technical details that have little to do with my day-to-day world.

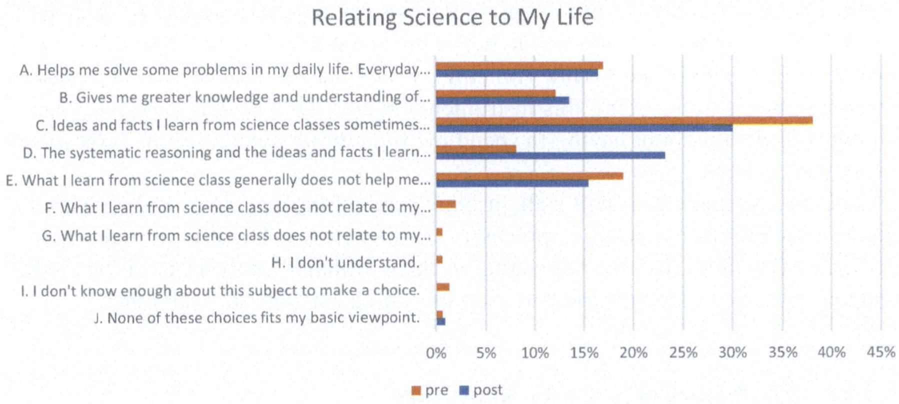

Fig. 5.18 Distribution of student responses, by percentage, to a question relating science to a respondent's life. (Note. This histogram shows both pre- (orange) and post-instruction (blue) student responses to the question)

G. What I learn from science class does not relate to my everyday life: my problems are solved by past experience or by knowledge unrelated to science and technology.
H. I don't understand.
I. I don't know enough about this subject to make a choice.
J. None of these choices fits my basic viewpoint.

Where students' perception of the usefulness of science in their everyday lives is technically not a facet of NOS, it can be, from the students' point of view whether science is something that is outside of or intimately woven into their everyday lives. Do they understand science as esoteric or practical? Figure 5.18 reflects that students mainly find science to be helpful in their lives. For pre-instruction respondents, scientific reasoning mostly appears to play a role in their everyday activities, such as cooking and exercise. This was also obvious in the qualitative analysis presented in Chap. 6. However, about 20% of pre-instruction respondents chose an answer stating that science played no role in their lives; at the post-instruction stage, few respondents chose such answers. In fact, 15% more post-instruction students responded that the systematic reasoning of scientific thought helped them a lot in their lives—option D—compared to students at the pre-instruction stage.

This sentiment is likely due to the emphasis during the class on the process of science, as opposed to just having students memorize a bunch of facts. It is reasonable to assume that knowing a load of geological facts would have little impact on the life of someone who is not a geology major. In contrast, emphasizing the processes of science, through having students experience first-hand the reasoned and creative process of rational thinking, helped them see its utility in their individual lives.

So, again, this is not about finding right or wrong answers in students' responses to the VOSTS; rather, my aim was to better understand and profile their thinking. If I observe a change in the profile over time, I can attribute that shift to this present teaching intervention. Using this instrument enables the researcher to gather large amounts of data without having to spend the time doing qualitative analysis for so many respondents.

The following sections will walk through some of the questions dealing with the nature of scientific knowledge, processes, and scientists. I will point out where I discerned the same patterns happening in the qualitative analysis as in the earlier chapters to underscore how the two sources of data corroborate each other.

5.4.2.1 The Nature of Scientific Knowledge

There were several questions that dealt mainly with the nature of scientific knowledge. I will pick a subset of those questions to highlight here. The first question focuses on the nature and use of models.

> (90211) Many scientific models used in research laboratories (such as the model of heat, the neuron, DNA, or the atom) are copies of reality. Your position, basically: (Please read from A to J, and then choose one.) Scientific models ARE copies of reality:

A. Because scientists say they are true, so they must be true.
B. Because much scientific evidence has proven them true.
C. Because they are true to life. Their purpose is to show us reality or teach us something about it.
D. Scientific models come close to being copies of reality, because they are based on scientific observations and research.
E. Scientific models are NOT copies of reality: because they are simply helpful for learning and explaining, within their limitations.
F. Scientific models are NOT copies of reality: because they change with time and with the state of our knowledge, like theories do.
G. Scientific models are NOT copies of reality: because these models must be ideas or educated guesses, since you can't actually see the real thing.
H. I don't understand.
I. I don't know enough about this subject to make a choice.
J. None of these choices fits my basic viewpoint.

Figure 5.19 shows that most students selected D—that models come close to being reality because they are empirically derived. This choice was popular for both pre-instruction and post-instruction. However, there was an uptick in the post-instruction responses indicating that models were indeed *not copies* of reality, but tools for understanding the abstract and therefore can change with time (choices E and F). This shows a shift from scientific realism to the notion that it is important that the model works to solve problems, rather than representing reality with fidelity. This problem-solving stance defines instrumentalism (Kuhn, 1962).

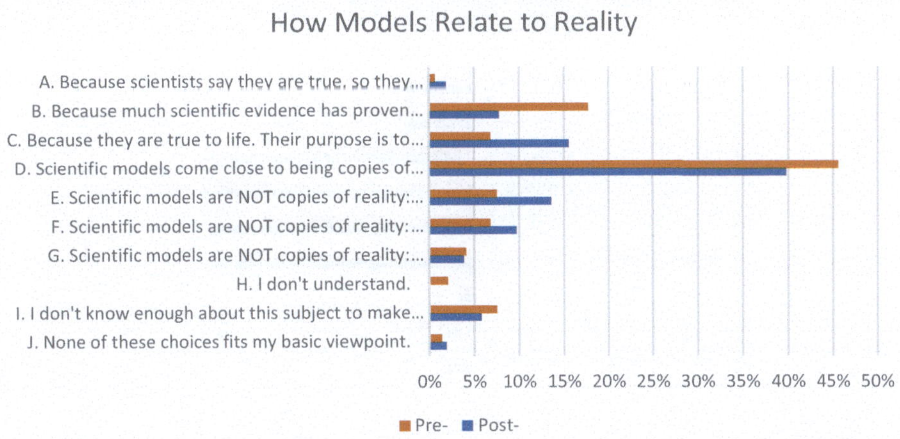

Fig. 5.19 Distribution of student responses, by percentage, to questions relating to how scientific models relate to reality. (Note. This histogram shows both pre- (orange) and post-instruction (blue) student responses to the question)

Related to this is the idea of scientists' models of classification and how those schemes relate to reality, which the following question takes up:

(90311) When scientists classify something (for example, a plant according to its species, an element according to the periodic table, energy according to its source, or a star according to its size), scientists are classifying nature according to the way nature really is; any other way would simply be wrong. Your position, basically: (Please read from A to I, and then choose one.)

A. Classifications match the way nature really is, since scientists have proven them over many years of work.
B. Classifications match the way nature really is, since scientists use observable characteristics when they classify.
C. Scientists classify nature in the most simple and logical way, but their way isn't necessarily the only way.
D. There are many ways to classify nature but agreeing on one universal system allows scientists to avoid confusion in their work.
E. There could be other correct ways to classify nature, because science is liable to change, and new discoveries may lead to different classifications.
F. Nobody knows the way nature really is. Scientists classify nature according to their perceptions or theories. Science is never exact, and nature is so diverse. Thus, scientists could correctly use more than one classification scheme.
G. I don't understand.
H. I don't know enough about this subject to make a choice.
I. None of these choices fits my basic viewpoint.

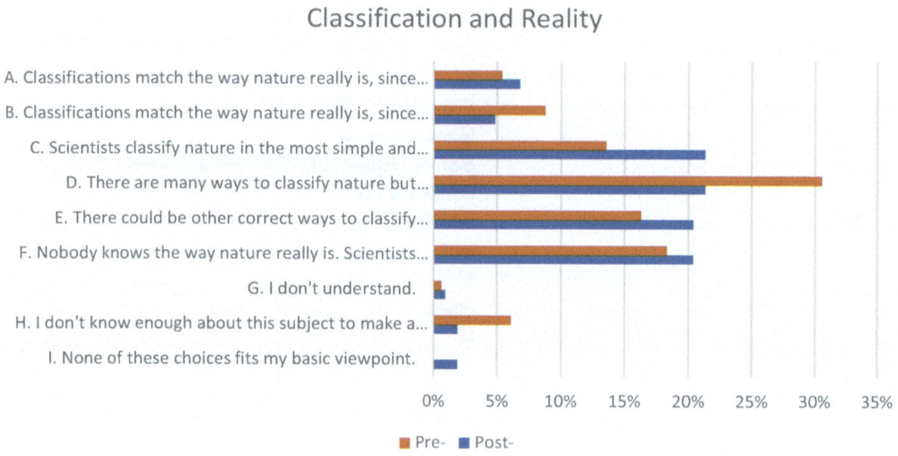

Fig. 5.20 Distribution of student responses, by percentage, to question relating to classification systems of reality. (Note. This histogram shows both pre- (orange) and post-instruction (blue) student responses to the question)

A large portion of the students, especially pre-instruction, took the realist view that any good classification system should be universal so as to diminish confusion among scientists (choice D). For post-instruction responses, students indicated that classifications were not fixed, but could vary depending on how the scientist views that reality (Fig. 5.20). It is not explicit in the answers, but this could be parallel to the instrumentalist view, where the best classification system is the one that works to solve the problem at hand. It also further emphasizes the construction of knowledge and the role of subjectivity as opposed to knowledge being already predetermined.

In class, we do not directly discuss the fact that scientists work within different theoretical frameworks. Evolution, plate tectonics, and relativity are all theoretical frameworks. They are all lenses through which to view the world. However, they are also sequestered to their own scientific discipline, looking at focused aspects of that world, so no sense of conflict exists from one framework to the next. In contrast, in what some refer to as the "social sciences"—sociology, anthropology, education, economics, political science and the like—investigators are free to choose a theoretical framework that best suits the questions they are asking.

For instance, in Chapters 6 and 7, I am using phenomenography and conceptual metaphor theories, respectively, looking at the same data and deriving different interpretations. I could have chosen feminist theory, queer theory, or critical race theory, for example, if my questions were different and revolved around power and opression, sexual orientation, or race, respectively. In GLGY 209, we discussed the implications of Lyell interpreting signs of the past using uniformitarianism as a theoretical framework versus Cuvier's catastrophism, Werner's Neptunism versus

Hutton's Plutonism, Dana's contractionism versus Wegener's horizontal displacement versus Samuel Warren Carey's earth expansionism. Each of these frameworks comes with assumptions, which determines *a priori* how an investigator will under stand the phenomenon of interest.

Another VOSTS question addressing student understanding of the nature of knowledge asks about whether things like scientific laws are found, as is, or invented.

> (91012) For this statement, assume that a gold miner "discovers" gold while an artist "invents" a sculpture. Some people think that scientists discover scientific THEORIES. Others think that scientists invent them. What do you think? Your position, basically: (Please read from A to I, and then choose one.) Scientists discover a theory:

A. Because the idea was there all the time to be uncovered.
B. Because it is based on experimental facts.
C. But scientists invent the methods to find the theories.
D. Some scientists may stumble onto a theory by chance, thus discovering it. But other scientists may invent the theory from facts they already know.
E. Scientists invent a theory: because a theory is an interpretation of experimental facts which scientists have discovered.
F. Scientists invent a theory: because inventions (theories) come from the mind—we create them.
G. I don't understand.
H. I don't know enough about this topic to make a choice.
I. None of these choices fits my basic viewpoint.

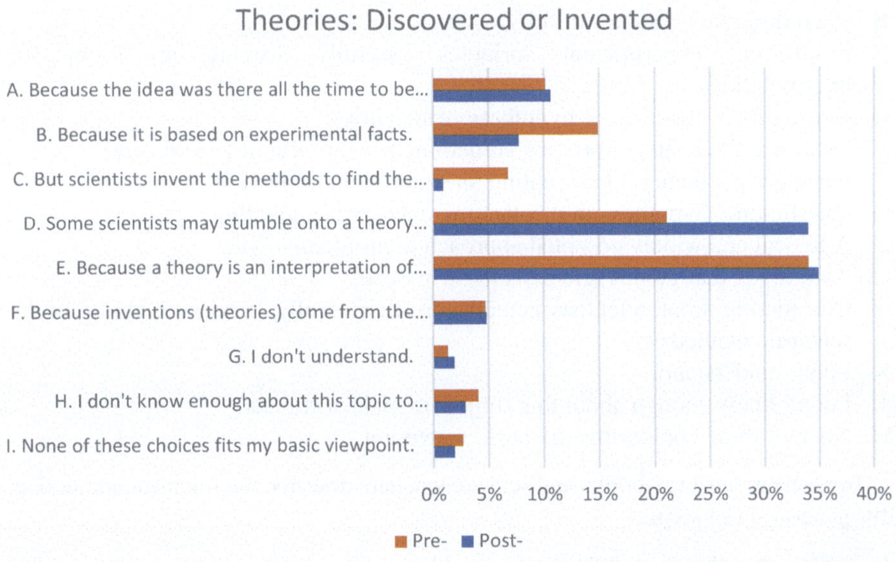

Fig. 5.21 Distribution of student responses, by percentage, to a question about the nature of scientific theories. (Note. This histogram shows both pre- (orange) and post-instruction (blue) student responses to the question)

According to Fig. 5.21, a large group of students (34% pre-instruction and 41% post-instruction) agreed that scientists invent theories (choices E and F). The other choices (A–D), which are more in line with theories as predetermined and just waiting for scientists to find them, were slightly higher in the pre-instruction responses (42%) compared to post-instruction (20%). In general, students understood scientific theories to be made from smaller facts, which are discovered or derived through experimentation. This parallels the analysis of the qualitative data that I discussed in the previous two chapters. Understanding how students view scientific knowledge can now set the stage for how they view the scientific process. The next section explores this description.

5.4.2.2 The Nature of the Scientific Process

In this section, I analyze the VOSTS focus on the process of science. The first point I address is the idea of a single, uniform scientific method. This is something that gets taught in almost all science classes from primary grades on up to introductory university level courses (for a broader discussion about this, see Chap. 6, Sect. 6.3.2.1). The following VOSTS question examines this idea:

> (90611) When scientists investigate, it is said that they follow the scientific method. The scientific method is: Your position, basically: (Please read from A to M, and then choose one.)

A. The lab procedures or techniques; often written in a book or journal, and usually by a scientist.
B. Recording your results carefully.
C. Controlling experimental variables carefully, leaving no room for interpretation.
D. Getting facts, theories, or hypotheses efficiently.
E. Testing and retesting—proving something true or false in a valid way.
F. Postulating a theory, then creating an experiment to prove it.
G. Questioning, hypothesizing, collecting data, and concluding.
H. A logical and widely accepted approach to problem-solving.
I. An attitude that guides scientists in their work.
J. Considering what scientists actually do, there really is no such thing as the scientific method.
K. I don't understand.
L. I don't know enough about this subject to make a choice.
M. None of these choices fits my basic viewpoint.

In essence, most textbooks and science teachers describe the scientific method as this general set of steps:

1. Make some observations/discern a problem
2. Do some background research/ask a question
3. Make a hypothesis based on background research and make some predictions
4. Test the hypothesis (usually involving controlled experimentation)

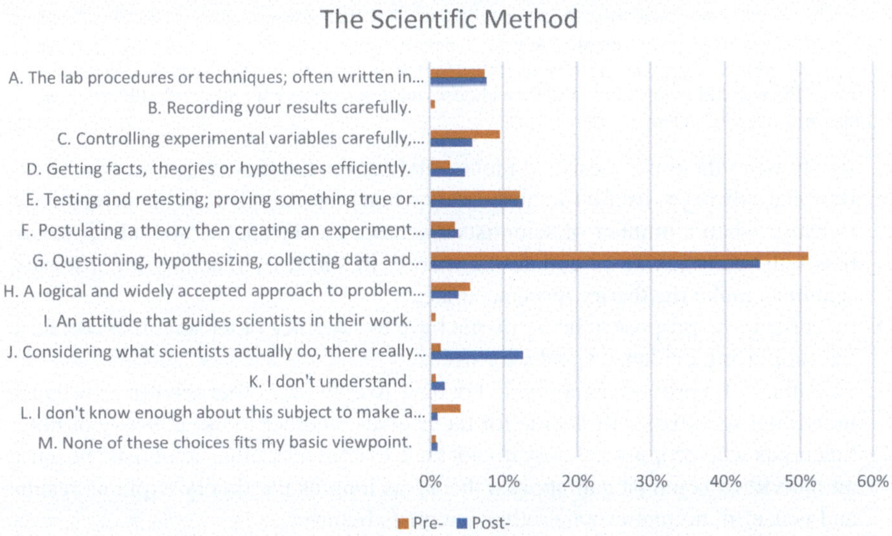

Fig. 5.22 Distribution of student responses, by percentage, to a question relating to the scientific method. (Note. This histogram shows both pre- (orange) and post-instruction (blue) student responses to the question)

5. Collect and analyze data to see if the prediction was accurate
6. Generate results to support hypothesis (if not, amend hypothesis and go back to step 3)
7. Communicate results

At issue here is that these steps are more of a convention for *writing up* research as opposed to actually doing it (see Chap. 6, Sect. 6.3.2.1 for a full discussion on the topic of the scientific method). To be sure, most students, both pre-instruction and post-instruction, identify the scientific method as the straightforward list of steps reflecting the dogma in science classes and textbooks starting from the early primary grades. However, post-instruction responses show a drop in the percentage of students choosing G (from 51% to 45%), and an increase in those choosing J (1% to 13%), which denotes that no single method exists (see Fig. 5.22).

This shift likely arises as a result of the narratives, in which none of the investigations adhere to the pattern of steps identified in the scientific method. Students reflected on the different narratives, and I challenged them to find an instance where the investigations followed such a series of steps. We then reflected on their search and discussed several historical cases to clearly show the progress of scientific investigations. This shift in understanding also occurred in the qualitative data analysis, with many more students in the post-instruction data explicitly stating that there was no single method scientists follow.

Once the investigations are concluded and scientists have formulated some explanations for the phenomena they've been studying, they need to report their findings to the larger community. This could involve consensus building, as portrayed in the following question:

(70231) When a new scientific theory is proposed, scientists must decide whether to accept it or not. Scientists make this decision by consensus; that is, proposers of the theory must convince a large majority of fellow scientists to believe the new theory. Your position, basically: (Please read from A to I, and then choose one.) Scientists who propose a theory must convince other scientists:

A. By showing them conclusive evidence that proves the theory true.
B. Because a theory is useful to science only when most scientists believe the theory.
C. Because when a number of scientists discuss a theory and its new ideas, scientists will likely revise or update the theory. In short, by reaching a consensus, scientists make the theory more accurate.
D. Scientists who propose a theory do not have to convince other scientists: because the supporting evidence speaks for itself.
E. Scientists who propose a theory do not have to convince other scientists: because individual scientists will decide for themselves whether to use a theory or not.
F. Scientists who propose a theory do not have to convince other scientists: because an individual scientist can apply a theory as long as the theory explains results and is useful, no matter what other scientists believe.
G. I don't understand.
H. I don't know enough about this subject to make a choice.
I. None of these choices fits my basic viewpoint.

Figure 5.23 shows that the majority of students, both pre-instruction and post-instruction, agreed that when proposing a new idea, there is a dialogical aspect to the process. The scientists talk and amend their own ideas to finally come to a consensus (choice C). The second most chosen response (A: 26% of pre-instruction responses and 15% post-instruction responses) suggests that the scientist just needs

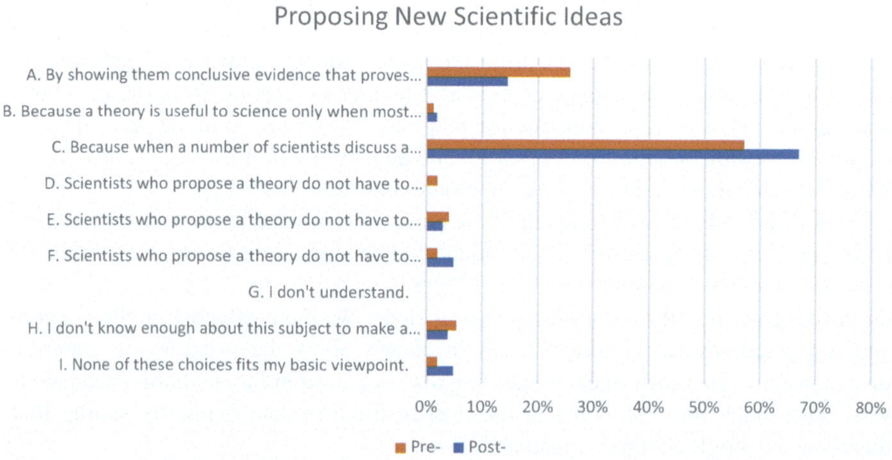

Fig. 5.23 Distribution of student responses, by percentage, to a question relating to proposing new scientific ideas. (Note. This histogram shows both pre- (orange) and post-instruction (blue) student responses to the question)

to prove the idea true to the community to reach consensus. The post-instruction students basically shifted from choice A to choice C. This shift may be attributable to the small-group work the students engaged in over the entire term. This work stressed reaching consensus, which involved students making arguments to the rest of their group and amending their ideas in response to others' arguments. Also, the narratives provided many opportunities for students to experience this process—as I presented data, asked for interpretations or explanations, then presented more data. This constant development of ideas through time also exemplifies the process of consensus building.

During the process of developing scientific knowledge, making assumptions must occur to allow such advancement. One VOSTS question explicitly focused on the role and nature of assumptions within the scientific process:

> (90521) When developing new theories or laws, scientists need to make certain assumptions about nature (for example, matter is made up of atoms). These assumptions must be true in order for science to progress properly. Your position, basically: (Please read from A to I, and then choose one.) Assumptions MUST be true in order for science to progress:

A. Because correct assumptions are needed for correct theories and laws. Otherwise, scientists would waste a lot of time and effort using wrong theories and laws.
B. Otherwise, society would have serious problems, such as inadequate technology and dangerous chemicals.
C. Because scientists do research to prove their assumptions true before going on with their work.
D. It depends. Sometimes science needs true assumptions in order to progress. But sometimes history has shown that great discoveries have been made by disproving a theory and learning from its false assumptions.
E. It doesn't matter. Scientists have to make assumptions, true or not, in order to get started on a project. History has shown that great discoveries have been made by disproving a theory and learning from its false assumptions.
F. Scientists do not make assumptions. They research an idea to find out if the idea is true. They don't assume it is true.
G. I don't understand.
H. I don't know enough about this subject to make a choice.
I. None of these choices fits my basic viewpoint.

For the role of assumptions in the scientific process (see Fig. 5.24), the majority of students, both pre- and post-instruction, agreed that for science to advance, the assumptions that are made need to be true (choice D), though instances exist of learning from false assumptions once an idea based on those assumptions has been proven wrong. A smaller percentage of the pre-instruction students (10%) maintained that any assumptions scientists make to progress with an idea must be true or they are just wasting time (choice A). Post-instruction responses show a gain (from 13% to 18%) towards stating that assumptions, correct or incorrect, are necessary for science to progress and that scientists can still learn from false assumptions (choice E).

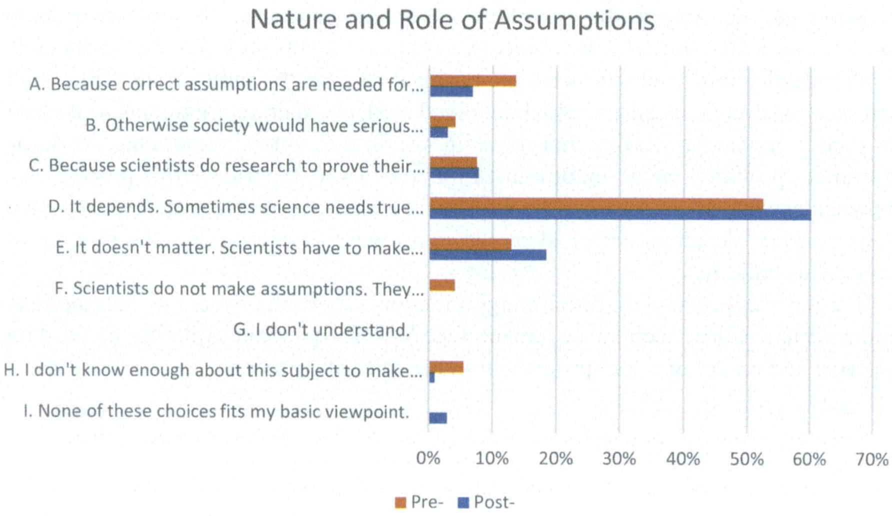

Fig. 5.24 Distribution of student responses, by percentage, to a question about the nature and role of assumptions. (Note. This histogram shows both pre- (orange) and post-instruction (blue) student responses to the question)

Through the multiple examples I presented in class, I explicitly bring up the role of assumptions within the historical narratives; students had to identify the assumptions that a particular scientist was making and then also evaluate the reasonableness of those assumptions. For instance, as I've discussed earlier, Darwin and Lyell both assumed a world operating under the rules of uniformitarianism, while William Thompson assumed many of the values he invested in his thermodynamic equations in the debate about the age of the earth (Stinner & Teichmann, 2003). Aristotle assumed that the structure of the rest of the earth was similar to the karst topography he experienced in his homeland of Greece. Nicholas Steno assumed that the surface of the earth was changing and that the signs and cross-cutting relationships he observed were a record of those changes.

These and other examples helped students to understand both the role of assumptions and how to judge the reliability of ideas based in those assumptions. Assumptions are often based on the subjectivity of the scientist; this last section will therefore discuss student understandings of scientist subjectivity as a factor that contributes to their scientific work.

5.4.2.3 Being a Scientist

Some of the VOSTS questions explored students' understandings about the nature of scientists, particularly where this fits into the complex relationships between reality, science processes, and science products. This first question starts to explore some of that space:

(90111) Scientific observations made by competent scientists will usually be different if the scientists believe different theories. Your position, basically: (Please read from A to H, and then choose one.)

A. Yes, because scientists will experiment in different ways and will notice different things.
B. Yes, because scientists will think differently, and this will alter their observations.
C. Scientific observations will not differ very much even though scientists believe different theories. If the scientists are indeed competent their observations will be similar.
D. No, because observations are as exact as possible. This is how science has been able to advance.
E. No, observations are exactly what we see and nothing more; they are the facts.
F. I don't understand.
G. I don't know enough about this subject to make a choice.
H. None of these choices fits my basic viewpoint.

The responses shown in Fig. 5.25 demonstrate a striking difference between pre-instruction responses and post-instruction responses. The choices can be separated into two groups:

- Scientist subjectivity should play no role in making scientific observations—choices C, D, and E.
- Scientist subjectivity does play a role in the observations a scientist makes—choices A and B.

Pre-instruction responses favour the idea that subjectivity plays (or should play) no role, at least for competent scientists, in the observations they make (44% of pre-instruction responses compared to 24% of post-instruction responses).

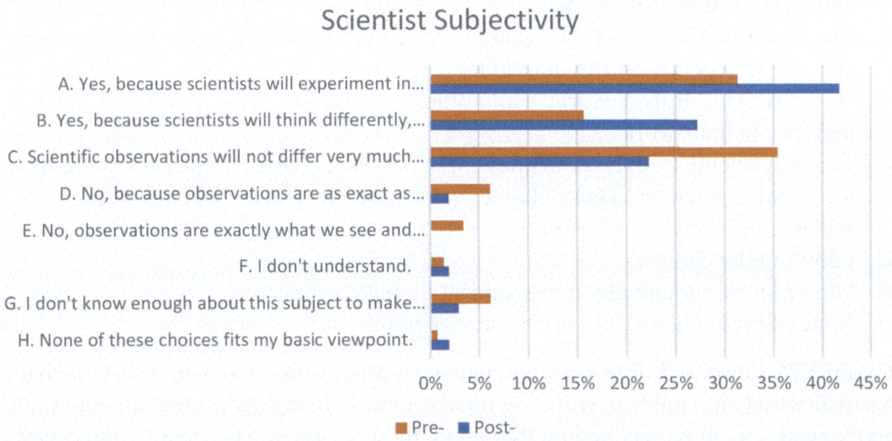

Fig. 5.25 Distribution of student responses, by percentage, to a question relating to scientist subjectivity. (Note. This histogram shows both pre- (orange) and post-instruction (blue) student responses to the question)

Post-instruction responses shifted to the idea that subjectivity does play a role in the observations scientists make (47% pre-instruction responses compared to 69% post-instruction).

This shift in understanding is also apparent in the qualitative data analysis in Chap. 6, where the acknowledgement of subjectivity was not only greater, but students also accepted it more as a part of science and not something that must be wiped clean from the process or avoided at all costs. I attribute this shift directly to the historical narratives, which situated the scientists of interest within their circumstances. For example, Mary Anning became an important source of fossil finds—having been trained by her father, she was able to "see" them when others who were not so trained could not discern them. Wegener was able to jump to the idea of horizontal displacement after seeing the map of the Americas and Africa because he was not enculturated in contractionism. Where Cuvier—adhering to catastrophism—saw mass extinctions, Lyell—a uniformitarian—saw long periods of little to no deposition, allowing for fossils to accumulate at one specific level. The list goes on.

Another question, one exploring the characteristics of scientists, asks this:

> (60211) The best scientists are always very open-minded, logical, unbiased, and objective in their work. These personal characteristics are needed for doing the best science. Your position, basically: (Please read from A to I, and then choose one.)

A. The best scientists display these characteristics, otherwise science will suffer.
B. The best scientists display these characteristics because the more of these characteristics you have, the better you'll do at science.
C. These characteristics are not enough. The best scientists also need other personal traits such as imagination, intelligence, and honesty.
D. The best scientists do NOT necessarily display these personal characteristics: because the best scientists sometimes become so deeply involved, interested, or trained in their field, that they can be closed-minded, biased, subjective, and not always logical in their work.
E. The best scientists do NOT necessarily display these personal characteristics: because it depends on the individual scientist. Some are always open-minded, objective, etc. in their work; while others can become closed-minded, subjective, etc. in their work.
F. The best scientists do NOT display these personal characteristics any more than the average scientist. These characteristics are NOT necessary for doing good science.
G. I don't understand.
H. I don't know enough about this subject to make a choice.
I. None of these choices fits my basic viewpoint.

Figure 5.26 shows a shift in students' understanding: from there being certain characteristics that are vital to performing good science—being open-minded, unbiased, and logical—to an understanding that while these characteristics can be important, they are not sufficient, and indeed, may not even be necessary for good science to happen. While many people have been taught an idealized version of how

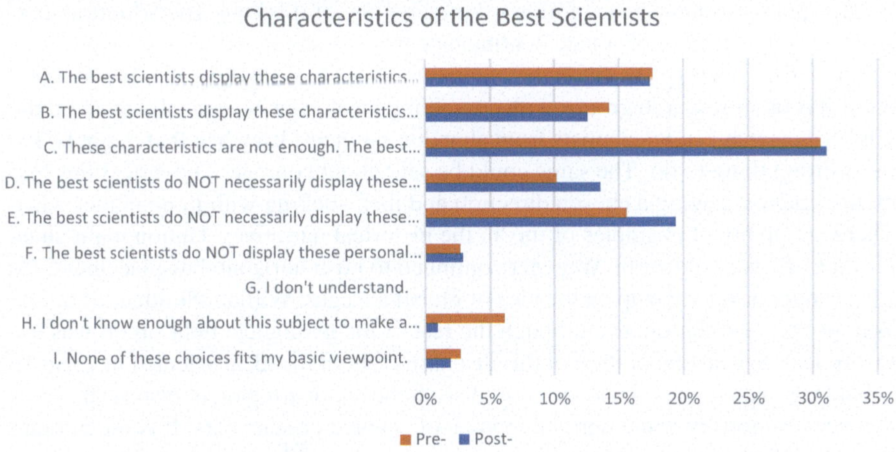

Fig. 5.26 Distribution of student responses, by percentage, to a question relating to characteristics of the best scientists. (Note. This histogram shows both pre- (orange) and post-instruction (blue) student responses to the question)

science gets done—the scientific method, which takes the human aspect out of the process—an idealized version of what makes a good scientist and who such a person is also persists. I will describe this idea in depth (Chap. 6, Sect. 6.3.2.3), but here is another angle. Take for instance, this list of qualities inherent to a good scientist (Chukwuemeka, 2022):

- Curious
- Detail-oriented
- Good communication skills
- Patient
- Logical and critical thinker
- Persistent
- Courageous
- Target-oriented
- Analytical
- Open-minded and free from bias
- Skeptical

These do indeed seem like the qualities that would make a good scientist. Yet, some consideration of the implications of each attribute makes it apparent that this "ideal" actually contains multiple contradictions: in particular, the notion that scientists must be simultaneously open-minded and skeptical is hard to reconcile. I feel it is important to emphasize this point because sometimes multiple critiques of a firmly held model are necessary to finally acknowledge its limitations. Skepticism is a form of bias, and it is certainly not having an open mind. Being target-oriented is to have a bias, as the scientist is not "aiming" just anywhere. They are aiming in a particular place, the place *they* think is correct.

The quality of persistence follows this same line of thinking. By definition, persistence is the "firm or obstinate continuance in a particular course in spite of opposition" (OED, 1994, p. 1315). If that opposition is lack of funding, that is one thing, but if it is other research groups with competing ideas, then we are talking about the kind of bias that keeps scientists from abandoning their idea when that might be the most fitting thing to do. The same could be said about courage. These qualities rely on the scientist having a chosen direction and then sticking with it, no matter what. There are plenty of examples of this in the historical narratives. Hutton maintained his idea of a very old earth. Wegener continued to push horizontal displacement. De Charpentier never gave up on the idea of global ice ages. William Smith used fossils as a tool for correlation, even though the rest of the geological field said it was too theoretical. See an elaboration of this discussion about the ideal scientist in Chap. 6. What I try to do in these instances is to disrupt the reconstruction of historical events that seek to portray some normative view of science or scientists. It is much more fertile for discussion to address situations as they more likely were, scientists in the dark, with their hands outstretched, trying to find a path forward but having no idea where that path will lead.

A dichotomy of thought about scientists is evident then. We attribute heroic qualities to them, such as persistence and courage, even as we paradoxically also mainly view scientific knowledge as some sort of independent truth. Such framing means knowledge or insights are not constructed by the scientist, where bias gets in the way or inhibits the seeing of reality. However, it seems that when a scientist is "right," that reveals persistence, stubbornness, and courage that paid off. When they are "wrong," then it is obvious that their persistence and stubbornness were simply biases preventing them from making true advancements; moreover, their courage was merely close-mindedness. In other words, we need to be careful how we interpret the actions of historical figures, knowing that we know how the story ends and they did not. We must not reconstruct history with the end already in mind, even though we already know the end. Someone like Alfred Wegener was maligned for his bias in seeking and presenting only evidence that supported horizontal displacement, then lionized for his courage in challenging the orthodoxy of the time. Closer to the truth, Wegener was passionate and inquisitive, very human in his devotion to his perspective.

Bringing these kinds of critical discussions into the classroom situate science and scientists in a way that helps students become critical consumers of the science that pervades their lives. In the age of "do your own research," we must be realistic that this is not possible. One cannot conduct their own experiments about climate change, COVID-19 vaccines, cigarette smoking, or any other controversy affecting public policy. One cannot read all of the literature on any of these topics to come to an "informed" position. But we can develop the ability to discern who speaks for science (Höttecke & Allchin, 2020), who is trustworthy, who is reliable—and then develop our understandings from there, while always keeping in mind that experts and authorities on a topic are also only human. The data here and in the next chapters indicate that my students did begin developing such a mindset.

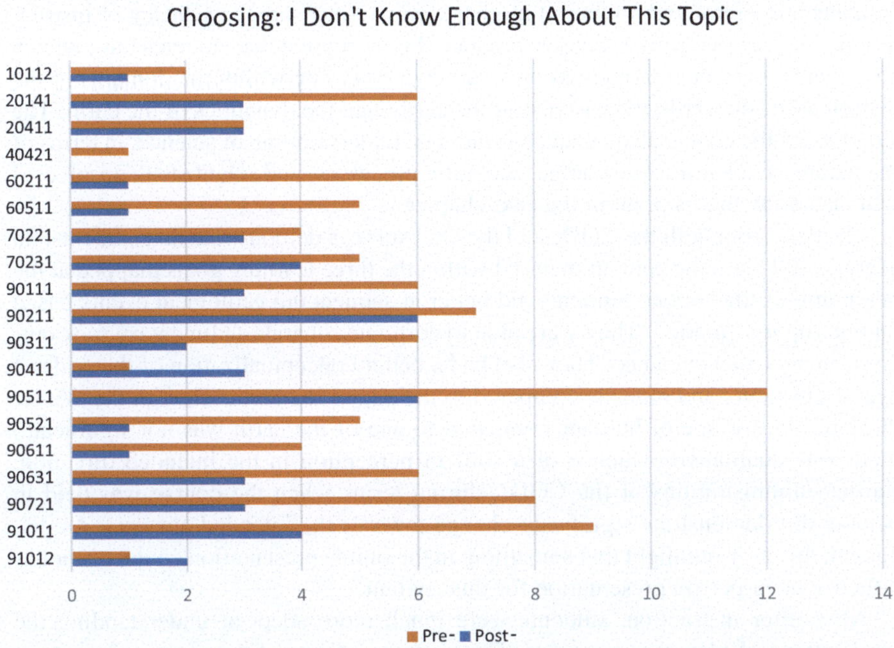

Fig. 5.27 Distribution of percentage of students choosing, "I Don't Know Enough About This Topic to Make a Choice," across all questions. (Note. This histogram shows both pre- (orange) and post-instruction (blue) student responses to the questions identified)

One other trend in the data that I think is important to point out is that each questions included the choice, "*I don't know enough about this subject to make a choice.*" The range of percentages of students choosing this option was 1% to 12%. Analyzing how many chose this option pre-instruction compared to post-instruction shows that with the exception of two of the VOSTS questions, students were more likely to select this option prior to instruction than post-instruction (see Fig. 5.27). Then, students showed a more nuanced understanding of science and were also surer of themselves when expressing their answers.

5.5 Conclusions

My goal with this chapter has been to evaluate the impact that historically contextualizing the geology content knowledge—in tandem with the other transformative teaching strategies—has on student learning, both for disciplinary content and within the context of NOS. Indeed, the students have successfully achieved my learning goals for the course. This is a very detailed analysis of the data collected. I hope for it to serve a couple goals. First, it is a demonstration of the efficacy of the teaching strategies, and second, it models a way to measure the efficacy of instruction that is part of a truly reflective practice. I have argued that students have shown advancements in their content knowledge understanding within the domains of the historicity of the earth, the vast age of the earth, and the dynamics of the earth. The structure of the course also enhanced students' understanding of science, in terms of the nature of scientific knowledge, scientific processes, and scientists themselves. I will elaborate on this point in the next chapter.

The data from both the CODs and the GCI surveys demonstrate that students, on average, did learn the content material within the three learning goals mapped at the beginning of the course. Students did better at sequencing geological events based on outcrop information. They were able to correlate separate sedimentary rock outcrops using fossil evidence. They also had a better conceptualization of the scale of age of dinosaurs and humans compared to the age of the earth. Though the Winter 2021 results for age of humans compared to age of the earth was not significant, there was qualitative evidence of a shift in perception in the intended direction. Earlier administrations of the CODs, during terms when the course was held in person, did demonstrate significant change towards the intended answers. As discussed, this may highlight that something in the online presentation context is not as effective as in-person presentation for this content.

Also, after instruction, students were much more adept at understanding the implications of planetary tectonics. They better understood the nature of isostasy, the structure of the earth, and the mechanics of plate tectonics. Their results also invited me to think about and further refine some of the questions, including how I might teach these concepts to facilitate students' productive model development to an even greater degree. This was especially apparent for the hot spot question (see Figs. 5.7 and 5.8), where I needed to be mindful of the orientational context I was using to teach the concept, versus that which I was using to evaluate learning. For the subduction zone earthquake question (see Fig. 5.9), it is possible that the image needs to be reworked so that the number of locations of possible correct answers is more limited. This is work that will take place moving forward. Where this might seem to be a modest goal for a geology course, in terms of the content breadth, it is important to remember that these concepts are not only expansive but also *foundational* to understanding geology. Taking time and developing understandings of these concepts in particular is critical if students will go forward to study the discipline, or if they will rejoin their own academic home while maintaining such foundational ideas.

When considering students' development of NOS understandings, the VOSTS questions largely support the interpretations I will make in the next chapter. Despite the inherent limitations of conventional multiple- or forced-choice questions, the VOSTS questions are stronger because the choices given were those based on qualitative data of Canadian students. This means that concerns about interpreting the questions or the choices decrease. Furthermore, the many choices per question represent a spectrum of opinions. Finally, the last three choices for each question allow students to bypass earlier choices they do not feel comfortable making. I only used the VOSTS survey during the last iteration of the course to check my prior interpretations of the qualitative data, which I'd been collecting throughout the previous five iterations of the course. I was pleased to observe that the results support these qualitative findings: students did show a shift in responses from pre-instruction to post-instruction in a direction parallel to the NOS learning goals.

In general, students were much more accepting of the role subjectivity and bias plays in the scientific process. These aspects were not something to be erased from the process or to be blamed when scientific knowledge ends up being wrong. They were seen as inherent to the process because they are inherent to humans. Another big shift in understanding concerns the constructed nature of science knowledge—that is, the act of constructing knowledge is an inherent aspect of knowledge development. As will be seen in the next two chapters, the notion of ideas being things is quite pervasive; however, post-instruction responses indicated a shift to a more constructivist stance. Classification schemes, and scientific laws, theories, and models are developed by scientists more for utilitarian purposes than as a true reflection of reality.

It is within this constructivist context that scientists' subjectivity and bias play a role. Also, students much more readily acknowledged the need and role of assumptions in the process. Finally, students began to see past the "ideal" portrayal of scientists for one that is more realistic. Scientists are not technicians or mindless automatons, nor are they removed from the temporal and social context in which they are situated. They are creative and maintain agency within the process of scientific knowledge development. Yes, they are limited by the data, but after that, almost anything goes. It is this description of scientists, which humanizes them, that was important to me. If we want to increase the number of students entering STEM, then making it welcoming to any human with a passion for creativity and solving problems seems to be a reasonable strategy.

The next two chapters will investigate the qualitative data collected over the past five iterations of the course. Using different theoretical frameworks in the analysis of those data will show themes within the context of the shifting of nature of science understandings from pre-course to post-course overlapping with many of the results presented above.

References

Aikenhead, G., & Ryan, G. (1992). The development of a new instrument: "Views on Science-Technology-Society" (VOSTS). *Science Education, 76*(5), 477–491.

Aikenhead, G. S., Ryan, A. G., & Fleming, R. W. (1989). Views on science-technology society: Form CDN.mc.7. In *University of Saskatchewan: Department of Curriculum Studies, College of Education*. University of Saskatchewan.

Bogdan, R., & Biklen, S. K. (2007). *Qualitative research for education: an introduction to theories and methods* (Vol. 5th). Pearson A & B.

Boulter, C., & Buckley, B. (2000). Constructing a typology of models for science education. In J. Gilbert & C. Boulter (Eds.), *developing models in science education* (pp. 41–58). Kluwer Academic publishing.

Caldwell, J. E. (2007). Clickers in the large classroom: Current research and best practice tips. *CBE–Life Sciences Education, 6*, 9–20.
Cheek, K. (2012). Students' understanding of large numbers as a key factor in their understanding of geologic time. *International Journal of Science and Mathematics Education, 10*, 1047–1069.
Cheek, K. A., Ladue, N. D., & Shipley, T. F. (2017). Learning about spatial and temporal scale: Current research, psychological processes, and classroom implications. *Journal of Geoscience Education, 65*(4), 455–472. https://doi.org/10.5408/16-213.1
Chukwuemeka, E. S. (2022). Characteristics of a good scientist: 8 essential qualities. *Scholarly Articles.*. https://bscholarly.com/characteristics-of-a-good-scientist/
Cowan, J. (2017). The potential of think aloud protocols for educational action research. *Active Learning in Higher Education, 20*(3), 219–232.
Czajka, C. D., & Mcconnell, D. (2018). An exploratory study examining undergraduate geology students' conceptions related to geologic time and rates. *Journal of Geoscience Education, 66*(3), 231–245. https://doi.org/10.1080/10899995.2018.1480826
Francek, M. (2013). Compilation and review of over 500 geoscience misconceptions. *International Journal of Science Education, 35*(1), 31–64.
Furey, E. (2022). *Standard deviation calculator*. Retrieved 20 May 2022 from https://www.calculatorsoup.com/calculators/statistics/standard-deviation-calculator.php
Höttecke, D., & Allchin, D. (2020). Reconceptualizing nature-of-science education in the age of social media. *Science Education, 104*, 641–666.
Kuhn, T. (1962). *The structure of scientific revolutions*. University of Chicago Press.
LaDue, N. D., & Shipley, T. F. (2018). Click-on-diagram questions: A new tool to study conceptions using classroom response systems. *Journal of Science Education and Technology, 27*(6), 492–507.
LaDue, N., Ackerman, J. R., Blaum, D., & Shipley, T. (2021). Assessing water literacy: Undergraduate student conceptions of groundwater and surface water flow. *Water, 13*.
Leighton, J. P. (2017). *Using think-aloud interviews and cognitive labs in educational research*. Oxford University Press.
Libarkin, J. C., & Anderson, S. W. (2005). Assessment of learning in entry-level geoscience courses: Results from the geoscience concept inventory. *Journal of Geoscience Education, 53*(4), 394–401.
Libarkin, J., & Geraghty Ward, E. (2011). The qualitative underpinnings of quantitative concept inventory questions. In A. Feig & A. Stokes (Eds.), *Qualitative inquiry in geoscience education research special paper 474* (pp. 37–48). United States Geological Society.
Libarkin, J. C., Geraghty Ward, E., Anderson, S. W., Kortmeyer, G., & Raeburn, S. P. (2011). Revisiting the Geoscience Concept Inventory: A call to the community. *GSA Today, 21*(8), 26–28.
McComas, W. F., Clough, M. P., & Nouri, N. (2020). Nature of science and classroom practice: A review of the literature with implications for effective NOS instruction. In W. F. McComas (Ed.), *Nature of science in science instruction: Rationales and strategies* (pp. 67–114). Springer.
MedCalc Software Ltd. (2022). *Comparison of means calculator (Version 20.109)*. MedCalc. Retrieved 20 May 2022 from https://www.medcalc.org/calc/comparison_of_means.php
New York State Education Department. (2013, January 24). Regents examination in physical setting/earth science. Question #19. https://www.nysedregents.org/earthscience/113/esci12013-exam.pdf
Niebert, K., & Gropengiesser, H. (2015). Understanding starts in the mesocosm: Conceptual metaphor as a framework for external representations in science teaching. *International Journal of Science Education, 37*(5–6), 903–933. https://doi.org/10.1080/09500693.2017.1025310
OED. (1994). *Oxford English Dictionary* (Vol. 2nd). Oxford University Press.
Powell, S., Straub, C., Rodriguez, J., & VanHorn, B. (2011). Using clickers in large college psychology classes: Academic achievement and perceptions. *Journal of Scholarship of Teaching and Learning, 4*, 1–11.
Preuss, P. (1992). Rhea's time. In *The ultimate dinosaur* (pp. 88–99). Bantam Books.

Rudwick, M. J. S. (1985). *The great Devonian controversy: The shaping of scientific knowledge among gentlemanly specialists*. University of Chicago Press.

Stinner, A., & Teichmann, J. (2003). Lord Kelvin and the age-of-the-earth debate: A dramatization. *Science Education, 12*(2), 213–228.

Stowell, J. R., Oldham, T., & Bennett, D. (2010). Using student response systems ("clickers") to combat conformity and shyness. *Teaching of Psychology, 37*(2), 135–140. https://doi.org/10.1080/00986281003626631

Sumy, D. F., Welti, R., & Hubenthal, M. (2020). Applications and Evaluation of the IRIS Earthquake Browser: A Web-Based Tool That Enables Multidimensional Earthquake Visualization. *Seismological Research Letters, 91*(5), 2922–2935. https://doi.org/10.1785/0220190386

van Someren, M. W., Barnard, Y. F., & Sandberg, J. (1994). *The think aloud method: a practical guide to modelling cognitive processes*. Academic.

Winchester, S. (2001). *The map that changed the world: William Smith and the birth of modern geology*. Viking Press.

Chapter 6
Students' Understandings of Science and Scientists

The English language is a form of communication! Conversation isn't just crossfire where you shoot and get shot at! Where you got to duck for your life and aim to kill! Words aren't only bombs and bullets—no, they're little gifts, containing meanings!

Philip Roth

Language is the blood of the soul into which thoughts run and out of which they grow.

Oliver Wendell Holmes

6.1 Introduction

The research reported in the previous chapter (Chap. 5) looked mainly at students' geology content knowledge gains, using different quantitative measures. The research presented in this chapter reflects my qualitative efforts to understand how my students understand the nature of science, in terms of its products and processes. This research is important for two reasons. First, I wanted to see if historically contextualizing scientific content knowledge, as I described earlier in Chap. 3, and implementing different pedagogical strategies that emphasize the scientific process, which I explored in Chap. 4, would lead to students developing a more refined understanding for the nature of science (NOS) as both product and process. Secondly, there has only been a handful of studies (8.2%) focused on NOS understandings in post-secondary students (Abd-El-Khalick & Lederman, 2023). This research will add to that small research base. With these reasons as a backdrop, I am asking these two research questions

- How do university non-science majors describe aspects of science, both products and processes?

G. Dolphin, *Teaching Geology Using the History and Philosophy of Science*, Science: Philosophy, History and Education,
https://doi.org/10.1007/978-3-031-68667-2_6

- How do students that participate in an introductory geology course, with historically contextualized content, explicit and reflective NOS instruction, and other transformative teaching practices describe aspects of science, both products and processes?

Developing students' NOS understandings continues to be a goal for science teaching, but valid and reliable measures have been a challenge. Many diverse instruments have been developed and implemented since the 1960s (McComas et al., 2020; see Table 6.1). However, due to the problematic nature of defining what NOS content is and what should be taught (Allchin, 2011; Erduran & Dagher, 2014; Lederman, 1992; Matthews, 2012; McComas & Clough, 2020; Osborne et al., 2003; Rudolph, 2000), operationalizing it is also problematic.

McComas et al. (2020) outlined five challenges standardized NOS instruments face:

1. Scored items on such instruments seem objective to the scorer, but to the respondent, the items are highly subjective—meaning they apply their own meaning for each item.
2. Blanket statements like "scientific knowledge is tentative" discount the contextual nuance that frames understanding of this query.
3. Little consensus exists about what the measurements mean: specifically, how they relate to real-world decision-making.
4. Little consensus exists as to which aspects, facets, and ideas about science are most important to teach—and therefore to assess.
5. The complexity of NOS understandings mean assessments are time-consuming—they take time to administer and require training to evaluate them well.

As I discussed in the last chapter, I do not find most forced-choice instruments to be particularly useful for assessment; moreover, I consider student development of NOS knowledge to be quite open to various instructional modalities, a view that diverges from the dominant perspective found in the related literature—as I examined in Chap. 1, Sect. 1.3. In light of this, I collected qualitative responses to two open-ended questions as a way to garner understanding about the student participants' understanding of the NOS. The responses were several sentences to a paragraph for each participant, making robust qualitative analysis of any one student's NOS understandings difficult. However, using phenomenographic theoretical framework (Hajar, 2021; Stokes, 2011), with the volume of responses I received

Table 6.1 Instruments used for measuring NOS understanding

Instrument	Acronym	Citation
Nature of science scale	NOSS	Kimball, (1967)
Nature of science knowledge scales	NSKS	Rubba, (1976)
Conception of scientific theories test	COST	Cotham & Smith, (1981)
Views of the nature of science	VNOS	Lederman et al., (2002)
Student understanding of science and scientific inquiry	SUSSI	Liang et al., (2008)
Students' ideas about nature of science	SINOS	Chen et al., (2013)

over the 6 years of the study, a was able to aggregate the data to look at the collective understandings of the group and then searched for corroboration with the VOSTS.

6.2 Methodology

The following section outlines the procedures salient to conducting the qualitative portion of this research. I will situate the research by describing the theoretical framework, the data collection and analysis procedures, and the demographic descriptions of the student participants. I also detail both my perspective as the researcher and clarify the verification strategies. This will help to give the context for the methodology and the analysis that the reader can take into account if they decide to incorporate similar evaluation strategies in their own transformative teaching.

6.2.1 Study Participants

Table 5.1 lists the number of students participating in the study for each of the five iterations over six years that I collected data. There was a total of 942 responses to the pre-course questions and 1266 responses recorded for the same questions post-course. General demographics of students in the course also appear in Chap. 5 (Tables 5.2, 5.3, 5.4, and 5.5).

6.2.2 Data Collection

For this research, I sought to expose student thinking about science by having them respond to two open-ended prompts at the beginning of the term (pre-instruction) and after the end of the final lecture (post-instruction). My aim here was also to elicit a more personal, experiential account of science from the students. I created the first prompt to elicit the participants' understanding of the process of science and the second question to focus on the products of science. The two prompts were the following:

- The last time I was acting like a scientist, I was doing… I know I was acting like a scientist because…
- When considering the development of scientific knowledge, it is most important to remember…

These unfinished statements provide students the freedom to present thoughts most salient to them to seek their "active" epistemologies as opposed to "professed" epistemologies (Sandoval, 2003). In other words, the goal was to get a more authentic portrayal of students' understandings (active) as opposed to answers based on what they thought I wanted to hear (professed). For instance, as I noted earlier, many introductory geology textbooks, declare in the beginning pages that science happens through the performance of the "scientific method"; indeed, many geologist, would stand by that proposition—a professed epistemology—when asked how to do geology. However, if told to describe a specific instance where they developed new scientific knowledge, these same geologists would most likely give a far different and more accurate description from their personal experience—an active epistemology.

Students completed the questions at a URL for the online platform, Padlet, where they read the prompt and provided their response. I collected data from students in the six separate offerings of the course between Winter 2016 to Winter 2021, resulting in 942 pre-instruction responses and 1266 post-instruction responses (Table 6.1).

The difference between the pre-instruction responses and the post-instruction responses is due to students giving informed consent at the end of the course rather than each time they responded to the prompts. While some students who gave pre-instruction responses dropped the course prior to granting informed consent, other students joined the class after the pre-instruction survey administration and so did not have pre-instruction responses to record. Further, because I was looking at the aggregate of responses for my analysis, I did not match responses by participant, from pre-instruction to post-instruction. I downloaded the responses from the Padlet platform into MS Word documents, which enabled me to quickly anonymize the data and then upload it into an online qualitative and mixed-methods data management system, Dedoose. Once in Dedoose, I began qualitative coding of the data (Salmona et al., 2019).

6.2.3 Theoretical Framework

I employed a phenomenographical approach with this research, which is an empirical methodology that frames the analysis in terms of how a *group of participants* understands a phenomenon—as opposed to each individual participant's understanding. In other words, phenomenography does not look at meaning-making at the individual level, as is often the product of qualitative research. This kind of analysis takes a lot of data from each of the participants. Because I only have several sentences to a paragraph of writing from each student, this kind of qualitative analysis is not appropriate. However, because I have responses from over 2000 participants, aggregating their answers should give an approximation of the average thinking of the group. Therefore, an expected change in that average thinking from pre- to post-course, supports the notion of curricular efficacy. Phenomenography is an appropriate framework to use because it begins with these key assumptions (Stokes, 2011):

- People have varying conceptions of the same phenomenon because they experience that phenomenon differently from other people.

- People can express their conceptions of the experienced phenomenon through verbal or written expression.
- A finite number of conceptions exist for any one phenomenon.
- Different conceptions can be organized hierarchically, from general to specific.

Starting with these assumptions, the researcher seeks to discern an account of how a group of people understands a phenomenon. For individuals, reality is based on how they experience or conceive of that reality and their descriptions follow this (even if this is not consistent with how reality may actually be). Instead, against this background, phenomenography aggregates the individual descriptions of reality held within an entire group to report a *holistic or collective* understanding (Hajar, 2021). The responses to the two questions were relatively short—a couple of sentences to a paragraph—making aggregating the approximately 2000 responses reasonable for gaining valuable insight into the beliefs of the whole group.

For teaching and learning, the product of phenomenography gives a picture of where students' collective understanding is, at the class level, as opposed to an individual level. Constructing a representation of reality, whether individually or through social negotiation, is based on continued interaction with that reality. In the context of teaching and learning science, a learner's conception of a notion or concept, as in NOS, can increasingly shift towards the scientific consensus view if the person is given greater exposure to that concept. This is so, even bearing in mind that the consensus view is itself a developing approximation of reality. As the individual group members' conceptions shift to more accurate approximations, so should the aggregate or collective understanding for the entire group shift—in this case, the class. The following analysis compares students' collective understandings of science (its processes and products) from pre-instruction to the collective understandings post-instruction.

As with all qualitative analyses, researcher subjectivity is critical in discerning and assigning importance to phrases, codes, themes, and hierarchies within the data. With this in mind, I did my best to curtail assumptions about the concepts I was researching by creating codes that were as descriptive and therefore depended on as little interpretation as possible. I also have included a subjectivity statement (Sect. 2.5). This presentation of my perspective makes transparent for readers how my subjectivity may affect the data analysis (Stokes, 2011).

6.2.4 Data Analysis

Phenomenography relies on a second order approach (Hajar, 2021), where analysis delineates how others (in this case, the students) understand a particular phenomenon (in this case, scientific products and processes). To keep my own prior knowledge and possible bias from influencing my analysis of the data, I developed codes directly from the words used with very little ingterpretation, a strategy known as bracketing the responses (Hajar, 2021). I practised discovery coding (Stokes, 2011), not imposing themes but discerning them naturalistically, as I read through the data. This is parallel to open coding, where one identifies words or phrases of interest with codes and then discerns larger scale themes by grouping the codes under broader thematic umbrellas.

Basically, I looked for phrasing where participants described something about science knowledge or the process of science. For instance, excerpts that followed the prompt, "The last time I was acting like a scientist I was…" usually described a scientific process, often an action such as "doing a titration," "doing a dissection," or had to do with the respondent's state of being: for example, "I was wearing goggles and a lab coat," or "I was in a science class." These direct codes that I then grouped into a larger theme, "situational," reflecting the notion that it was the situation the participant experience—being in science class, wearing science clothes, using science equipment—that seemed to relate them all. This process resulted in 103 individual codes at which point saturation was reached. I then categorized these codes into the four following themes:

- Nature of scientific knowledge—27 codes, such as *can change*, *discovered*, *falsifiable*, and *objective fact*
- Scientific processes—56 codes, including *classifying*, *collaborating*, *explaining*, *measuring*, and *predicting*
- Scientists—14 codes that addressed the qualities of being a scientist, including *maintaining objectivity*, *open-minded*, *creative*, and *must adapt*
- Putting science in everyday activities— 6 codes that included activities participants referred to often when describing the most recent time when they behaved like a scientist. This included *physical exercise*, *mechanics*, *mixing drinks*, *cosmetics*, *pool testing*, and *cooking*

Dedoose records each time a code gets applied and keeps a tally of how many applications of each code take place in the entire data set. I used these sums as a proxy for salience of the concepts within the aggregated data, looking for discernable patterns within the data that would inform how the group, on average, understood the concepts of science processes, science knowledge, and scientists (Hannah & Lautsch, 2011).

While the qualitative research field is divided concerning the usefulness of counting codes as part of the analytical process (Maxwell, 2010), code counts derived from reviewing a large number of responses to a couple of open-ended prompts do provide an overview of important ideas within the aggregate data. This is especially pertinent when assessing the shift in counts from pre-instruction responses to post-instruction responses.

In some cases, as I further refined my interpretations of student responses, it made sense to assemble related codes into more general groups (see Fig. 6.1). For instance, there were several codes describing scientists (Curious, Open-minded, Patience, Dedicated, and Creative) which seemed to all be describing intrinsic or "born with" qualities of scientists, in contrast to Being objective, or Needing to adapt. I therefore assembled these qualities into a single group, called Internal characteristics. There were other codes such as Bias is a limitation, and

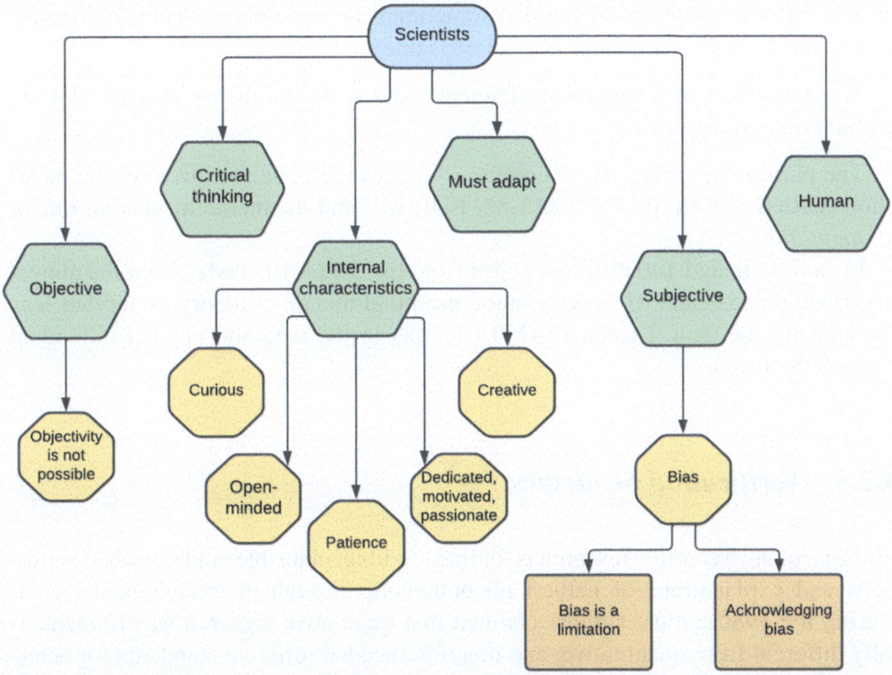

Fig. 6.1 Hierarchy of coding themes. (Note. Shows levels of hierarchy from most encompassing (blue oval) to most specific (orange rectangle), for the theme "Scientists.")

Acknowledging bias, that I assembled under a more general code of bias. Since bias is a more extreme form of subjectivity, I placed it subordinate to and within the more general group of subjectivity, all within the descriptions of scientists.

Metaphorical language appeared throughout the data. Examples of such include the following:

1. Progress sometimes requires *taking one step back* in order to take *two steps forward.*[1]
2. It is important to combine our knowledge from *different fields*.
3. It is important to remember the *foundation* from which these theories develop.

With this observation, I began a deliberate search for metaphor (Steen, 2007), coding with a method for identifying metaphor-related words, based on the Metaphor Identification Process Vrije Universiteit (MIPVU; Steen et al., 2010). I describe this process in detail in Chap. 7. It is important to note that analyzing the data through the hierarchical lens I described above regarding science processes or products as

[1] Inset, numbered sentences are direct excerpts from the data.

well as this metaphor lens meant that sometimes the same phrase had two different codes, one for each. The statement below is one such example:

4. It is important to remember that science and scientific theory in particular are built on assumptions.

The phrase, "are built on assumptions" received a code for invoking the metaphor SCIENCE KNOWLEDGE IS AN EDIFICE and the hierarchical code *role of assumptions*.

In sum, I created 140 different codes from the data—103 codes from the above-described process and 37 codes using conceptual metaphor theory as another lens for viewing the data. I isolated 4259 excerpts in the data and applied individual codes 8082 times.

6.2.5 Verification Strategies

The goal of any scientific research is finding credible, plausible, and reliable descriptions and explanations of natural phenomenon, through methodological rigour. During the 1980s, some authors claimed that qualitative research was fundamentally different from quantitative, and therefore needed different standards for determining rigour (Guba & Lincoln, 1981, 1982; Lincoln & Guba, 1985). The standards put forth during this time used post hoc methods for determining and ensuring "trustworthiness." In other words, the strategies involved—such as member checks, audit trails, triangulation, and interrater reliability—were carried out only after the data were collected. This approach does not consider any of the decisions responsible for making sure the collected data are solid and appropriate in the first place (Morse et al., 2002).

In contrast, Morse et al. (2002) argued that the post hoc approach takes the responsibility for rigour out of the hands of the researcher, while also downgrading the perception of the veracity of qualitative research. Instead, they proposed several verification strategies to ensure methodological rigour, along with the validity and reliability of research outcomes. In particular, they emphasized the necessity of implementing verification strategies *within* the qualitative design, as opposed to strategies external to the qualitative design. As such, they proposed the following strategies, which I consider below in light of this project:

- ***Investigator responsiveness:*** This means that the researcher has awareness of and reflects upon the progress of the research so they can make any appropriate changes in methodologies when the circumstances demand that. I did this during the analysis phase, described in the next point.
- ***Methodological coherence:*** This means aligning the research questions and the methods, which could entail changing the methods in response to changing questions. In this research, for instance, when I began to see the pervasive use of metaphor within student responses, I responded by including in the methodology

a way to look explicitly for the scope of metaphor use, which then became part of the analytical design.

- ***Theoretical sampling and sampling adequacy***: In this research, theoretical sampling happened within the context of the developing theory. Sampling, analysis, and theorizing were not discreet steps, but continuous and part of an iterative process. I maintained sampling adequacy by choosing appropriate participants for my area of inquiry, along with ensuring an appropriate number of responses. I offered students some extra credit points as a reward for responding to the prompts. I also encouraged them, both verbally and in writing, to write as much as they could when responding to the prompts. This also helped to fulfill the data requirement.
- ***Active analytic stance:*** This stance was reflected in my methods, wherein I begin analysis as the data were coming in. Awareness of the emerging themes helped me in my teaching, which in turn helped guide my future analysis.
- ***Saturation:*** This refers to saturation of the data, which means that by the end of the coding process, no new themes are coming into view, and that the respondents have expressed all the different ways of experiencing the phenomenon. This was indeed the case, as the final three documents of the total 20 documents required only a few codes to be added per document.

6.2.6 Researcher's Perspective

As qualitative research demands that I situate myself in relation to this research project, I'll begin by stating that I'm a geologist by training. After some work in the groundwater industry, I switched careers to become a public-school teacher, teaching earth science to ninth grade students in Central New York. That was where I became interested in teaching using the history of geology and in helping students to better learn how science happens. This focus came about after reading the groundbreaking book by Michael R. Matthews (1994), *Science Teaching: The Role of the History and Philosophy of Science*. My introduction to understandings of NOS were those of Lederman's group and the seven tenets (see Chap. 1, Sect. 1.3.1). As I developed more curriculum, more inquiry activities, and read more about the history of geology, I found the seven tenets to be inadequate for how I wanted students to understand the nature of scientific process and science knowledge.

It became more important to me to have students analyze specific instances of development in geology, rather than have them categorize them as examples of, say, theory-laden NOS or socially embedded NOS. I wanted them to see the complexities of the situation and try to understand how those competing influences could have influenced the direction of geological development. I thus pursued a more open process of analysis, in contrast to a mere categorization of instances into pigeonholes. My approach shifted to highlight NOS-in-the-making as opposed to ready-made NOS—which is to say, constructivism as opposed to realism.

During those years as a high school teacher, I developed about half of what I have gone on to teach in GLGY 209, an earth science course at the university level (Dolphin, 2009). Having become very interested in understanding how students learn and how to use that knowledge to develop more effective teaching, I went back to graduate school and received my PhD in science education. My dissertation focused on using the history of the development of plate tectonics to understand how students learn about plate tectonics. Eventually, I accepted my current position at the University of Calgary (U of C), in the Department of Geosciences, where I conduct geoscience education research and teach (mainly) introductory geology courses.

It was at U of C where I further developed my understanding of the history of geological concepts and also the philosophy and methodologies of geology. I combined this new knowledge with the instruction I had been developing for the previous decade (Dolphin & Benoit, 2016; Dolphin et al., 2018). I also connected with Robert Kelly (an art professor and art educator), who helped me to outline the design constraints and engineer solutions to the problems I discussed in Chap. 2. I have since also developed an upper-level geology course focused on the history and philosophy of geology. My research in student learning now expands out from this project to include the use of metaphor and its influence on student learning as well as the use of virtual outcrop models, which are computer-based, three-dimensional landscape models.

6.3 Findings

This section will describe findings in terms of students' understandings of science products, science processes, and scientists, along with any observed changes from pre-instruction responses to post-instruction responses, based on student responses to the open-ended prompts noted above in Sect. 2.2 (reminder: *The last time I was being like a scientist I was...I know I was being like a scientist because...*; *When considering the development of scientific knowledge, it is most important to remember...*). I will also describe some of the observed themes and apparent shifts in students' responses from pre-course to post-course.

6.3.1 Similarities Among Responses

Three codes stood out as the most applied, for both pre-instruction responses and post-instruction responses (Fig. 6.2): *science is situational* (409 applications), *science is the phenomenon* (278 applications), and *desirable or expected results* (206 applications). Although application of these first three codes happens more often in the pre-instruction responses, they were still among the dominant codes in the post-instruction responses as well; therefore, I take them to be strongly held, common

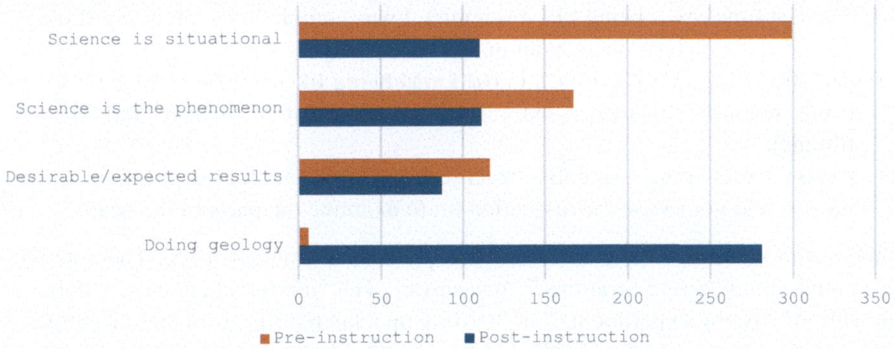

Fig. 6.2 Most commonly applied codes. (Note. The bar graph shows the amount of the four most commonly applied codes within the data, separated into pre-instruction (orange) and post-instruction (blue) cohorts)

understandings or data themes. The following subsections describe these three most pervasive codes in more detail. A fourth code, *doing geology* (287 applications), happened considerably more in the post-instruction responses and so will be described in the section on differences between pre-instruction and post-instruction responses (see Sect. 3.2).

6.3.1.1 Science Is Situational: Location, Location, Location

This code was the most applied of all. Excerpts so tagged indicated that what counted as doing science were the circumstances under which the science activities occurred. For most, this meant merely being in a science class or science lab. Especially for pre-instruction responses, the science courses that student referred to were chemistry, biology, and then physics, with the vast majority being high school sciences, as follows:

5. I was in Biology 30 in my grade 12 year.
6. I know I was doing science, because first of all it was a science course.
7. There was a lab section associated with the class, where we made individual labs on our own and labs with other peers in the class.
8. I know I was doing science because I was enrolled in Chemistry 30, and we did labs with chemicals.

For many, being in a science course or lab was enough to acknowledge they were doing science. Other situational conditions included wearing scientific apparel or using scientific equipment, as notes in these responses:

9. I know I was being like a scientist because it included wearing the scientific lab gear like the lab coat and googles.

10. The last time I was being like a scientist, I was completing a lab in my Biology 30 class in which we were examining and identifying various types of microorganisms using microscopes. I know I was being like a scientist because I was using scientific measures and scientific equipment to analyze and identify [things].
11. I know I was being a scientist because I had to wear safety goggles, an apron, gloves, and got to use the dissection kit to examine the parts of the heart.

Others elaborated about the activities that qualified as doing science. These activities could entail "doing titrations," "measuring pH," "mixing chemicals," "doing a dissection," "doing experiments," or "writing up a lab report," as in these examples:

12. I know I was being like a scientist because I did dissections and made observations about what I dissected. I also had to create a lab and write up everything I did and everything I observed.
13. The last time I was being like a scientist was in high school when we did acid base titrations in Chem class. Something about handling acids and looking at formulas really made me feel like that's what real scientists do.
14. The last time I was being like a scientist was throughout my high school labs and experiments. This included chemistry titrations and biology dissections.
15. I know I was being like a scientist because I was doing scientific experiments that included mixing different solutions to form something new and collecting data and writing reports on the data I found from the experiment.

Science is situational is commonly associated with such other codes as *developing knowledge/study* (31 co-applications), *dissection* (12 co-applications), *mixing chemicals/ingredients* (19 co-applications), and *performing experiments* (33 co-applications) Notably missing from the responses—especially the pre-instruction responses—is any reference to the more cognitive scientific practices outlined by the NGSS (NGSS Lead States, 2013) or other global science standards documents (see Table 4.3), such as asking questions and defining problems, developing and using models, constructing explanations, engaging in argumentation, and evaluating and communicating information. Importantly, many of these codes—*modelling, argumentation*, and *making explanations*—did appear in the post-instruction responses (see Sect. 3.2.1).

Students included the notion of performing experiments as part of the definition of what they thought counted as doing science. They most often related experimentation with *mixing chemicals* together (15 co-applications) or *testing hypotheses* (48 co-applications). Students also had occasion to reference other activities not typically associated with experimentation, as in this excerpt:

16. I know I was being like a scientist because we were conducting research to answer a hypothesis and performing experiments to learn concepts of the course, like dissection.

Oftentimes, they related experimentation to following strict procedures, mixing chemicals, and achieving some desired result, as in these comments:

17. I was acting like a scientist because I was following an experiment that required the gathering of materials and then following step-by-step procedures.
18. I was doing science because of the class name, but also the work that was required to do, such as following instructions or steps to get a certain outcome.

6.3.1.2 Science Is the Phenomenon: Views on Natural Processes

This code was the second most applied to the data (278 applications) and also occurred more often in pre-instruction responses (167 applications) compared to post-instruction responses (111 applications). I applied this code to excerpts where students described an actual natural phenomenon that they witnessed or caused, and which they viewed as doing science. The following excerpts will help clarify this:

19. I know I was doing science because when I was cooking the chicken it went from frozen to thawed to cooked. Seeing bubbles form on the pan and seeing the steam come off the chicken indicated that a chemical reaction had happened. Even though cooking chicken is a simple thing when you look at it, you know that science is involved as it involves reactions.
20. I know I was doing science because I observed the physical changes of the ice on the paths, melting from a solid to a liquid state. This shows the connection which everyday life has with science. Being human is science as well because of the complex systems in our bodies and cells that we are all made up of.
21. I know I was doing science because the process of breathing in oxygen and releasing carbon dioxide, using your lungs and your airbags to keep breathing. That said, I believe that you are doing science every second of the day, even without realizing it. Though the natural science of the body is indeed science, we are not always knowing that we are doing it.
22. I was doing science when I was driving this morning. I know this is science because in order for the car to get energy, the fuels in the car are burnt. And so, this is an example of conversions of heat energy to electrical energy to mechanical energy.

In the cases coded as *science is the phenomenon*, students always picked some everyday occurrences as opposed to doing something in science class or lab.

6.3.1.3 Desirable/Expected Results: Conflating Science and Technology

Students' responses frequently portrayed science knowledge in terms of some desired or expected result of a mainly experimental process. Students often commented that the outcomes scientists were in search of were usually something they already had in mind—an expectation. The following excerpts illustrate this:

23. In the case of baking, it is the active ingredients which chemically react, so that the baked good rises appropriately and bakes within a certain time at a certain heat. All these measurements and factors (such as temperature and time), when done correctly, scientifically play a role in the success of the overall product.
24. I knew it was science because I experimented with different techniques and solutions to find out which works best for which plants.
25. I know I was doing science because the electricity previously didn't charge my phone but after fiddling around with it and fixing the wires, it started to charge.
26. I know I was doing science because I was aiming for a smooth paste texture, and the rich properties in both the substances come together to hydrate the skin and provide nutrition.

Excerpts tagged with this code were characterized by the students' notion that science is about maximizing some result; having better baked goods, growing healthier plants, getting the phone charger to charge, better hydration of the skin. Added to this is the student notion that if the experiment does not work out as desired, if the expected outcome did not materialize, that meant the scientist had failed. These statements were tagged with the subcode *human error* (68 applications), as with these examples:

27. However, I know I was being like a scientist because I had to analyze both my qualitative and quantitative data to understand what my sources of error were and how I could prevent making the same mistakes in future labs.
28. We should also be aware that experiments are not 100% accurate and there is always a possibility of error whether it be human error or a calculation error.
29. We had to do many labs for chemistry and as real scientists, we had to be precise and careful as everything counts including mistakes.

The subcode, *trial and error* (80 applications), is also incorporated within the excerpts coded *desirable/expected results*. I applied this code when the phrase showed up explicitly, or the excerpt implied a situation where the process of science included multiple attempts before finally getting it right, as in the following:

30. It is hard to 100% agree with one scientist's opinion. It is about trial and error, and it takes a long time or maybe forever to perfect something.
31. The answer is not always clear and can take many years filled with trial and error before you get to the correct end result.
32. I know it was science because we were using physics laws to build motorized planes. To see if we did it properly, we went through multiple testing phases.

In summary, the code applications of *science is situational*, *science is the phenomenon*, and *desirable/expected results* were applied more than other codes in the aggregated data. However, they pertained considerably more to pre-instruction responses than post-instruction responses (65.1).

6.3.2 *Differences from Pre- to Post-Instruction Responses*

In this section, I discuss some of the less dominant codes, compared to those I discussed at length above. Nonetheless, they are important, and I will describe how I used them to characterize the students' NOS understandings and ultimately compare those characterizations from pre-instruction responses to post-instruction responses. I could thus document any observed shifts in how students understand science and how the course may have informed changes from their pre-instruction to post-instruction responses. I structured the characterizations to address how responses described the scientific process, the nature of scientific knowledge, and scientists.

6.3.2.1 Descriptions of the Scientific Process

Within the pre-instruction responses I found many references concerning the process of science or the activity of doing science. The first prompt (I knew I was being like a scientist because…) elicited such an answer explicitly, but such references to the process of science occurred in response to both prompts. In response to the first prompt, many students related a daily activity to doing science (Fig. 6.3). Notably, for all but one of these daily experiences (*physical activity*), no difference was observable from pre-instruction responses to the post-instruction responses. Excerpts 33–38 serve as examples of each of the codes:

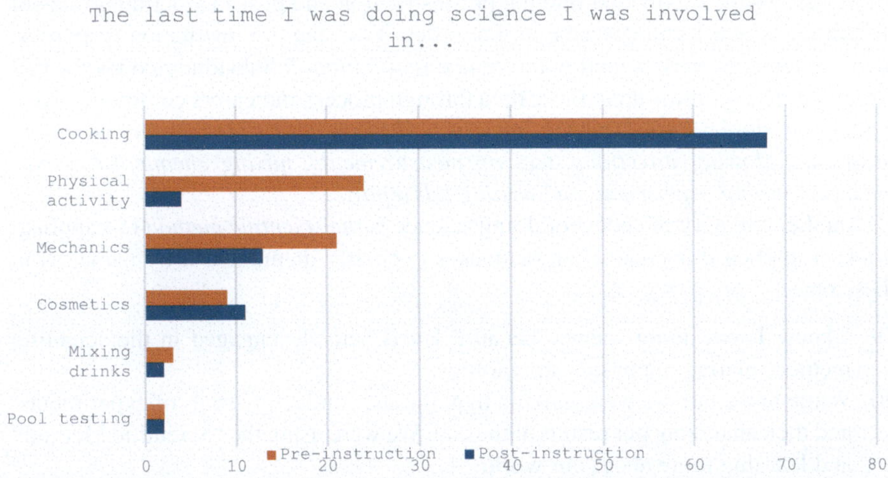

Fig. 6.3 Codes for "The Last Time I was Doing Science". (Note. The bar graph shows the amount of codes and type of activity in which students equated their daily activities with doing science, separated into pre-instruction (orange) and post-instruction (blue) cohorts)

33. I was making eggs the other day and focusing on controlling my variables in order to make the perfect sunny-side up egg for my hash browns.
34. I was experimenting with different company products, and I hypothesized through test and trial that the one that showed the most visible results in a short amount of time must be the best treatment for my skin.
35. I was cleaning things at work with different chemicals, and I used things like grease remover and sanitizer water to see the different chemical reactions that occur, such as bubbling and discoloration of a dirty place when you use different chemicals.
36. I was setting Grand Marnier on fire. I was doing it for different lengths of times to figure out which amount of time would "burn the glass with the alcohol" to make a specific drink.
37. I was observing and testing how hard I need to kick a soccer ball in order for it to reach the goal.
38. I did a water test at the pool where I work. I knew I was doing science because I had to take samples from the environment and add chemicals to conduct a test.

In the above excerpts from the data, responses repeatedly affirm the importance of "experimenting" or "testing" to try to achieve some desirable outcome. Responses often included language like hypothesizing, controlling variables, and mixing chemicals or ingredients, very similar to the responses associated with *science is situational,* described above. The students comment on their everyday activities but overlay these science-related activities onto them. Mostly, these are problems to be overcome, not questions to be answered.

Continuing the theme of scientific processes, Fig. 6.4 lists numerous codes applied to the data and the frequency of each code for activities related to scientific processes. These are divided into the pre-instruction and post-instruction groups of responses, which I will describe in that order. As in the pre-instruction responses, many statements were tagged with *science is situational.* Subsidiary codes for that larger theme also show up in these data through process indicators commonly associated with doing school science, such as *collecting data*, *doing dissection*, *doing titration*, *following directions*, *making measurements*, *mixing chemicals/ingredients*, *performing experiment*, and *writing lab reports*.

Another category of codes for doing science is *the scientific method* (54 applications). I applied this code when responses explicitly mentioned it by name, as in these ones:

39. I knew I was doing science because I was actively engaged in the scientific method relating to physics and biology.
40. While in the lab, we were making hypotheses, conducting trials of experiments, and then analyzing our results at the end. We were using the "Scientific Method" and learning more about our world.

I also applied this code when the process was detailed in the response, as in this excerpt:

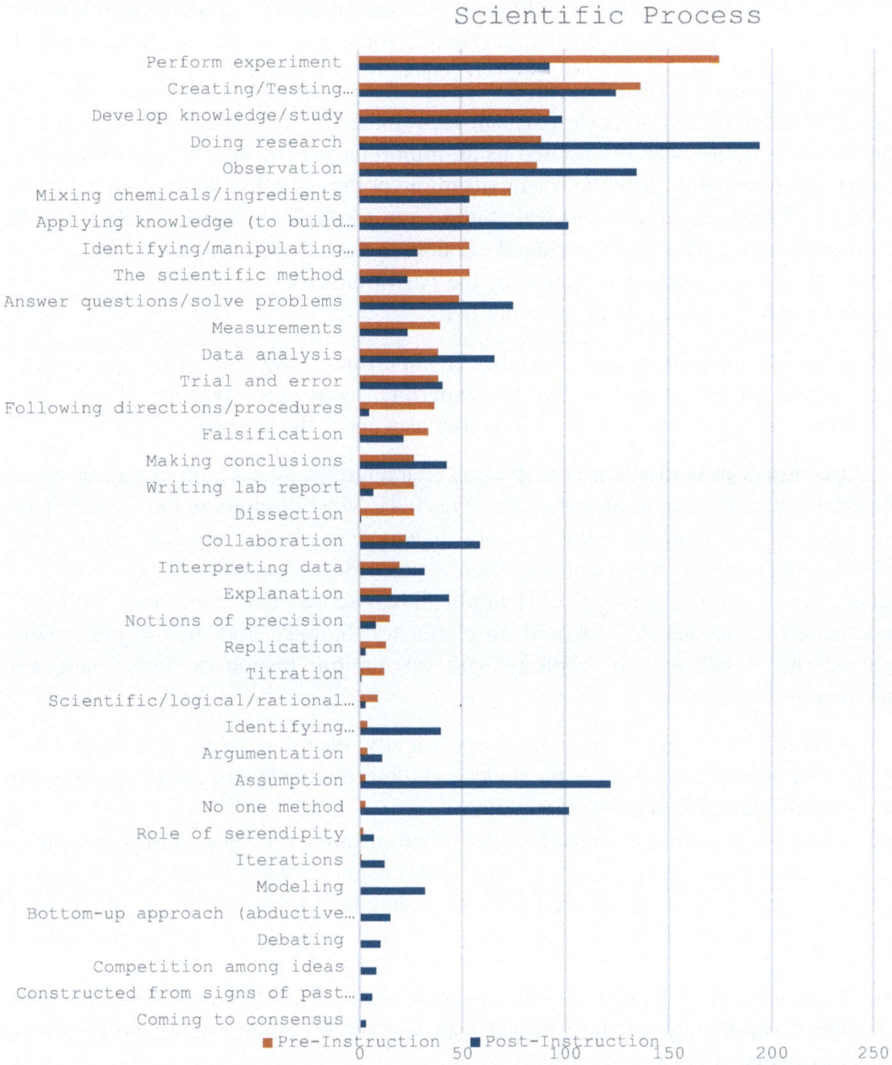

Fig. 6.4 *Codes for the "Scientific Process"*. (Note. The bar graph shows the amount of codes that reference scientific processes, divided into pre-instruction (orange) and post-instruction (blue) responses)

41. I know I was doing science because I created a research problem, formulated a hypothesis, tested my hypothesis with the manipulation of variables and data collection, and finally came to a conclusion with the data I had gathered.

In a few cases, the notion of performing the scientific method was presented not just as a framework for structuring scientific investigations, but as a rigid procedural tactic, as follows:

42. We had to find a problem and had to provide an answer or a conclusion to the problem by doing tests in lab and going step-by-step with the scientific method.

Another common code I applied to pre-instruction responses related to the role of experimentation. The code *perform experiment* (174 applications) was fitting here, as this action was often cited as an important part of the process of science, important for testing, proving, or falsifying hypotheses. Often associated with the process of *perform experiment* was doing so to *test/prove a hypothesis/theory* (40 co-applications). This is an abridged version of the scientific method. The importance of doing science was rooted in the performance of an experiment to prove some knowledge claim, as pointed out here:

43. I was testing a hypothesis, running an experiment, had control groups, experimental groups, had to ensure the experiment was being performed with reliability and validity, and all of the other aspects of the scientific method.

In contrast, post-instruction responses characterized science processes in much broader and more nuanced terms (see Fig. 6.4). Most obvious is the switch from *science is situational* to *doing geology*. It was gratifying to see that some post-instruction responses referred more often to geology, both in a *science is situational* sense and in a *doing geology* (281 applications) sense. This contrasted with pre-instruction responses that focused on chemistry, biology, and then physics, with almost no reference to geology (pre-instruction responses only saw six applications).

44. I knew I was doing science, more specifically related to geology, because I had the opportunity to observe the rocks and [determine] if these rocks were similar on different places of the mountain.
45. I know I was doing science because diagrams are a crucial part of demonstrating geological findings, since experiments cannot always be carried out in real life, and this diagram showed how the mountain mass lowers due to erosion off the top, in which isostasy pushes the crustal roots up to replace the eroded material.
46. I was observing the layers of the cut-up mountains in Algonquin Park, trying to figure out the age of the national park and the events that gave the layers its formation.

Excerpts 44 to 46 exemplify the *doing geology* code. They represent actual skills in geology, specifically correlation (excerpt 44), modelling (excerpt 45), and relative dating (excerpt 46). In addition to references to these geological process skills, students also referred to behaviour that I coded as *doing research* (72 co-applications). This code was common in the pre-instruction responses (89 applications), but much more so in the post-instruction responses (194 applications), as in these excerpts:

47. I was researching alternative truths to some data for the final project. I know I was doing science because I was challenging a hypothesis through reasoning and research.

48. I was researching Alfred Wegener's theory of continental drift as well as other theories that explained the reason for the movement of tectonic plates.

Numerous other codes are associated with *doing geology,* further demonstrating its difference from *science is situational,* which was more common at the pre-instruction stage. These include *applying knowledge* (102 applications, 28 co-applications), *collaboration* (59 applications, 30 co-applications), *developing knowledge/study* (99 applications, 31 co-applications), *earthquake assignment* (52 applications, 36 co-applications), *identifying patterns/relationships in data* (40 applications, 15 co-applications), and *modelling* (32 applications, 18 co-applications). Apart from *developing knowledge/study*, these codes are very limited or non-existent in the pre-instruction responses.

49. I was in my Geology 209 class while working on the seismic activity assignment, [and] I used the concepts learned in class to establish data that needed to be used in the assignment, such as finding out what variables to use and creating a hypothesis that could be adequately used throughout the investigation.
50. I was working on my long-term project with my group to complete a letter written from Aristotle discussing his contributions to earthquake theory.
51. I was recording data and looking through the seismic activity maps in which I had to interpret the events and write down what I saw and explain why it was happening.
52. I was analyzing data from multiple resources to try and answer questions concerning seismic activity in regions of Japan and Chile.
53. I know I was doing science because I was identifying the different types of weathering each rock underwent, and the evidence for said weathering, as explained in the lectures, when the professor explained numerous methods of weathering and how to determine them in rocks.
54. We learned about volcanic hotspots and with the demonstration from J___, I was able to further understand the topic by replicating that demonstration and comprehend it from my perspective.

They also practise the social aspects of *collaboration* (59 applications), *argumentation* (11 applications), and *debating* (10 applications). Another code, *assumptions*, also showed a big change from the pre-instruction stage to the post-instruction one, as is clear in these remarks:

55. I know this is science because, just like the play we saw in class regarding geological theories, scientists work hard to accumulate a claim based on evidence and assumptions, which they represent proudly as they should.
56. That is an assumption though [as] they could have been covered in fur; however, there is no way for us to prove or to know whether they did or not because we cannot go back in time to get the perfect information.
57. Furthermore, it is crucial to remember that assumptions do not limit science but rather enhance it, as it provides a starting point to any investigation.

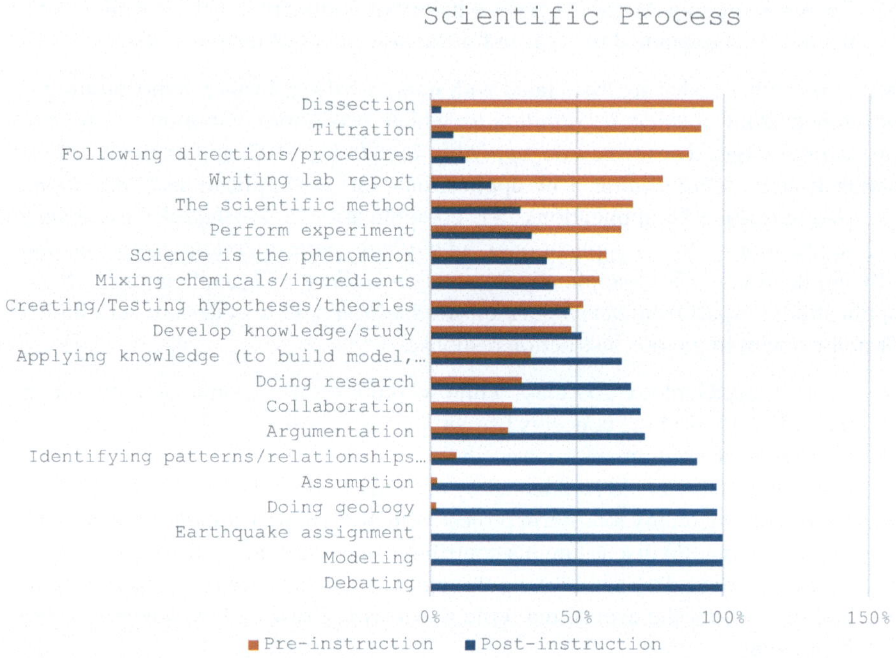

Fig. 6.5 Shifts in student descriptions of scientific processes. (Note. This histogram shows the frequency of applications of each code in pre-instruction (orange) and post-instruction (blue) descriptions of the scientific process. Counts have been normalized to the total number of applications for each code, calculated as a percentage)

In the pre-instruction responses, the code *assumptions* was applied only three times, compared to 122 applications in the post-instruction responses. In the above responses, students indicate that scientific knowledge is evidence-based (see the section on scientific knowledge, below). Nevertheless, assumptions also play an important role in the developmental process of scientific knowledge (excerpt 55), assumptions are like best guesses about something out of our reach for verification (excerpt 56) or they are simplifications, which are necessary for the progress of the investigation (excerpt 57).

6.3.2.2 Descriptions of Scientific Knowledge

Excerpts tagged with codes related to this theme reveal how the students understand NOS knowledge, thereby revealing their ontological stance (Fig. 6.6). As with the previous section on scientific processes, I will begin with pre-instruction responses. The two most applied codes (for both pre-instruction and post-instruction responses) suggest two different notions about scientific knowledge. The first code implies that scientific knowledge is *always changing/evolving* (147 applications pre-instruction, 152 applications post-instruction), evident in the next few statements:

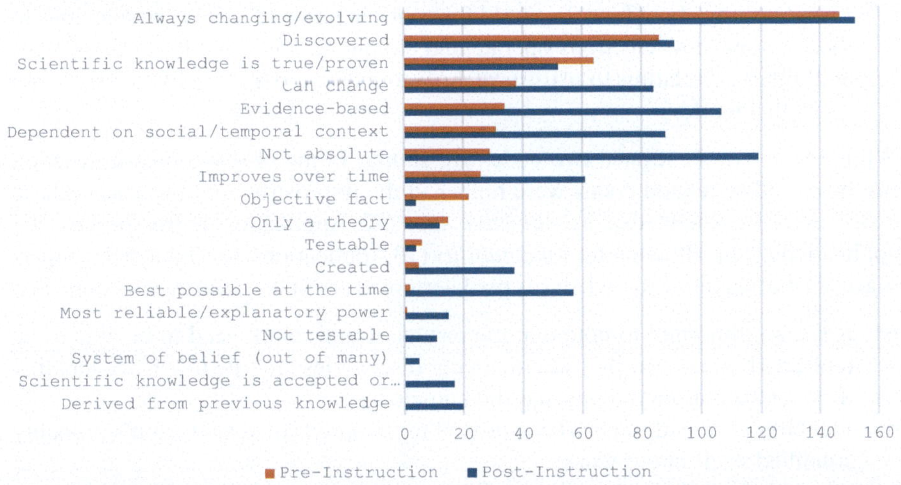

Fig. 6.6 Nature of scientific knowledge. (Note. The bar graph shows the amount of codes that reference scientific knowledge, divided into pre-instruction (orange) and post-instruction (blue) responses)

58. The primary things to keep in mind is that science is dynamic and is constantly changing, but it is also important to recognize that the foundation of science relies on tested hypothesis.
59. It is important that people don't latch themselves on to the idea that something written 50 years ago is still completely, unquestionably factual. Knowledge evolves rapidly.
60. It is most important to remember that what we know and understand in regard to science is constantly evolving; therefore, it is beneficial to consistently question our knowledge to further develop these ideas and our understanding of different scientific theories.

Excerpts 58 to 60 support the notion that students understand scientific knowledge to arise from the scientific processes of testing hypotheses, performing experiments, and following the scientific method. However, these results are temporary and dynamic—constantly and rapidly changing and evolving. Of course, one need only look at the history of science to see evidence that scientific knowledge has indeed changed over time.

Importantly, the code of *always changing/evolving* appears to set up a dichotomy with the second most applied code, that scientific knowledge is *discovered* (177 total applications). The following statements reflect the discovery viewpoint:

61. It's okay to fail in science because that means you just discovered something that is missing in order to prove that scientific knowledge.
62. It is important to remember no discovery is too small. Small discoveries can often lead to something huge and even if it does not, trial and error are the best ways of learning and can lead to new discoveries you never went out looking for in the first place.

63. I was learning about how the world functions in terms of forces and reactions to those forces, how electrons operate and atomic theories, and lastly derivatives and integrals relating to a form of how to express what we have discovered about this world.

While the above-mentioned two codes are similar in the pre- and post-instruction responses, other related codes were higher in the pre-instruction responses. These codes describe knowledge as *objective fact* (22 Applications), *true/proven* (64 applications), and allowing for *repeatability* (13 Applications, see Table 5.4), implying again that scientific knowledge is predetermined and static, as in these statements:

64. It is also important to note that the results of your work need to be able to be replicated in order to [be] sure that your discoveries are true to science and that they can be considered as scientific knowledge.
65. I feel like if something has been proved for hundreds of years then it's probably solidified itself in our world.
66. Scientific knowledge is gathered through a variety of means and must pass rigorous testing to prove its validity and consistency before it is taken as "fact."

Any unexpected outcomes from these predetermined truths are blamed on *human error* (38 applications), as follows:

67. I knew it was about science because the rocket did not meet the expected result because of error.
68. Once the lab was complete, we calculated our percent error to explain where we, as scientists, must have allowed room for error. This prevented us from repeating the same mistakes in subsequent labs.

It is noteworthy that this characterization of scientific knowledge as unchanging runs counter to the students' expressions that science is *always changing/evolving*. In fact, the two are incompatible. This seeming incompatibility also often appears in the very same excerpts:

69. It is important to remember that some things are always constantly changing, which means we should keep an open mind when it comes to new theories or new discoveries.
70. It is important to remember that what we believe to be true is always changing as we discover new things.
71. It is important to understand that science is constantly changing and evolving. Theories that were once studied have been changed greatly, or added on [to], so that it relates to the world we live in now. Scientific discoveries are always occurring and being analyzed.

Turning to the post-instruction responses, I applied about twice as many codes overall for these responses (848) as I did for the pre-instruction responses (475; see Fig. 6.5). The codes applied to these post-instruction statements often do align with many of the ideas I have discussed above regarding the pre-instruction responses. As evidence, the codes *always changing/evolving*, *discovered*, and *true/proven* were the most similar in number across pre- and post-instruction responses.

However, some codes for the post-instruction responses stand out, indicating some important shifts in students' understandings of the nature of knowledge. For instance, students qualified their stance with respect to the dichotomy between knowledge that is *always changing/evolving* to knowledge that is *true/proven*, giving it some nuance. This was revealed through statements I tagged with the codes *can change* (84 applications), *not absolute* (119 applications), *evidence-based* (107 applications), and *dependent on social/temporal context* (88 applications), some of which are shown below:

72. Many aspects of science are not debatable or up for interpretation, but it's crucial to keep in mind that some concepts may change as more information is learned about them.
73. There is no way to know for sure, but we can use combinations of theories and assumptions from the past and present to come up with the closest thing we can get to the truth (given the technologies we have at the time).
74. It is important to remember that science is dynamic and rarely definitive and what is definitive is just the most supported idea at the time.
75. It is important to remember all of the factors which can influence our observations and findings. What we may take as scientific fact can be influenced by political influence, funding, sample, [and] researcher, among others.
76. It is most important to remember to take into account the external factors that may affect the analysis, experimentation, or case study of a specific theory or topic.

According to the pre-instruction responses, change in scientific knowledge was something that always happened, while post-instruction responses qualify this change as something that happens only under certain conditions—increased information, revised assumptions, better explanatory power, or effective problem-solving. Other pertinent codes reflected the nature of the change when it does happen. The post-instruction responses showed understandings that scientific knowledge is *derived from previous knowledge* (20 applications), *improves over time* (61 applications), and becomes the *best possible at the time* or is *most reliable* (72 total applications), as in these excerpts:

77. Even scientific laws are not 100% correct, but they are simply the best possible answer we have developed so far. Scientists will always be working to add new pieces of information to old or new processes to make them more accurate and reliable.
78. It is important to remember to keep an open mind, as science is constantly developing, and what ideas, theories, and/or hypotheses we may now believe are the only explanations for certain aspects of science, [but they] may actually end up being revised, added on to, or completely abolished in the future.
79. Much of science involves making observations, compiling evidence, and coming up with the best explanation for that phenomenon.
80. In geology, the best theory is the one that has the most explanatory power, seeing that not all geologic hypotheses can be tested.

The reliability must always be enhanced based on the available evidence at that time in history, which determines whether the *scientific knowledge is accepted or rejected* (17 applications), as follows:

81. When a theory becomes accepted as the best theory, it is just that, and only that. The best theory is one that best explains all of the evidence we currently have relating to the theory.
82. It is most important to understand that science is imperfect. The majority of the population thinks that the scientific method is how science is done, [when] in fact, Plummer goes as far as to say all scientific knowledge evolves through the scientific method in his textbook. In reality, science doesn't follow a specific process. Knowledge evolves through a convoluted process that doesn't follow any specific path. Ideas aren't easily "rejected" and are usually revised to suit the data, assuming the data is in line with what the scientists want to hear.

In the above two excerpts, scientific knowledge is subject to a judgment by scientists. Does the knowledge change in such a way to better explain the data? If so, it is usually accepted. This shift in the post-instruction responses to a conditional changeability for scientific knowledge, with it being subject to acceptance or rejection, parallels another shift in understanding—that knowledge is *created* (37 applications), a construction. This is opposed to something predetermined and just waiting around to be found, a persistent idea in the pre-instruction responses.

83. Doing research for projects in this class among different scientists throughout time and identifying how all of their observations and ideas impacted one another and no one individual was completely right. And taking little pieces from a bunch of information to create a better, well rounded correct idea is when I was acting like a scientist.
84. One should keep in mind that patterns we might find are human constructs and often rely on assumptions made.
85. Also—assumptions are not only necessary but human. However, we have to acknowledge that they are just that—assumptions. The patterns we see in the world are imposed, not inherent.

Student responses showed an understanding that knowledge was created and based on empirical evidence. This shows a definite shift from realism to constructivism.

6.3.2.3 Descriptions of Scientists

In the descriptions above, post-instruction responses showed shifts in understandings that became apparent through the changes in the type and number relevant codes from the pre-instruction stage to post-instruction stage. Further, I developed many new codes for the post-instruction responses, indicating some new and possibly more nuanced understandings. To account for how students characterized scientists, from pre-instruction to post-instruction responses, I almost doubled the number of codes I applied: 228 applications in pre-instruction responses compared

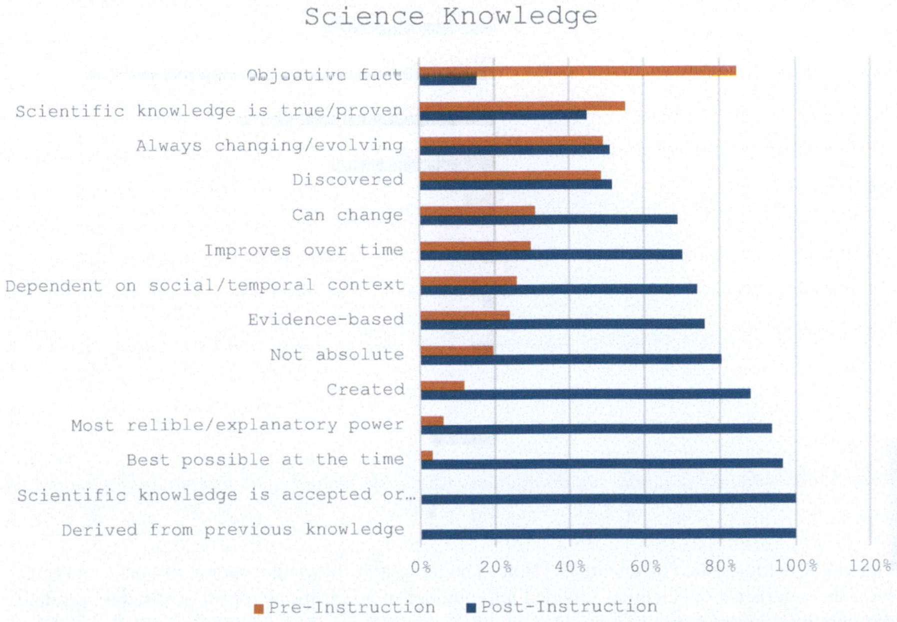

Fig. 6.7 Shifts in student descriptions of scientific knowledge. (Note. This histogram shows the frequency of applications of each code in pre-instruction (orange) and post-instruction (blue) descriptions of scientific knowledge. Counts have been normalized to the total number of applications for each code, calculated as a percentage)

to 425 applications for post-instruction. However, I did not develop any new codes in this theme for the post-instruction responses, although undoubtedly some shifts occurred in the type and amount of code I applied at this stage (see Fig. 6.8).

In general, pre-instruction responses characterize scientists as needing to be *open-minded* (74 applications) but also able to exhibit *critical thinking* (35 applications), as in these examples:

86. Current scientific knowledge can be disproven with new research and discoveries, but the only way to advance is to accept new information and build on that to further your understanding, which is not possible when you don't keep an open mind.
87. One discovery could be changed with new information, and without an open mind it can be worthless.
88. It's necessary to question and critically engage with the information provided—even if correct, searching for other sources or researchers that might counter the original claim.
89. It is also important to be highly skeptical of even one's own findings; if you act as your own harshest critic, it is more likely that others who attempt to find fault in your findings will have difficulty doing so since you've already attempted to find the flaws in your own theory.

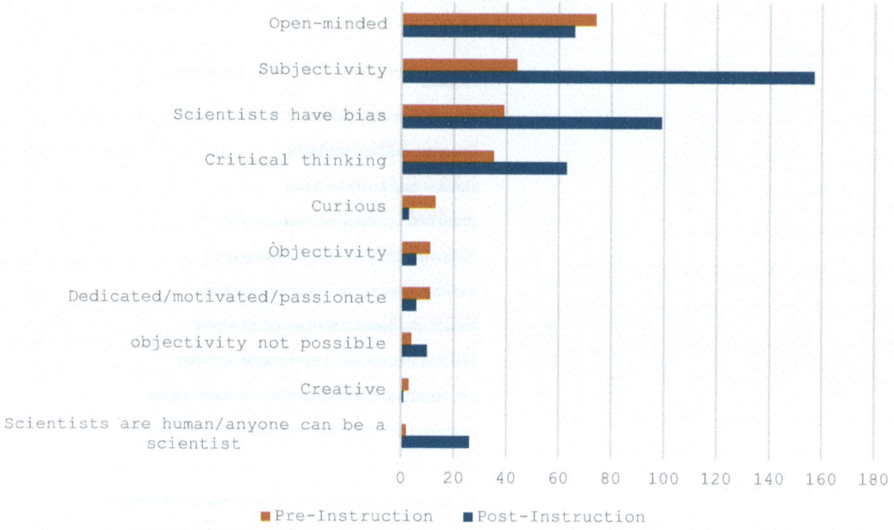

Fig. 6.8 Characteristics of scientists. (Note. The bar graph shows the amount of codes that reference characteristics of scientists, divided into number of applications for pre-instruction (orange) and post-instruction (blue) responses)

Though none of these responses (excerpts 86 to 89) explicitly mention the word scientist, I have interpreted the subjects in these statements implicitly to be someone doing science: a scientist. These excerpts underscore the need for scientists to be open to a new possible explanation for a phenomenon, even if it might contradict their previously held explanation. However, in addition to the need for this open-mindedness, scientists must also keep a critical mind. They must be critical of any new data or findings, even if such findings support their original idea.

Also incorporated in this description of scientists is the acknowledgement of researcher *subjectivity* (44 applications), and *bias* (39 applications), and that scientists are or should strive for complete *objectivity* (21 applications). The following excerpts illustrate these descriptions:

90. Everyone comes from different backgrounds and thus may interpret things in varied ways.
91. It is important to understand the general laws of chemistry, biology, and physics, to further understand the research you are performing. It is also important to understand that it is okay to have a different perspective than others because science is based on hypothesis and testing.
92. Knowledge often consists of multiple sides of interpretation and must be considered without bias.
93. The biases that have and will exist may scramble what we hope to be pure, truthful science.

94. The development of scientific knowledge calls for an unbiased approach as we develop a theory or experiment.
95. This is why the scientific process is an approach free of bias, and we, as scientists, need to remember to approach each experiment with an open mind regarding what will happen.
96. Objectivity is a staple in scientific development, but as humans we struggle to accommodate new information when our vision is clouded by the old. Scientific development is as much an open state of mind as it is a process in which humanity furthers itself technologically and academically.

Student responses acknowledge that different scientists are prone to diverse interpretations of data given their unique background experiences and knowledge, and that bias can disrupt the purity of science. To remedy this, scientists should approach a problem, free from bias, and also use testing to mitigate any bias. Students' pre-instruction responses drew attention to scientists' subjectivity and bias as limitations and as aspects that should be mitigated to maintain the purity of the knowledge. Responses also contained some intrinsic characteristics of scientists, such as being *curious* (13 applications), and being *dedicated/motivated/passionate* (11 applications). Though the post-instruction responses did not add any new codes to the description of scientists, I observed a notable shift towards understanding them.

The post-instruction responses, as with the pre-instruction responses, favour the idea that scientists need to be *open-minded* (63 applications) while also possessing *critical thinking* (63 applications). However, post-instruction responses showed a decided increase in the notion of *subjectivity* (157 applications) and *bias* (99 applications) in scientists, marking it as a concern. This matches the increase in responses that emphasize scientific knowledge as *dependent on social and temporal context* (88 applications), as the excerpts below demonstrate:

97. There are always going to be technological advancements and new points of view for everything, so looking at something in a different way does not make it incorrect.
98. We categorize and create symbols for communicating knowledge, but everyone has different perceptions of the world around them (which creates biases and different understandings) so, aspects of an individual contribute to how they perceive the world.
99. This is OK. If you are interested in a career in science, acknowledge that your personal opinions may guide your theories and research directions, but try to keep open-minded about other possibilities.
100. It is important to remember that science is not entirely related to facts and figures. There is also a bunch of guesswork and bias involved.

These four excerpts acknowledge that scientists have different points of view or perceptions of the world and even that scientists will be biased for or against some ideas. This perspective, however, is much more prevalent in the post-instruction responses, while also revealing some acceptance of subjectivity and even bias as part of the process.

At post-instruction, student responses demonstrating the notion that all scientists are human comes with two implications (26 applications): first, they are not perfect; and second, any human can therefore be a scientist. The following excerpts show these views:

101. It is important to remember that scientists are human too.
102. Understanding that science is not perfect is not necessarily a bad thing, though! It's okay that scientists are humans with biases and certain beliefs that affect their work—this is actually central to how science works.
103. It is not an idea that one person comes up with and that is that. Since everyone has their own interpretation of a topic, people will have their own perspective on a topic, leading to varying opinions on a phenomenon, but this is important to the process of expanding scientific knowledge.

These three excerpts express the idea that scientists are human, which is attended by the awareness that they are not perfect. So, if a scientist is just a regular human, then any human could also become a scientist, as these remarks reveal:

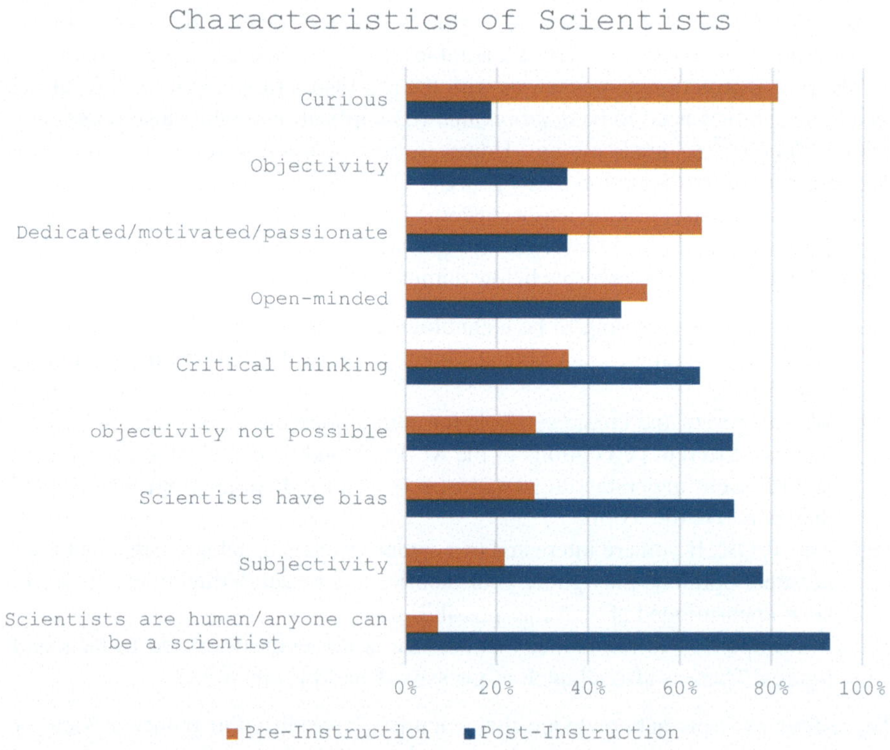

Fig. 6.9 Shift in student descriptions of scientists' characteristics. (Note. This histogram shows the frequency of applications of each code in pre-instruction (orange) and post-instruction (blue) descriptions of characteristics of scientists. Counts have been normalized to the total number of applications for each code, calculated as a percentage)

104. Science doesn't have to be some great discovery. Science is something that anyone can take part in.
105. Students often limit their true potential as they think their ideas are of less value or cannot compete with those that have already been proven.

While students did not often express this idea in their post-instruction responses, even its occasional presence does represent a shift from the pre-instruction responses.

6.4 Interpretation: How Students' Thoughts About Science Changed

In this section, I will develop meaning for the responses iterated above, starting with the most common assigned codes in the data and then why there are differences from pre-course to post-course. The final part of this section will put the data in a broader context and use course strategies—mainly the historical aspects of the course— to explain students' observed shifts in expressed understanding.

6.4.1 From Science Is Situational to Doing Science

The most applied code, *Science is situational*, reflected students' expressions that doing science was intrinsic to the situation or state of being of the student. They were in science class, or they were wearing science-related clothing, or they were working with scientific instrumentation. They seldom, if ever, mentioned activities like questioning, problem-solving, modeling, explaining, or arguing. It could be that students did not mention such activities because they did not have the language to identify these activities, or they did not realize that these were also scientific activities. That the students did mentions these activities much more in the post-course responses, indicates that they either developed such language and/or learned that these activities also engender doing science.

The paucity of such cognitive activities in the pre-course responses is not surprising, given how little these kinds of processes happen in typical science classrooms or labs. For most of the students, being in a science class, participating in science-associated procedures, and wearing science-related apparel were actions that meant they were doing science.

In excerpt 16, the student relates "performing experiments…like dissection." Performing experiments (252 applications), then, to many students seems to have the very broad meaning of any activity that occurs within a lab setting. Students have little experience asking questions in a science class and designing experiments to answer their questions. Rather, they are used to activities for confirming what has been taught in class (American National Research Council [NRC], 2005) and thus would find it difficult to appreciate the structure and purpose of experimentation.

The code *following directions* (42 applications), as part of doing science, was recurrent within the findings, and reflects the typical student experience when doing school science lab work. The key aspect during school activities was to follow directions, thereby confirming a particular principle or concept, with everyone getting the same results (NRC, 2006). If the results varied, it was attributed to human error (See excerpts 67 and 68).

Closely associated with *science is situational* is the code *science is the phenomenon.* This was the second most applied code. It identified students describing their observations of natural phenomena, which they then related to doing science by describing the phenomenon as they would have learned about it in science class, using scientific language. They were making "reactions" happen (excerpt 19), causing ice to experience a "physical change" from solid to liquid (excerpt 20), "using lungs ...[as] airbags to keep breathing" (excerpt 21) and causing "conversions of heat energy to electrical energy to mechanical energy" (excerpt 22). According to the excerpts tagged with this code, everyone is a scientist and science is happening everywhere and all the time, because the students understand science to be the actual phenomena, as opposed to constructing explanations of those phenomena. The notion that science involves creating the descriptions and explanations of the phenomena, to understand those phenomena, is not within their definition of science.

If we again ponder the traditional science class or science lab, it is apparent that the students heavily privilege learning disciplinary content knowledge—the products of science—without much discussion of the process of science or how that process relates to the creation of the product. Instead of teaching, "This is what we know and why," or even better, "This is what we observed, and this is the meaning we made from those observations, how we explain those observations," we more often teach the list of facts, as if they were a discovered part of the real world.

Just as the students associate the activities they do in science class and lab, such as titrations, dissection, and mixing chemicals, with science, so do they equate constructs with reality and therefore consider the natural phenomenon as also science. They. Are in science class, therefore what they learn (the phenomena) must be science.

In the above excerpts (23–32), the students envision the goal of science as achieving perfection of knowledge or being able to cause some desired outcome (like flying a plane). This notion of scientific knowledge being predetermined is not what we traditionally evaluate as the process or goal of science. Thus, situations in which a scientist's error prevents them from achieving the "correct end result," or where "fiddling around" or "trial and error" is necessary to perfect the final answer, do not exemplify the process of science.

I have therefore interpreted these excerpts as students' conflating science and technology. This observation shows up in the literature and is related to some of the issues of science teaching noted above. However, it primarily appears rooted in the inadequate attention to explicitly outlining the relationship among science, engineering, and technology (Rau & Antink-Meyer, 2020). Despite the ongoing promotion of science, technology, engineering, and mathematics (STEM disciplines) to students, precise definitions for science, and especially technology and engineering, have been obscure. Nonetheless, some have described them along these lines (Rau & Antink-Meyer, 2020):

- ***Science***—is about asking questions about the natural world, then collecting data to discern a causative relationship and construct explanatory models.
- ***Engineering***—is about defining a problem and testing alternative solutions, discerning the best possible solution.
- ***Technology***—encompasses both science and engineering activities, by being a product of the scientific investigation or engineering process. It is a product that generally fulfills human needs or desires.

Others have commented on students' conflation of science and technology (Jadrich & Bruxvoort, 2013), describing the purpose of science as examining data from the natural world so as to develop descriptive or explanatory models (see also Harkema et al., 2009). The purpose of technology is then to employ engineering strategies to produce a corresponding solution or desired result. Multiple circumstances contribute to students conflating these various aspects—but especially the concreteness attributed to scientific and technological products, a popular culture focus on technological products over scientific ones, and the kinds of experiences students have in science laboratories.

When considering the products of both science and technology, it should be obvious that a major difference between the two is that for science, the product is a model or explanation illustrating a causal relationship. In short, it is abstract. Natural selection, plume theory, kinetic molecular theory, relativity, climate change, and the like are explanations constructed to explain our observations of the natural world. They are not something that people can put their hands on or manipulate in any physical way. In contrast, the products of technology are precisely that— concrete objects that we can and do use to help alleviate some difficulty. Science provides the model that technology then deploys to create an instrument or procedure to facilitate some desired outcome.

Examples abound. There is a burgeoning technology that relies on clustered, regularly interspaced, short palindromic repeats (CRISPR-Cas9) like a molecular-level set of scissors that cuts a target DNA strand in a designated location, removes some DNA, and inserts a new (Vidysagar & Lanese, 2021). The technology is based in the theory that cellular DNA is a molecule that maintains the codes that cause cells to act in a certain way.

During the writing of much of this chapter, news of the progress of the James Webb Space Telescope was commonly positioned in the "science" sections of much of the news media. Within the science news of U.S. News & World Report (usnews.com/science/news) several articles appeared (Dec 25th, 2021; Jan third, 2022; Jan fifth· 2022; Jan eighth, 2022) describing the progression of the mission to send the telescope out to space and to unfold the intricate protective umbrella that would keep the very sensitive telescope cold enough to make its observations. The telescope will undoubtedly be a tool for doing groundbreaking science. However, it is just that—a tool, not science. This conflation is a very common and further muddies the boundary between science and technology.

In popular media, the conflation is also very common. The villain in many science fiction movies is the "mad scientist"; however, what usually gets the focus of attention is not the science the villain does, but the "doomsday" machine, chemical, or drug the villain develops and uses to threaten the world (Jadrich & Bruxvoort, 2013). In other words, most attention is paid to the concrete product—the technology—the villain will use to achieve their ends. In these popular narratives, the abstract explanatory model that the antagonist would have used as a basis for developing the technology in the first place is overlooked. Certainly, this emphasis has been evident in the several excerpts presented in this section: students focused on the concrete over the abstract and the notion of science providing a desired outcome, with its success or failure judged by whether this sought-after outcome was achieved.

Another important factor influencing students' conflation of science and technology very likely comes from their previous experiences in science class, especially with science experiments or science labs. During both the primary and middle years of schooling, science is not usually taught by discipline experts, meaning that teachers often have very little background in science, which therefore negatively affects how they teach science (Harlen, 1997; Kolbe & Jorgenson, 2018; Peters-Burton & Hiller, 2013). In general, teachers with little science background knowledge are also less likely to implement inquiry-based activities. However, Peters-Burton and Hiller (2013) found that primary grades teachers were more inclined to implement hands-on activities, but they were mainly variable manipulations and trial-and-error activities. These activities typically included some desired concrete outcome but avoided science content. The teachers were more concerned with students being actively engaged than with them learning the science. Of course, the lesson the students learn from such activities is that doing science is about "fiddling" with variables until they get the desired result. No attention is paid to asking questions, collecting data with the intent on formulating explanations for studied phenomena. Sadly, student experiences are not much different in secondary school science environments. Of course, in secondary science, the focus is on the concrete results and propositional knowledge; the desired outcome is the expected outcome and any variation from the expectation is seen as error on the part of the student who has made a mistake somewhere during the procedures and must then go back and do it correctly (Campbell & Bohn, 2008).

Multiple circumstances reinforce students' descriptions of science and their conflation with technology, which all contributed to my formulation of these response codes as *science is situational*, *science is the phenomenon*, and *desirable/expected results*. The products of technology are more concrete, making them easier to experience than the abstract products of science (explanatory models). Media representations, both fiction and non-fiction, often underline the technology that is related to the science as opposed to the science itself. Finally, students' K–12 science experiences commonly stress processes of trial and error to achieve a desired result—normally the concrete, propositional knowledge—with little to no attention on actual science inquiry processes, like framing questions, designing investigations, or constructing explanations in response to collected data.

The descriptions of science I had observed in the responses tagged with these codes did not match the consensus view. My interpretation here is that the students' responses are not surprising as they reflect their experiences with much of popular media, along with less-than-adequate learning experiences during science classes in their K–12 schooling, the science education environment where most of these students have recently been.

Student responses reflect what school science and popular media (to some extent) portray: strict sets of procedures student must follow (*science is situational*) concerning propositional knowledge (*science is the phenomenon*), and demonstration or verification of that propositional knowledge (*desirable/expected results*). Lab experiences also often invoke engineering strategies, like *trial and error*, which accentuate the importance of the *desirable/expected results,* while simultaneously downplaying the important role of inquiry, arguing, and knowledge construction.

The literature is clear on recommendations for science teaching, including laboratory experiences to foster more robust understanding of NOS (Jadrich & Bruxvoort, 2013; NRC, 2006). Structurally, science should do the following:

- Develop learning outcomes, to
  - enhance mastery of disciplinary content knowledge
  - develop scientific reasoning
  - understand the complexity and ambiguity of empirical work
  - develop practical skills
  - cultivate interest in science and in learning science
  - develop teamwork abilities
- Integrate activities with regular instruction
- Build in opportunities for reflection on process and products of science, including the
  - development and assessment of models
  - existence of multiple models simultaneously
  - acknowledgement and acceptance of bias, error, and similar

Table 6.2 shows the strategies established in GLGY 209 to address these recommendations, along with how I used them to explain the characterization of the post-instruction responses when compared to the pre-instruction responses. It is important to remember this course did not have an associated laboratory period. Students engaged in all activities, at least partly, during regular class time, which facilitated integrating the lecture with the activity.

Where students have used their daily activities as example of doing science (see Fig. 6.3), the activities they recount are almost never questions to be answered but some problem to overcome, creating a recipe for a better meal, fixing an engine, making a better tasing drink. This means they are more along the lines of engineering rather than science, as I have also talked about above (Jadrich & Bruxvoort, 2013; Rau & Antink-Meyer, 2020).

Table 6.2 Teaching recommendations compared with instructional strategies implemented in GLGY 209

Recommendations	Instructional strategies
Developed learning outcomes	All activities have rubrics aligned with learning outcomes (see Chap. 2, Sect. 2.5; Fig. 2.5)
Enhancing mastery of disciplinary content knowledge	Lecture and activities: Extended Seismic Activity, Seafloor Data Activity, Geologic Timeline Scale Demonstration
Developing scientific reasoning	Inquiry activities: Extended Seismic Activity, Seafloor Data Activity, Age of the Earth Activity, Models of Earth Dynamics Activity
Understanding the complexity and ambiguity of empirical work	Extended Seismic Activity, Seafloor Data Activity, Pangaea Jig-saw Puzzle Activity, Bathymetry Mapping Activity, Age of the Earth Dramatization and Activity, Models of Earth Dynamics Activity
Developing practical skills	Visible Geology Activity, Extended Seismic Activity, Seafloor Data Activity
Cultivating interest in science and in learning science	Historical and sociological contextualization: Women in Geology Wikipedia Activity, Age of the earth Dramatization
Developing teamwork abilities	Permanent small groups for all graded work, ITP Metrics Team Feedback Evaluations
Integrate activities with regular instruction.	Interrupted Case Study Method, Braiding History, Inquiry, and Model-Based Learning
Build in opportunities for reflection on process and products of science	Unit 4—The What and the How of Science, Interrupted Case Method
Development and assessment of models	Age of the Earth Activity, Models of Earth Dynamics Activity, Extended Seismicity Activity, Seafloor Data Activity
Existence of multiple models simultaneously	Models of Earth Dynamics Activity, Age of the Earth Activity, Visible Geology Activity, Bathymetry Mapping Activity, Interrupted Case Method (where multiple models are presented that the student must evaluate based on provided data)
Acknowledgement and acceptance of bias, error, and similar	Historical Cases (emphasizing bias, error, and similar), Multiple Models of Earth Dynamics Activity, Age of the Earth Dramatization and Activity

Note. These recommendations, from the NRC (2006), and Jadrich and Bruxvoort (2013), aim to enhance students' understandings of NOS content during laboratory activities. I developed my own instructional strategies for GLGY 209

Importantly, these are not metaphors for science, although I will demonstrate where students often did use metaphor when referring to science process and product (Chap. 7). Though DOING SCIENCE IS COOKING is a metaphor that scientists often use to make sense of the scientific process (Harwood et al., 2005), this is not how students were using it in the above examples. Though the activities are similar to the activities described for *science is situational* above, these responses are different in that they reflect how students transferred *their conception of doing science*—like following directions, mixing chemicals/ingredients, and trial and error to get a desirable outcome—into their daily activities, as opposed to in science class. It is not just being in a particular location, such as the science lab, but actually replicating the actions that defines doing science, even when this occurs in a non-lab context. Each of these activities, *collecting data, doing dissection, doing titration, following directions, making measurements, mixing chemicals/ingredients, performing experiment*, and *writing lab reports* is certainly done in a laboratory environment and would count as developing practical skills recommended by the NRC (2006).

6.4.2 From Routine to Creative

My interpretation of the excerpts referring to science process (36–43) is that students expressed understanding of the scientific process as a routine activity, with the implication that scientists act more like technicians performing such routines. A technician is "a person qualified in the practical application of one of the sciences or mechanical arts, especially a person whose job is to carry out practical work in a laboratory or to give assistance with technical equipment" (OED, 1994, p. 2018). The coding of the pre-instruction responses brings into focus understandings that strongly align with this definition. In particular, the responses offer little to indicate that scientists need to be creative—only that they should follow directions and collect data (from experiments) that prove (or disprove) some hypothesis. Certainly, scientists do take part in these activities and science does get done as a result of them, but this portrayal does not cover the range of cognitive activities, processes, and people that constitute scientific fields of study and endeavour.

In contrast to the pre-course descriptions of the process of science, my interpretation of post-course responses indicates a shift in students' understanding of "doing science"— from just being in the place of a scientist by being in a lab, wearing science clothes, or manipulating science equipment, and following directions or technical procedures, to *doing geological activities*, as excerpts 44–54 demonstrate. To achieve this code, it was not enough to say, "I was in geology class." Instead, I applied this code when the participant described what they were doing in a meaningfully scientific (in this case it was geologic) way. They were trying to figure out the stratigraphy of their location (excerpt 44), using models (excerpts 45 and 46), and researching alternative explanations (excerpts 47 and 48).

Based on the post-course responses, students demonstrated a shift in their understanding of the process of science reflective of what appears in the *doing geological activities* code, only these responses were more general about science and not just about geology. In other words, the post-instruction responses reflect the scientific process as being a more cognitive undertaking, as opposed to technical/procedural one.

Excerpts 49–54 are representative examples affirming a shift in which students, post-instruction, understand science differently than they did based on their pre-instruction responses. In contrast to being in a situation or routine that was situated in some way to science, they recounted cognitive activities of doing geology, such as collaborating (excerpt 50), identifying and categorizing (excerpt 53), modelling (excerpt 54), and explaining (excerpt 51). Many of these activities parallel what the implemented course strategies emphasize, such as the talking explicitly and implicitly about the following:

- the processes of geology (see Chap. 4, Sect. 4.8.2)
- the role of models and modelling (see Chap. 4, Sect. 4.8.7)
- observations and identification of patterns (see Chap. 4, Sect. 4.8.2)
- the application of prior knowledge to new data as a way of testing the prior knowledge (see Chap. 4, Sects. 4.8.1 and 4.8.4)

Students also practiced many of the social aspects in the process of science. These social aspects became explicit in the recounting of the different controversies highlighted in the course and implicit within the small-group structure of the course. In their groups, they collaborated, developed arguments for activities, and debated (especially during their exams, which they took in groups).

The role of assumptions also stood out in the post-course data as compared to the pre-course data (excerpts 55–57). This is not surprising since we spent a lot of time discussing the role of assumptions in the scientific process. Many of the short-term assignments asked students to analyze models with respect to the assumptions that went into making them, as is clear in their remarks.

In summary, in comparing the pre-instruction responses to the post-instruction stage, I noted that students' understanding of the scientific process showed a shift away from a passive approach, one that does not even require *someone* to be doing science. This is what I characterized as *science is the phenomenon*: if the phenomenon is happening, then science is happening. If a scientist is involved, they might only have to be in an appropriate location (a laboratory) or wear the appropriate clothing (lab coat, goggles) or use the appropriate equipment (microscopes) to be doing science. Somewhat more active, but still cognitively passive, is doing the "routines of science" (*following procedures*, *doing titrations* or *dissections*, *mixing chemicals, performing the scientific method*, *performing experiments*, and *writing up lab reports*).

According to pre-instruction responses, some cognitive activity does happen when *trial and error* has the person focusing on a *desirable outcome*. In this case, the person acting like a scientist is fiddling with variables until they achieve some maximized or predetermined result, which is analogous to solving an engineering problem and developing associated technology. These understandings reflect students' experiences with school science, as I have already discussed. However, based on the post-instruction responses, students' understandings of science portrayed the process as more cognitively active, as they described *doing geology,* accompanied by *doing research* and doing the *earthquake assignment*, an inquiry-based short-term activity.

Students also acknowledged the role of *assumptions* in the process of science as well as the social aspects, such as *argument*, *collaboration*, and *debate*. This shift towards process is emblematic of a shift in how the students understand the nature of scientific knowledge—that is, their ontological stance. When the creative aspect of the process of science is not acknowledged, as was predominant in the pre-instruction responses, scientific knowledge must be static and absolute. Scientists are not responsible for creating scientific knowledge, only proving or falsifying hypotheses, using controlled experiments and the scientific method. Based on post-instruction responses though, students revised their understandings to incorporate more creative aspects into their NOS knowledge. I will explore these assertions in the next section. Figure 6.5 gives a graphical representation of the codes described above, showing the shift in the descriptions present in pre-instruction responses to post-instruction responses, and by proxy, a shift in the students' understandings of the scientific process.

6.4.3 Scientific Knowledge: From an Always Changing Truth to Grades of Reliability

Based on the responses students describe science knowledge as something that is found, discovered as is. This implies that scientific knowledge is already part of the natural world, rather than something involving cognitive processes. Excerpts 61–63, portray scientific knowledge as small pieces of a larger knowledge; something that scientists find, whether inadvertently or by deliberately looking for it. In Chap. 7, I will return to this aspect to address the related metaphors, IDEAS ARE OBJECTS and THEORIES ARE PUZZLES, (Chap. 7, Sects. 7.3.4 and 7.3.6, respectively). That said, student responses also depict science knowledge as always changing or evolving. With these two codes (*Scientific knowledge is discovered* and *Scientific knowledge is always changing*), I explored how students do not reconcile the apparent dichotomy of a static truth that gets discovered or proven with the notion that knowledge is ever-changing, ever-evolving.

Both ideas remain in circulation and work in tandem: the latter idea is reflected in the few stories they hear in science concerning how we used to think this way and then we thought that way (*changing/evolving*), due to some *discovery*. The discovery aspect undoubtedly derives from the presentation of science as a list of facts to be memorized, without any appreciation for the constructive process leading to the initial development of those ideas or that the continuous accumulation of data could lead to thinking differently about some phenomenon. While the number of applications for these two codes is similar from pre-course to post-course, other related codes do indicate a development of more nuance from pre-course to post-course.

For instance, in the pre-course responses, there is a greater emphasis on scientific knowledge being *true* or *proven* (excerpts 64–66). They portray true scientific knowledge as replicable and static in the world, something that has been proven consistently to be valid. This characterization can certainly emerge from a number of sources, most notably the confirmatory lab activities that most students encounter in science class (NRC, 2006), where students prove the already known concept through testing and repeatability.

The implicit and explicit message from confirmatory labs is that responsibility for any deviation from the expected outcome of an experiment lies with the scientist as "human error" (excerpts 67and 68) preventing them from exposing the unchanging knowledge until they learn from their mistakes and avoid any future ones.

The tension in the responses arises when trying to reconcile fixed, true knowledge with ever-changing knowledge. Some responses support the belief that smaller "truths" are *discovered* and become part of a bigger scientific knowledge that is *constantly changing/evolving*. It implies that scientific knowledge is merely the accumulation of facts, discoveries, truths (Excerpts 69–71). It does not imply the notion of paradigm shifts (Kuhn, 1962), nor does it take into account constructivist notions of knowledge (Carey, 2009; Giere, 1988; Nersessian, 2008; Thagard, 2012). Importantly, this dichotomous view of science revealed in the several examples above exaggerates the notion that scientific knowledge is both durable and tentative.

Nonetheless, the scientific community will continue to use consensus scientific knowledge (its durability) until it accumulates too many difficulties and more reliable knowledge comes to light (its tentative nature). For instance, Newtonian Mechanics has existed for hundreds of years. It has not changed, and we continue to teach it in high school and undergraduate physics classes. What has changed is our understanding of wider parts of the universe, such that Newtonian Mechanics is not the best tool for solving many problems, and Relativity and Quantum Mechanics do a better job. These were not additions to Newtonian Mechanics, but completely different systems of thought.

We met this same idea in GLGY 209 when, for instance, multiple models for mountain building existed at the same time. Proponents for each model needed to decrease the difficulties their model faced, while poking holes in the competing models. It was not about discovering some predetermined right answer, or even proving one model over the other. It was about creating an argument that one model was more reliable than the other at solving the problems posed to it. The pre-course coded excerpts seem to eliminate the continuum space between the ephemeral and absolute understandings of scientific knowledge, such that scientific knowledge is both unchangeable fact and always changing.

However, post-instruction responses do support the argument that some important shifts in students' understandings of the nature of knowledge took place. Notably, student responses showed more nuance. Excerpts 72 to 76 support the idea that scientific knowledge has conditional stability: It is not absolute, and it *can* change. That change depends on various things, but usually comes as a result of information accumulating, which leads to the ideas most supported by the evidence, or that seem closest to the truth. Reponses also reveal that dominant ideas could result from influence by social, political, economic, or other factors.

Based on excerpts 77 to 80, students understand that scientific knowledge does not just change as a matter of course but begins with past knowledge and changes in response to new information. Change, instead, is qualified. It happens when it makes previous knowledge more reliable or has a greater explanatory power. This change may come as an addition to or subtraction from the original explanation, or it may mean entirely rejecting the earlier idea for another one. The key here is that it is subject to the quality of the explanation. Responses emphasized that reliability must always be enhanced based on the available evidence at that time in history, which determines whether the *scientific knowledge is accepted or rejected* (excerpts 81 and 82). It is these subtle changes in wording that indicate a more thoughtful understanding of the nature of knowledge.

Excerpts like 83 to 85 also support this shift from a more realist epistemology, pre-instruction—a stance in which the world is as we perceive it and knowledge of the world is not influenced by perspective—to a post-instruction stance, wherein knowledge is constructed in the mind based on prior knowledge and current context. They also reflect the implicit message of the three historical narratives. Each historical interlude sets up a problem with new data, followed by an interlude that describes how the original idea had to be amended to accommodate those data. There was no example of "discovering" some new truth, only a narrative that traced an idea's human-guided developmental trajectory, situated in a specific social and temporal context, and responding to new evidence. The idea of scientists creating the knowledge they then used leads to the discussion of the next section, which will outline how the pre- and post-instruction responses frame scientists.

Summarizing the shift in descriptions of scientific knowledge from pre-instruction to post-instruction, Fig. 6.7 shows that students' pre-instruction descriptions about science maintain a seemingly contradictory aspect. They describe scientific knowledge as being both true and proven facts, or as something that is discovered, while simultaneously also describing it as always changing or evolving. This may arise from how much of the information presented in a science class is the stuff of science, presented as facts to be memorized. However, at the same time, instruction acknowledges that ideas we have had in the past are not the consensus view of today. Humanity is always learning, and therefore scientific knowledge must always be changing, in an additive manner, which is understood to be *normal science* (Kuhn, 1962). There was no mention of changing paradigms in the responses, or switching between better or worse concepts, only the implication that new knowledge is correct and old knowledge is wrong.

In contrast, the post-instruction descriptions of scientific knowledge qualified the change to something that *could* happen, not that it necessarily must happen. Plus, the change could be dependent on new knowledge, but it could also be due to social and temporal context changes. They acknowledged that present knowledge is built from past knowledge and improves over time to be the best conceptually at the time. Student responses showed an understanding that knowledge was created and based on empirical evidence. This shows a definite shift from realism to constructivism.

6.4.4 Scientists: From an Idealized to Humanized

Excerpts 86–89 paint a picture of scientists as both critical, yet at the same time, open-minded. This seems like a nice ideal, but as with all ideals, reality falls a bit short. We idealize scientists, in part by claiming they should always be trying to disprove their own ideas, to be their own worst critic (for instance, see Nance & Murphy, 2016, p. 7; Chukwuemeka, 2022; D'Augostino, 2017). This does not reflect reality, but is the idealization we have communicated to students, which they have dutifully learned.

In reality, new ideas in science need a champion, someone, usually the developer of the idea, who promotes the idea over others (Greene, 2015); Someone who continues to address the limitations of the idea, while simultaneously exposing the limitations of competing ideas (Frankel, 2012).

Instead of asserting that scientists must be free from bias, or that bias would undermine pure science (see excerpts 90–96), many post-course responses make room for subjectivity and bias, as given human traits: indeed, it is something to make allowances for (excerpts 97–100). This is not to say that bias is desirable in science, or that we should not try to strive for objectivity, but that scientists are human, and humans have biases. This more realistic view affords space for the ideal scientist to act as a fierce proponent for their ideas to gain traction against competitors (Frankel, 2012).

Scientists' subjectivity, and indeed, even bias, have played an important role in the development of scientific knowledge. How does a scientist decide on what they want to research, what they find interesting to investigate? However, subjective/biased is not what constitutes the common projection of the ideal scientist, and students' acknowledgement of this leads to the last difference between students' pre-instruction and post-instruction responses that describe scientists. In particular, the code *scientists are human/anyone can be a scientist* acknowledges that scientists are not, or do not, need to be free of their subjectivity and bias.

Humans have biases and will make mistakes within the process of doing science. However, students also acknowledge that these biases are "okay." It "is important to the process," "central to how science works." As the many in-class examples made clear, scientists have continued to advocate for their explanations, even in the face of seemingly contradicting evidence. Students attended to these narratives and began to understand the human aspects of science and the scientific process.

This "humanization" of scientists also carries another implication: the notion that scientists are not God-like or superbeings, but just regular people with a passion for solving problems or answering questions. So, if a scientist is just a regular human, then any human could also become a scientist (see excerpts 104 and 105).

Pre-course responses characterized scientists as either free of bias, needing to strip themselves of bias, or potentially displaying their humanness, framing it as a weakness or limitation and not as an integral part of who they are. This is also in line with the notion of a scientist as a technician, like an automaton, just going through the motions, with no use or need for subjectivity or bias. The post-instruction shift reveals some embrace of this humanness as part of the process—that we can expect subjectivity and even bias to be present, and that these aspects are not to be stripped away, but merely accounted for in the final product. This perspective thus reassures anyone who might be interested in science that nothing like this has to hold them back from participating.

Most people hold in their mind an idea of who makes a good scientist. They are curious, detail-oriented, creative, persistent, open-minded critical thinkers who are free of bias (D'Augostino, 2017). These are the very qualities that showed up in the codes applied to the students' pre- and post-instruction survey responses (Fig. 6.9), though they were especially prevalent in the pre-instruction responses. The post-instruction responses, however, focused on the joint notion that subjectivity and bias were unavoidable characteristics of scientists—but were also characteristics *necessary* for doing science. This acknowledgement allows scientists to be human, and therefore encourages any human to become a scientist. To be clear, we still do need transparency so as to manage such subjectivity and bias. I presented some of that, with respect to my own subjectivity, at the top of this chapter in Sect. 2.6 (on the researcher's perspective). We also, of course, need controls to manage bias, which is what peer-reviews can provide. This is a reliable, though not fool-proof, means to uncover a scientist's potential blind spots, which could negatively influence their research.

6.5 Discussion: Outcome Space and the Shift in Understanding

For phenomenography, the outcome space is a graphical representation of the categories of description (Stokes, 2011). The outcome space I have created for this analysis shows a shift in understandings of science—from an epistemological stance of naïve realism to one of constructive realism. Figure 6.10 illustrates these descriptive categories in a hierarchical manner, with two separate groups of hierarchies arranged side by side to illustrate the aggregate shift of the students' pre-instruction (left) to post-instruction (right) responses. The pre-instruction responses foreground the two major themes of *science is situational* and *science is the phenomenon*.

- ***Science is situational.*** This view relies on the notion that merely being in a science lab (or class), wearing science-related clothing, and/or working with science-related equipment means you are doing science. Additional emphasis is on application over inquiry, aiming for some expected or desired result, a perspective that conflates science and technology, and the positioning of scientists as technicians. Students never claimed this orientation outright, however their descriptions of what scientists do are sympathetic to this notion. Further, students portrayed scientific processes as rigid and quite procedural. Science happens by following "the scientific method." It is primarily about testing hypotheses via making measurements, doing experiments (mainly titrations and dissections), mixing chemicals according to strict directions, and then writing up a lab report describing what was done and how it worked. If the results were not as expected, then human error was to blame.

- ***Science is the phenomenon.*** This understanding basically equates science with the particular phenomena that scientific practices are used to describe and explain, such that any kind of cause and effect, any kind of action in nature, is science—like melting ice, driving a car, or breathing out carbon dioxide. With this way of thinking, everyone is a scientist because everyone is partaking in the abundance of cause-and-effect relations in nature. This contrasts with assertions made in post-instruction responses where all scientists are human; further, any human *can* be a scientist because they have the mental capacity and are not disqualified based on maintaining a subjectivity. *Science is the phenomenon* essentially objectifies scientific knowledge, making it tangible, concrete. Students' responses also support this objectification by referring to scientific knowledge as something that gets *discovered*, is *proven/true*, or is *objective fact*.

With scientific knowledge objectified, the job of the scientist is straightforward. No creativity is mandated or involved, no guesswork needed. Scientists uncover, find, or discover the scientific truths, as they already exist. They may do experiments, mixing chemicals and following directions, but only to verify or falsify what they have already found. It is all very prescriptive because everything is already settled. Scientists, according to this account, are more like technicians, wearing a particular uniform and trained to work with specialized equipment, with the goal of rigidly performing a task or tasks having definite and expected outcomes resulting from a "trial-and-error" approach. These views reflect the scientific activity that most students have experienced in much of their school science curriculums.

Based on these responses, students are expressing their epistemological stance along the continuum of naïve realism and scientific realism. Naïve realism is an epistemological stance asserting that there is an actual true reality, one that is independent of our minds; moreover, our perceptions of that reality are not mediated by other entities, like our senses. For instance, a red tomato is actually red (Fish, 2017), rather than a characteristic that we perceive based on our physiological sensitivity to a narrow segment of the electromagnetic spectrum.

Similarly, scientific realism also posits that a mind-independent reality exists. However, within this framework, the scientific claims about that reality—even about things that are unseen, like atoms—are at least approximately true descriptions of that reality, even though we might not be able to perceive some objects or everything about that object (McCain, 2016).

I have interpreted the responses claiming that scientific knowledge is *always changing/evolving* as conceptually situated within this scientific realism. This was a logical choice as these comments clarified that the change occurred because we learned more about the concept, made more discoveries, and basically added to the knowledge base; these factors then led to knowledge change. This is to say, then, that students framed these changes as primarily due to the accumulation of knowledge, as opposed to a different way to interpret old knowledge. I frequently applied this same code to both pre- and post-instruction responses, but shifts in several other post-instruction code applications indicate that students' aggregate understanding of science did change by the post-instruction stage.

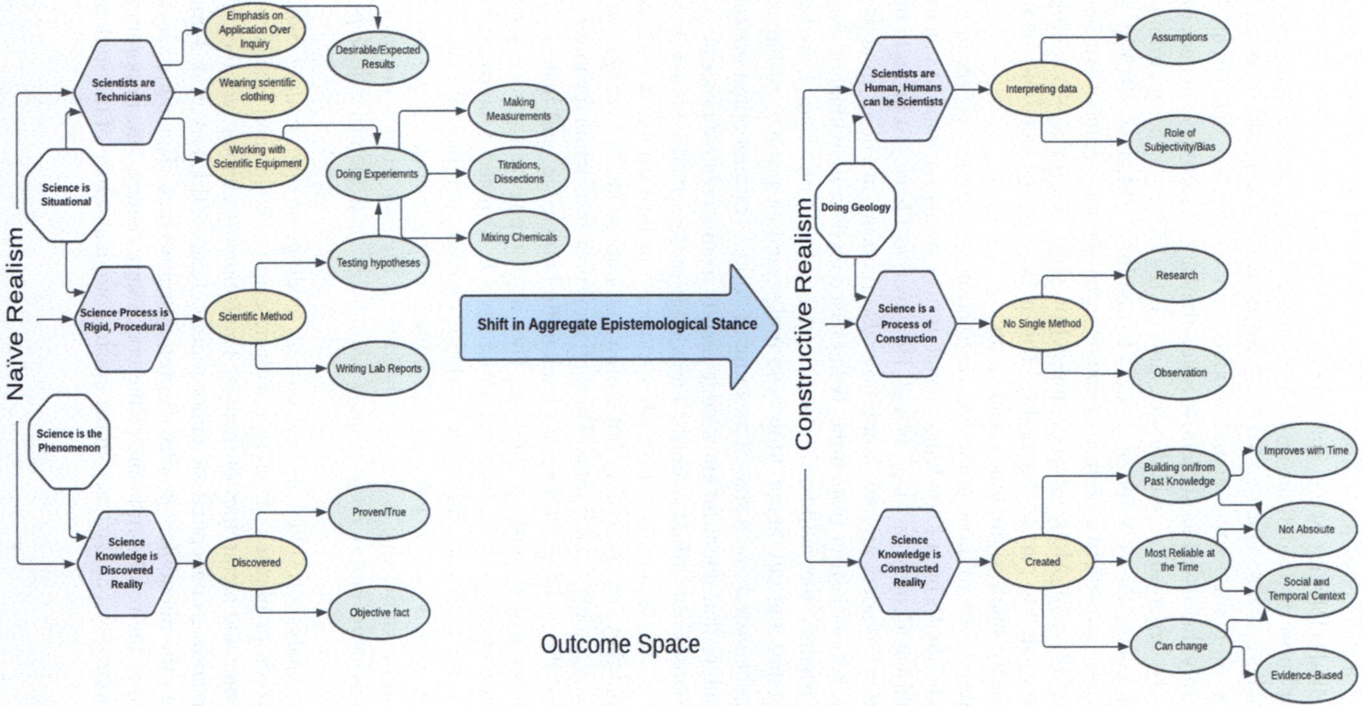

Fig. 6.10 Phenomenographic outcome space. (Note. A graphic illustrating my interpretation of the shift in students' understanding of science: pre-instruction responses show a naïve or scientific realism while post-instruction responses indicate a shift to constructivism)

Notably, a shift in students' understandings of science and scientists is apparent when looking at their post-instruction responses. To be clear, objectified data and prescriptive methodologies were still highly expressed in the post-instruction responses, but I observed a definite move away naïve realism and towards constructive realism (Giere, 1988; Madzia, 2013). Constructive realism holds that our conceptualization of reality—our mental models—are "organism dependent, [but] cannot be simply dismissed as subjective since they are objectively related, on the one hand, to the organism's embodiment, and on the other, to the environmental structures that enable the very living process to continue" (Madzia, 2013, p. 652). Put another way, constructive realism maintains that the environment we interact with is objectively real. Moreover, our interaction with that reality—via embodiment—is also related to part of that reality. This is because the same biochemistry takes place during smell, sight, touch, and so on.

The meaning given to those embodied interactions are subjective because they rely on the individual's prior knowledge, which is necessarily personal. As an example, the animal we identify as "dog" is real. However, as I have just brought up the concept of "dog," readers are, no doubt, now also thinking of "dog," though their own conception(s) will differ from mine, being based on their individual prior experiences. Nevertheless, they will be similar enough and based on a reality, so that we could get together and talk about "dog" and understand each other, perhaps even negotiating and constructing a shared meaning of "dog." This embodied interpretation acknowledges that scientific knowledge results from a creative process.

Fewer responses objectified scientific knowledge or described scientists in ways that made them like technicians. Instead, responses put forward that *scientists are human* and that they *interpret data,* but within that interpretation *assumptions* and *subjectivity/bias* can play major roles. Through their post-instruction responses, students indicated the constructed nature of scientific knowledge, and that *no single method* exists that would constrain thinking. They also emphasized a process involving more open-ended activities, such as *doing research*. As noted, I interpreted their responses to indicate that scientific knowledge is *created*, and thus a construct; but it *can change,* as it is *evidence-based* and influenced by the *social and temporal context*. Scientific knowledge is the *most reliable at the time,* so it is *not absolute*. Scientific knowledge *builds on/from past knowledge*, and therefore it *improves with time*.

The above analysis resulted from discovery coding, or open coding, where I designated words or phrases of interest with descriptive codes and then grouped them into themes and hierarchies of themes. The following chapter will look at the same data but adding conceptual metaphor theory (Jelec, 2014; Lakoff & Johnson, 1980, 1999) to the phenomenography theoretical framework. This added layer of analysis changes the results in terms of the codes and themes, but, importantly, it will also demonstrate a consistency in the epistemological stance of the students.

References

Abd-El-Khalick, F., & Lederman, N. G. (2023). Research on teaching, learning, and assessment of nature of science. In S. K. Abell & N. G. Lederman (Eds.), *Handbook of research on science education* (Vol. III, pp. 850–898). Lawrence Erlbaum Associates.

Allchin, D. (2011). Evaluating knowledge of the nature of (whole) science. *Science Education, 95*(3), 518–542.

Campbell, T., & Bohn, C. (2008). Science laboratory experiences of high school students across one state in the U.S.: Descriptive research from the classroom. *Science Educator, 17*(1), 36–48.

Carey, S. (2009). *The origin of concepts*. Oxford University Press.

Chukwuemeka, E. S. (2022). Characteristics of a good scientist: 8 essential qualities. *Scholarly Articles.*. https://bscholarly.com/characteristics-of-a-good-scientist/

D'Augostino, T. (2017). *What makes a good scientist?* Michigan State University. Retrieved 18 May 2022 from https://www.canr.msu.edu/news/what_makes_a_good_scientist

Dolphin, G. (2009). Evolution of the theory of the earth: A contextualized approach for teaching the history of the theory of plate tectonics to ninth grade students. *Science Education, 18*(3–4), 425–441.

Dolphin, G., & Benoit, W. (2016). Students' mental model development during historically contextualized inquiry: How the 'tectonic plate' metaphor impeded the process. *International Journal of Science Education, 38*(2), 276–297.

Dolphin, G., Benoit, W., Burylo, J., Hurst, E., Petryshen, W., & Wiebe, S. (2018). Braiding history, inquiry, and model-based learning: A collection of open-source historical case studies for teaching both geology content and the nature of science. *Journal of Geoscience Education, 66*(3), 205–220. https://doi.org/10.1080/10899995.2018.1475821

Erduran, S., & Dagher, Z. (2014). *Reconceptualizing the nature of science for science education* (Vol. 43). Springer.

Fish, B. (2017). *Naïve realism. Oxford Bibliographies in Philosophy.* https://www.oxfordbibliographies.com/view/document/obo-9780195396577/obo-9780195396577-0340.xml.

Frankel, H. (2012). *Wegener and the early debate* (Vol. 1). Cambridge University Press.

Giere, R. N. (1988). *Explaining science: A cognitive approach*. University of Chicago Press.

Greene, M. T. (2015). *Alfred Wegener: Science, exploration and the theory of continental drift*. Johns Hopkins University Press.

Guba, E. G., & Lincoln, Y. S. (1981). *Effective evaluation: Improving the usefulness of evaluation results through responsive and naturalistic approaches*. Jossey-Bass.

Guba, E. G., & Lincoln, Y. S. (1982). Epistemological and methodological bases of naturalistic inquiry. *Educational Communication and Technology Journal, 30*(4), 233–252.

Hajar, A. (2021). Theoretical foundations of phenomenography: A critical review. *Higher Education Research & Development, 40*(7), 1421–1435. https://doi.org/10.1080/07294360.2020.1833844

Hannah, D. R., & Lautsch, B. A. (2011). Counting in qualitative research: Why to conduct it, when to avoid it, and when to closet it. *Journal of Management Inquiry, 20*(1), 14–22.

Harkema, J., Jadrich, J., & Bruxvoort, C. (2009). Science and engineering: Two models of laboratory investigation. *The Science Teacher, 76*(9), 27–30.

Harlen, W. (1997). Primary teachers' understanding in science and its impact in the classroom. *Research in Science Education, 27*(3), 323–337. https://doi.org/10.1007/BF02461757

Harwood, W. S., Reiff, R. R., & Pillipson, T. (2005). Putting the puzzle together: Scientists' metaphors for scientific inquiry. *Science Educator, 14*(1), 25–30.

Jadrich, J., & Bruxvoort, C. (2013). Confusion in the classroom about the natures of science and technology. In M. P. Cough, J. K. Olson, & D. S. Niederhauser (Eds.), *The nature of technology: Implications for learning and teaching* (pp. 405–420). Sense Publishers.

Jelec, A. (2014). Are abstract concepts like dinosaur feathers? Conceptual Metaphor Theory and the conceptualisation strategies in gesture of blind and visually impaired children.

Kolbe, T., & Jorgenson, S. (2018). Meeting instructional standards for middle-level science which teachers are Most prepared? *The Elementary School Journal, 118*(4), 549–577. https://doi.org/10.1086/697540
Kuhn, T. (1962). *The structure of scientific revolutions*. University of Chicago Press.
Lakoff, G., & Johnson, M. (1980). *Metaphors we live by*. University of Chicago Press.
Lakoff, G., & Johnson, M. (1999). *Philosophy in the flesh: The embodied mind and its challenge to western thought*. Basic Books.
Lederman, N. (1992). Students' and teachers' conceptions of the nature of science: A review of the research. *Journal of Research in Science Teaching, 29*(4), 331–359.
Lincoln, Y. S., & Guba, E. G. (1985). *Naturalistic inquiry*. Sage.
Madzia, R. (2013). Constructive realism: In defense of the objective reality of perspectives. *Human Affairs, 23*(4), 645–657. https://doi.org/10.2478/s13374-013-0155-z
Matthews, M. (1994). *Science teaching: The role of history and philosophy of science*. Routledge.
Matthews, M. (2012). Changing the focus: From nature of science (NOS) to features of science (FOS). In M. S. Khine (Ed.), *Advances in nature of science research: Concepts and methodologies* (pp. 3–26). Springer.
Maxwell, J. A. (2010). Using numbers in qualitiative resaerch. *Qualitative Inquiry., 16*(6), 475–482.
McCain, K. (2016). *The nature of scientific knowledge: An explanatory approach*. Springer.
McComas, W. F., & Clough, M. P. (2020). Nature of science in science instruction: Meaning, advocacy, rationales, and recommendations. In W. F. McComas (Ed.), *Nature of science in science instruction: Rationales and strategies* (pp. 3–22). Springer.
McComas, W. F., Clough, M. P., & Nouri, N. (2020). Nature of science and classroom practice: A review of the literature with implications for effective NOS instruction. In W. F. McComas (Ed.), *Nature of science in science instruction: Rationales and strategies* (pp. 67–114). Springer.
Morse, J. M., Barrett, M., Mayan, M., Olson, K., & Spiers, J. (2002). Verification strategies for establishing reliability and validity in qualitative research. *International Journal of Qualitative Methods, 1*(2), 1–19.
Nance, D., & Murphy, B. (2016). *Physical geology today*. Oxford Univeristy Press.
National Research Council. (2005). *America's lab report: Investigations in high school science*. The National Academies Press.
National Research Council. (2006). *America's lab report: Investigations is high school science*. The National Academies Press.
Nersessian, N. J. (2008). *Creating scientific concepts*. MIT Press.
NGSS Lead States. (2013, May 25). Next generation science standards: For states, by states. National Academies Press. Retrieved July 2 from https://www.nextgenscience.org/
OED. (1994). *Oxford English dictionary* (Vol. 2nd). Oxford University Press.
Osborne, J., Collins, S., Ratcliffe, M., Millar, R., & Duschl, R. (2003). What "ideas-about-science" should be taught in school science? A Delphi study of the expert community. *Journal of Research in Science Teaching, 40*(7), 692–720.
Peters-Burton, E. E., & Hiller, S. E. (2013). Fun science: The use of variable manipulation to avoid content instruction. *Journal of Science Teacher Education, 24*(1), 199–217.
Rau, G., & Antink-Meyer, A. (2020). Distinguishing science, engineering and technology. In W. F. McComas (Ed.), *The nature of science in science instruction rationales and strategies* (pp. 159–176). Springer.
Rudolph, J. L. (2000). Reconsidering the 'Nature of Science' as a curriculum component. *Journal of Curriculum Studies, 32*(3), 403–419.
Salmona, M., Lieber, E., & Kaczinski, D. (2019). *Qualitative and mixed methods data analysis using Dedoose: A practical approach for research across the social sciences*. Sage Publications, Inc.
Sandoval, W. A. (2003). Conceptual and epistemic aspects of Students' scientific explanations. *The Journal of the Learning Sciences, 12*(1), 5–51.

Steen, G. J. (2007). *Finding metaphor in grammar and usage* (Vol. 10). John Benjamins Publishing Company.

Steen, G. J., Dorst, A. G., Herrmann, J. B., Kaal, A. A., Krennmayr, T., & Pasma, T. (2010). *A method for linguistic metaphor identification* (Vol. 14). John Benjamins Publishing Company.

Stokes, A. (2011). A phenomenographic approach to investigating students' conceptions of geoscience as an academic discipline. In A. D. Feig & A. Stokes (Eds.), *Qualitative inquiry in geoscience education research* (Geological Society of America special paper 474) (pp. 23–35). Geological Society of America.

Thagard, P. (2012). *The cognitive science of science: Explanation, discovery, and conceptual change*. MIT Press.

Vidysagar, A., & Lanese, N. (2021). What is CRISPR? *Live Science*. https://www.livescience.com/58790-crispr-explained.html

Chapter 7
Students' Metaphors for Science

> *Metaphor lives a secret life all around us. We utter about six metaphors a minute. Metaphorical thinking is essential to how we understand ourselves and others, how we communicate, learn, discover and invent. But metaphor is a way of thought before it is a way with words.*
>
> James Geary

> *An idea is a feat of association, and the height of it is a good metaphor.*
>
> Robert Frost

7.1 Introduction: Metaphor and Thinking

As I noted in Chap. 6, while coding of student responses to the open-ended questions given at both the beginning and end of the course regarding students' thoughts about scientific processes and products, I started to notice patterns in their use of metaphor within the responses. Importantly, considering students' use of metaphor actually opens a metaphoric window into their thinking, so I also document those occurrences of metaphor in my coding process. For instance, a student might say that observations are the *foundation* of science knowledge. In this case, I would identify the excerpt with the *Science knowledge is evidence-based* code, and also *Science knowledge is an edifice* (metaphor) code. This chapter will explore only the metaphorical analysis of the data, after which I link this interpretation to the interpretations from the open coding that I presented in Chap. 6.

These incidental metaphors–meaning I did not ask them to give metaphors of science, but they just expressed their ideas metaphorically–give insight into how the students conceptualize ideas like scientific knowledge and scientific processes and can also help instructors to choose words and metaphors that will more likely reinforce useful thinking and disrupt less useful thinking.

G. Dolphin, *Teaching Geology Using the History and Philosophy of Science*, Science: Philosophy, History and Education,
https://doi.org/10.1007/978-3-031-68667-2_7

7.1.1 Diving Deep into Conceptual Metaphor Theory

Literature on metaphor, metaphor use, and reasoning spans multiple disciplines, such as cognitive science, neuroscience, psychology, philosophy, and linguistics. Though most people have considered metaphor to be merely an artistic device in creative writing, research in these various disciplines portrays a much different story. Conceptual metaphor theory (CMT) explains how our thinking—and with it our language—is, indeed, mostly metaphorical (Lakoff & Johnson, 1980, 1999). CMT asserts that most of our abstract or conceptual thinking (love, understanding, theory, and so forth) is done using metaphor. In essence, making sense of abstract concepts entails generating meaning for them by relying on concepts derived from more concrete experiences. This sense-making happens by aligning constituent aspects of the concrete experience (the "source" concept or domain) with the analogous aspects of the abstract concept (the "target" concept or domain) in a process known as "mapping."

CMT explains how one can use aspects of the source domain to logically reason out the implications in the target domain. For instance, starting with the metaphor, IDEAS ARE OBJECTS, learning ideas would be *taking possession* of those things; communicating those ideas to others would be *transferring* those things to others (Reddy, 1979). These reasoned implications of metaphors are identified as "entailments" (Lakoff & Johnson, 1980).

The following excerpts from the student responses to the open-ended questions demonstrate entailments for the metaphor, IDEAS ARE OBJECTS (also see Sect. 3.4):

1. It is most important to remember that *acquiring* it [scientific knowledge] stems from the curiosity of man.[1]
2. It is important to remember that everyone learns through different techniques, and you should not get discouraged if someone else *picks up a topic* faster than you.
3. It is important to remember that integrity is very important, and someone else's scientific discovery should never be *stolen* and presented as that of someone else.

The process of interpreting the abstract in terms of the concrete is one of objectification, which is "a process through which non-physical concepts [the target] acquire the properties of physical objects [the source] in the mental lexicon which permits us to imagine, manipulate and talk about them as if they were concrete 'things'." (Jelec, 2014, p. 63). This process is the basis of embodied cognition (Barsalou, 1999; Clark, 2011; Shapiro, 2011), wherein we come to understand our environment through our concrete, embodied experiences with it. We then apply that concrete meaning to abstract concepts to give them meaning.

---

[1] As in Chap. 5, numbered, inset statements are direct quotes from the student responses and have been lightly edited (some punctuation and spelling to facilitate clarity of meaning).

This attribution of meaning is most often unconscious and automatic but can certainly influence how we reason about the target concept; specifically, through the dynamic of "metaphorical framing" (Donaldson & Allen-Handy, 2019), our mental organization of the concrete source concept influences our mental organization of the abstract target concept. Thibodeau and Boroditsky (2013), for instance—showed that people reasoned about crime prevention differently, depending on the metaphors used by media to portray crime.

A group of participants in their study were more inclined to take an enforcement approach to crime prevention when reading about crime portrayed by the metaphor, CRIME IS A BEAST ATTACKING. However, the other group was more likely to take a reform approach after reading an article using the metaphor, CRIME IS A VIRUS INFECTING. Similarly, Hauser and Schwarz (2019) showed that the metaphor SURVIVING CANCER IS SURVIVING A BATTLE caused cancer patients to perceive cancer treatment as more difficult to endure and cancer as less controllable, interpretations that could potentially cause poorer health outcomes. Metaphor use influences how we perceive our environment (like a monster attacking or a virus infecting), we will react in line with that perception—again, unconsciously and automatically. This phenomenon has been identified as a cascade of associations (Kahneman, 2011), and has also been described as an "experiential gestalt" (Lakoff & Johnson, 1999).

The importance here is that students' metaphors about science can at once inform on how they are thinking about science, but also reinforce that framing. Because metaphors are models, they are limited representations of a target concept. They necessarily highlight certain aspects of the target concept and hide other aspects (Lakoff & Johnson, 1980). Identifying both what is highlighted and what is hidden by a metaphor is important to help people avoid misunderstanding the target concept. An example from the class is that students' interpretations of the tectonic plate metaphor. They imagined plate, like a ceramic dinner plate–their most common embodied experience–to represent the earth's crust. This differs from scientists who conceptualize the earth's lithosphere as being broken into separate parts (plates) that move independently of each other: this aspect is what the metaphor highlights. However, the students missed what the metaphor hides: while dinner plates are separate entities stacked on a shelf or set on a table, the earth's lithosphere is really a solid whole with no distinct, continuous separations or spaces between them. This idea, then inhibited them from understanding how earthquakes happen (Dolphin & Benoit, 2016).

7.1.2 Metaphor in Science and Science Education

This same influence on perception appears in scientific fields as well. Allchin (2013) described an example this way: "The *language* of 'laws of nature' is a metaphor that can powerfully shape thinking. That is, all of the familiar meanings of laws *in the human political realm* are transferred *to nature*" (p. 140; emphasis original). As a result, we view scientific laws not as empirical descriptions of natural phenomena, but as *controllers* of natural phenomena. The positive aspect here is that scientific metaphors can advance fields of study (Kuhn, 1993; Palma, 2018), as metaphoric framing of the target can direct a scientist to think in a particular way (Haack, 2019; Nersessian, 1985, 2002; Reynolds, 2021).

However, metaphoric framing while assisting scientific thinking may also hinder scientific advancements. For instance, the PROTEIN AND SUBSTRATE ARE LOCK AND KEY metaphor, though useful early in its use, was so powerful that it inhibited the molecularization of genetics for decades (Müller-Wille & Rheinberger, 2012). Further, Zarkadakis (2015) maintains that achieving true artificial intelligence will be impossible until researchers abandon the BRAIN IS A COMPUTER metaphor. Without question, metaphor plays a crucial role in science, influences how teachers teach scientific content (Dolphin, 2016; Dolphin & Tillotson, 2015; Tobin & LaMaster, 1995) and how students think (Dolphin & Benoit, 2016).

Metaphor use can also inhibit effective teaching. For instance, I asked an instructor teaching in a traditional, lecture only manner—but who was convinced they were actively engaging their students—to describe their teaching. Within those descriptions I discerned their use of two main metaphors for teaching (Dolphin, 2016):

- The PUZZLE MASTER metaphor, in which the instructor holds all the pieces of the puzzle—in this case the puzzle was the theory of plate tectonics—and doles them out for students to put together.
- The FIELDTRIP LEADER metaphor, in which the instructor expressed that they were pulling students along, covering a lot of ground, and not wanting the students to get lost.

If we consider the nature of learning as being students constructing concepts from the ground up (Carey, 2009), using information in the environment beginning with concrete experiences and building and projecting to more abstract concepts (Nersessian, 2008; Shapiro, 2011), then the important role of the instructor is one of creating a learning environment rich with experiences that students can then leverage readily to construct the intended meaning, where students have the freedom to reason while still constrained by the cognitive resources provided.

In contrast, the PUZZLE MASTER and FIELDTRIP LEADER metaphors I documented above imply that the instructor considered their role as either the one delivering the ready-made knowledge or as the key figure guiding students around the landscape of ready-made knowledge. In both cases, within this metaphoric framing, the instructor envisioned themself having the access to ready-made knowledge and it was their responsibility alone to distribute or lead students to that knowledge, an orientation that leaves little room for student autonomy or for students to take much responsibility for their own learning. Therefore, consistent with these implications, it only made sense for the instructor to give direct, lecture-based instruction, incorporating no real student autonomy, no opportunity for students to miss a piece or get lost in the content. This perspective follows Reddy's (1979) conduit metaphor in relation to communication noted above, which emphasized that knowledge gets transferred from communicator to listener—from teacher to student during a lecture. This conceptualization places undo responsibility for communication on the communicator, sustaining the need to send exactly what they want the listener to understand; correspondingly, it also frees the listener of much active responsibility for understanding, as they merely need to "get it."

Other research has shown how epistemological stances of teachers inform their teaching. The IDEAS ARE OBJECTS metaphor, for example, has been invoked by early-career science teachers: they sought to give, or in some way transfer, key concepts to their students. When asked explicitly how they thought learning happened, they responded by saying students constructed their own knowledge—a constructivist philosophy— but their incidental language instead supported the metaphor of IDEAS ARE OBJECTS, which also paralleled their transactional teaching practice (Dolphin & Tillotson, 2015). I introduce these examples of metaphor and teaching because it is the instructor that controls how students experience their learning (though language and action) by either reinforcing or problematizing particular metaphors of learning.

A final example again illustrates how metaphor is involved in thinking and understanding. Students have a difficult time understanding elastic rebound theory as the mechanism of earthquakes, because they relate the process to tectonic plates. For these novices, the word *tectonic* had no meaning, but *plate* did. They invoked their concrete experiences with ceramic dinner plates; they did not think of these objects as having elastic properties, but rather breaking brittlely when dropped and crashing when pushed together. Their powerful past experiences with ceramic dinner plates—that they are separate and brittle—made them think of the crust, which consists of tectonic plates, in those same terms. This imagery thus made it difficult for them to build a mental model that incorporates the gradual building and eventual release of elastic potential as the mechanism for earthquakes. In this light, they described earthquakes as happening when tectonic plates smash together and break. As I have commented elsewhere, this metaphor use happens automatically and unconsciously (Kahneman, 2011). Understanding the use and influence of metaphors on thinking is thus crucial to effectively teaching science content.

The link between metaphor and thinking in science also raises the issue of metaphor and thinking about the nature of science. In analysing students' responses to the two writing prompts:

- The last time I was acting like a scientist, I was doing… I know I was acting like a scientist because…
- When considering the development of scientific knowledge, it is most important to remember…

their extensive use of metaphors was obvious. I thus sought to formally investigate what students' use of metaphors conveyed, and how this compares to the findings and discussion regarding students' NOS thinking addressed in Chap. 6.

7.2 Methodology

I used the Metaphor Identification Procedure Vrije Universiteit (MIPVU) method for metaphor identification (Steen et al., 2010). This methodology first analyzes text, word by word. For this study, I was interested in metaphor use within the context of how the participants talked about issues regarding the nature of science, so only focused on words associated with a science process or science product. The process begins by analyzing each word in a sentence. The sentence provides the contextual meaning, and the dictionary is consulted to discover if a more basic definition of the word exists, one that is "more concrete, specific, and human-oriented" (Steen et al., 2010, p. 35). "Human-oriented" means within the context of human experiences, the bases of experientialism (Lakoff & Johnson, 1980). If a significant difference exists between the contextual and the basic meanings, the word is tagged as metaphor, and then as either a direct or indirect metaphor.

7.2.1 *Direct Metaphors*

Direct metaphors are those where the speaker (or writer) makes an explicit comparison to science (the target concept) from a more concrete concept (the source concept). For instance, Flannery (2001) advanced the quilting metaphor—DOING SCIENCE IS QUILTING—to emphasize the more feminist aspects of the process of doing science. Thus, Flannery was making a direct and explicit comparison between the scientific process and quilting. Quilting is not something that is often associated with doing science and might otherwise be considered incongruous with the narratives and writing about science. Nonetheless, Flannery integrated the concept of quilting, using the comparison to express particular aspects of the science process.

This quilting metaphor is relevant as it represents the process of creating order out of a multiplicity of pieces, just as science continues to construct the order that underlies the multiplicity of phenomena confronting us daily. A research project often involves ideas and information that have been around for years but that may be used in a new context in the present work. Such a project is thus a patchwork of techniques and pieces of information that may have been gathered at very different times and in very different contexts—but happen to fit together as part of the solution to a problem at hand. Scientists and quilters both spend their time trying to fit pieces together to make a pleasing whole; often, this involves playing with the pieces, rearranging them to make them fit and allowing them to be used most effectively.

Flannery's comparison is not literal, but involves a cross-domain mapping from quilting onto science, where she then points out specific source-target connections:

- Some ideas have been around for years, like the material pieces used to make a quilt.

- An investigation is a patchwork of techniques, like the pieces of a quilt.
- The ideas may have been gathered through time for different contexts, just like the pieces of a quilt have been collected from various times and different sources, and for different reasons.
- Science ideas should fit together to create a "pleasing whole," just like the finished quilt should be an aesthetically appealing cumulate of many smaller pieces of material.

In my analysis, a word would get the label, "metaphor-related word (MRW) direct" (see Fig. 7.1). if the lexical unit meets these criteria: (a) incongruous meaning with rest of the text, (b) integrated with text through means of comparison, (c) non-literal comparison, and (d) direct mappings from the source domain to the target domain. However, no examples of direct metaphor use were present in the student data.

7.2.2 Indirect Metaphors

The main difference between MRW-direct and MRW-indirect is whether the metaphor is used deliberately and explicitly–like the quilting metaphor just described–or whether the comparison is implicit. As it happened, all the metaphor examples from student data were labelled as MRW-indirect. Discerning the indirect MRWs follows a similar set of steps, described above, for ascertaining direct metaphors (see Fig. 7.1). Let's look at an example from the student data to unpack the difference between MRW-direct and MRW-indirect:

4. It is important to combine Our knowledge from different *fields*.

"Fields" is the lexical unit of interest. Within the context of the sentence, *field* implies an area or division of an activity, subject, or profession (Meriam-Webster Dictionary, n.d.). However, a more basic, concrete definition is an open land area free of woods and buildings (Meriam-Webster Dictionary, n.d.). These understandings are different enough to require two distinct definitions in the dictionary listing—the basic, more concrete definition is meaning one, and the contextual, more abstract definition is meaning two. The basic (concrete) meaning easily maps onto the contextual (abstract) meaning, helping to make sense of the target (contextual meaning) in terms of the source (basic meaning). Doing work in a field, like farming or herding livestock, maps onto the work (research, synthesis, and so on) that one does in a subject or type of work.

A person can be unfamiliar with a field and set about exploring it, just like someone might be unfamiliar with a particular subject and do some research to explore and learn more about that subject. So, *field* is labelled MRW-indirect since the meaning is implicit; I also assigned to it the code SCIENCE KNOWLEDGE IS A LANDSCAPE. Another implicit instance would be the use of the phrase "*pushing the boundaries* of a scientific discipline," which I would therefore also tag as MRW-indirect. Additionally, metaphors that occur in the text *incidentally*—as in, "combine

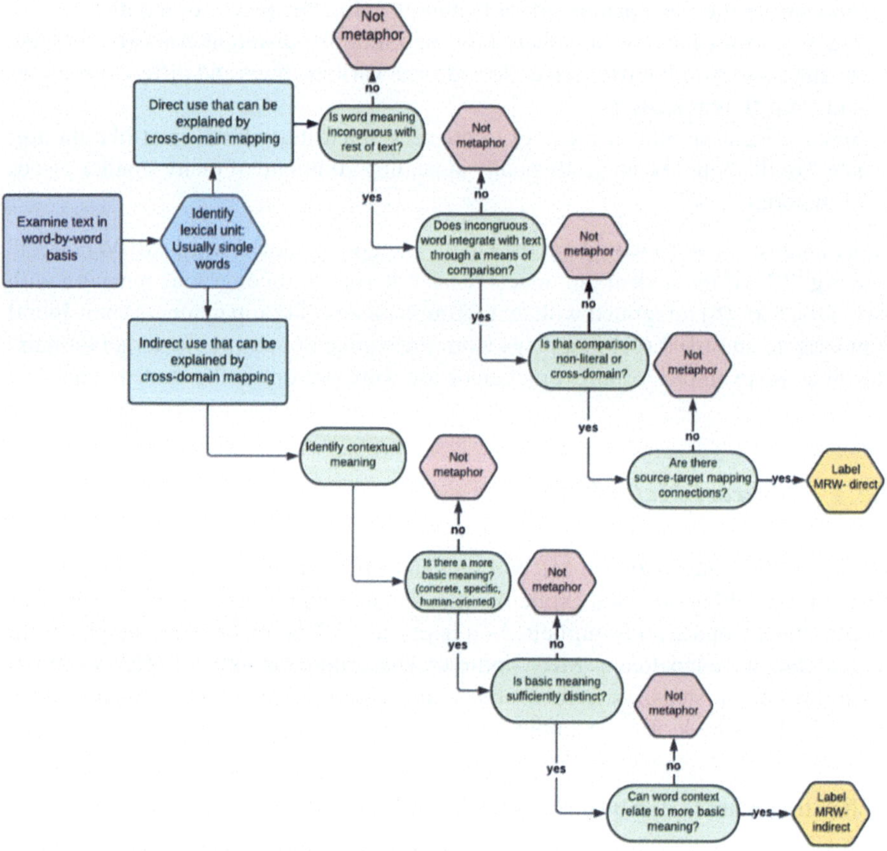

Fig. 7.1 Identifying metaphor-related words. (*Note.* A flow chart showing the process of identifying metaphor-related words, both direct (top chain) and indirect (bottom chain))

our knowledge from different fields," from above—would receive the MRW-indirect designation. In contrast, deliberately using a metaphor to clarify or explain an abstract concept—like the quilting metaphor for science (Flannery, 2001) I discussed above, or deploying acquisition and participation metaphors for learning (Sfard, 1998; Wegner & Nückles, 2013)—would receive an MRW-direct designation.

7.3 Findings

After coding all the data, I started relating MRWs to larger themes, usually primary metaphors; those that are "very plausibly and directly motivated by experience" (Grady, 1999, p. 6) and have very simple mappings. As well, individual codes might be complex metaphors—built from the combination of primary metaphors—or metaphor entailments, which are, as noted, the logical and thoughtful mapping of characteristics of the source domain onto aspects of the target domain (Lakoff & Johnson, 1980). Figure 7.2 shows the common metaphor code applications that I generated. Be The following section will document the metaphors I identified in the student data, loosely following the sequence shown in Fig. 7.2. Then, I will demonstrate that they are related through how the students, as an aggregate, spoke about—and therefore understand—both scientific products and processes.

7.3.1 Science Knowledge Development Is Moving Along a Path

Figure 7.2 shows that references to the distinct, identified metaphors were similar in both pre- and post-instruction responses. As with the code applications described in Chap. 6, the metaphors represent both the processes of science and the products of science. The most identified metaphor was SCIENCE KNOWLEDGE DEVELOPMENT IS MOVING ALONG A PATH (204 applications).

5. Scientific knowledge is a matter of finding something and *running with it*, with the hopes of *never stopping*.
6. When we are *closed off* to new knowledge, we cannot allow for science *to move forward*.
7. I was measuring amounts of matter, I had a hypothesis surrounding this concept and I was looking for an answer, I also took the time to study and analyze *the steps needed to come to a conclusion*.

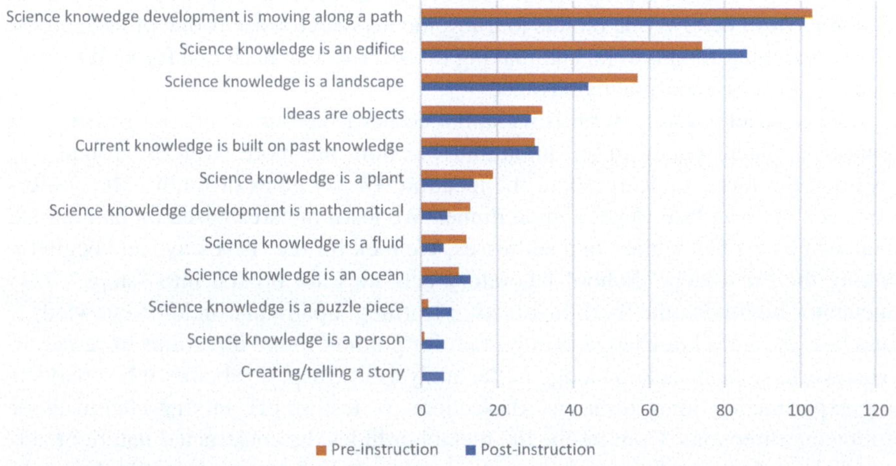

Fig. 7.2 Common metaphor code applications. (*Note.* Most common applications of metaphor coding for pre-instruction (orange bars) and post-instruction (blue bars) data, sequenced from most applications to least)

8. I was simply *looking for the best way* to explain the topic to myself, in order to *further my own understanding*.
9. Some scientists and studies in the past have been overlooked so I think it's very important to celebrate *scientific milestones* and those who *took the first step*.
10. The *only way to advance* is to accept new information and build on that to *further your understanding*, which is not possible when you don't keep an open mind.
11. We should use that information as a *stepping-stone* to help *further* develop newer ideas.

In each of the above excerpts (5–11), the concept of progress in science or understanding science is represented by taking steps, running, progressing, or furthering the participant's position in the best possible way or along the best path, paralleling the conceptual metaphor of LONGTERM PURPOSEFUL ACTIVITIES ARE JOURNEYS (Lakoff & Johnson, 1980). This metaphor has numerous entailments, such as the following:

- States are locations
- Purposes are destinations
- A route is a means for achieving a purpose
- Difficulties are impediments
- Stages in progress are landmarks
- Progress is distance travelled

This metaphor encompasses concepts that relate travelling with learning. Maybe one is leading the way, blazing a trail, or paving the way for others. Perhaps multiple ways exist to get to a destination. There may be obstructions that close ways off or make forward progress difficult, so one might have to take one step back before making further forward progress. It could be an uphill climb the whole way, or all downhill from here. While on this journey, the pervasive sense is one of looking for clues, searching for answers, and making discoveries that push one forward to continually progress, continually advance.

The experiential basis for this metaphor comes from our experience of being on journeys. When we are in an unfamiliar location, the more we look around and explore, the more we learn about the location. We set goals of finding the conference centre or a place to get a great dinner. We learn the area based on landmarks, intersections, street names, and addresses. We look for the "best way" and begin by taking the "first step." School laboratory activities are broken into "steps." This metaphor highlights the work involved in learning a particular field of knowledge, and how previous knowledge is important for guiding future directions of research. This work, students acknowledge in the analysis in Chap. 6, whether it be conducting experiments, doing titrations, dissections, or testing pH, mixing chemicals or following directions. Conversely, the metaphor hides the constructed nature of scientific knowledge by framing knowledge as already laid out, ready-made: It is only up to the investigator to explore and find it.

7.3.2 *Science Knowledge Is an Edifice*

The second most referenced metaphor is SCIENCE KNOWLEDGE IS AN EDIFICE (160 applications), parallel to THEORIES ARE CONSTRUCTED OBJECTS (Lakoff, 1994c). The metaphor maps the built structure of scientific knowledge: a foundation supports the structure and bricks are incorporated into the structure. Within it, rooms and windows and doors can be opened to provide different viewpoints. The following excerpts illustrate how this metaphor presented in participant responses:

12. I was listening to my geology professor's lecture *about the structure and foundations of science and geology.*
13. We must never forget about the inquisitive nature of our ancestors, who *laid the foundations upon* which our current scientific and technological *knowledge rests.*
14. It is also important to understand that *a lot* about science is not always about finding a *groundbreaking* discovery but *adding onto* or improving a hypothesis to help others better understand a certain topic.
15. We must remember that it [science knowledge] is never complete and is subject to change and enhancement, [and] as time moves forward and technology improves, *new windows of information open.*
16. Developing it is about *building on to existing* scientific knowledge, much like *laying a brick in a wall.*
17. One important concept for learning science is our understanding of the nature and *structure of scientific knowledge* and the process by which it is *developed.*
18. It is important to remember that this scientific knowledge is *not concrete* and is subject to change.
19. Science can *build on itself* overtime and *can shift* between perspectives and environments.
20. Science is about repeatability and for your work to *withstand* scrutiny.
21. By having an idea, you *open up doors* for others to either prove or disprove your theory.

Excerpts 12–21 show the breadth of use of this metaphor, which includes the following entailments:

- The mind is a builder
- Thinking is building
- Changes in thinking shift buildings
- Historical knowledge is a foundation
- Ideas are building blocks
- Unchanging ideas are objects set in concrete
- Exploring new knowledge is opening windows or doors

This metaphor and its entailments derive from our extensive experiences living and working in buildings and witnessing them being built (if not building them ourselves). Primary initial steps to build a structure are breaking ground and structuring a foundation. It is this foundation that supports the rest of the building, which is usually made of constituent pieces (bricks, blocks, or building stones). The building usually has different rooms to explore or use, with doors that allow access and windows that afford different views and perspectives. This metaphor highlights the constructed and durable nature of scientific knowledge and the fact that newer knowledge is often derived (renovated) from older knowledge. It also highlights the "craft" or artisanship of theory building (Rivard, 2020). However, it does hide the iterative nature of theory construction (see Chap. 4, Sects. 4.5 and 4.8.2, and Rivard, 2020).

7.3.3 Science Knowledge Is a Landscape

The next most common metaphor code is SCIENCE KNOWLEDGE IS A LANDSCAPE (101 applications). The landscape metaphor was one of the first used in data visualization, giving spatial meaning to information that is not inherently spatial (Fabrikant et al., 2010). The metaphor emphasizes our experiences with various landscapes to give meaning to the concept of scientific knowledge. The following excerpts from student responses illustrate this usage.

22. It is important to combine our knowledge *of different fields* in order to look at things from a different perspective and with different ideas.
23. I had a *hypothesis surrounding this concept* and I was *looking for* an answer, [and] I also took the time to study and analyze *the steps needed* to come to a conclusion.
24. An inquiry would *lead to exploration* of possible causes, related issues, and other correlations between *differing fields*, people, or information.
25. It is important to know that science is always changing, and in some sense, science is never definitive because someone will always be able *to dig deeper into* the true meaning.
26. Studying the processes that early scientists used to *stretch the boundaries* of what was known at the time, to develop tools and medicines, fundamentally shaped the world we live in today.
27. It is important to understand that each person learning [this material] comes from a different educational *background*, with *different levels* of strength in *every scientific field*.
28. New answers to old questions are being *unearthed* everyday.

The entailments included with this metaphor are as follows:

- Disciplines are bounded fields
- Previous knowledge is in the background
- Inquiry and/or learning are traversing the surface (see Sect. 3.1, above), exploring, looking for, digging deeper or below the surface, unearthing

- Making new knowledge or having insights is pushing the boundaries, breaking ground, developing, finding, discovering

Within this metaphor, scientific knowledge is a landscape, with different disciplines being different bounded fields (excerpts 22, 24, 26, and 27). Enhancing one's knowledge in a specific discipline would be finding the best way to reach the destination (excerpts 23 and 24), which comes from learning in the discipline; increasing knowledge about the discipline is pushing the boundaries and broadening the field (excerpt 26). Maybe a person has only a superficial understanding of a topic, or they are just scratching the surface, and must dig deeper to enhance their understanding (excerpts 22 and 25). This metaphor highlights that we are usually familiar with some parts of a discipline, but not with other parts—just as we would be with a field or an outdoor section of land. Work involves digging in, trying to enhance our understanding; the ideas, like buried treasures, are valuable (Demirbilek & Korkmaz, 2021).

The metaphor also hides the constructed nature of scientific knowledge, just as with SCIENCE KNOWLEDGE DEVELOPMENT IS MOVING ALONG A PATH, noted above. Knowledge is not a physical thing that one can hold, walk on, or simply explore—nonetheless, this metaphor effectively objectifies knowledge, simultaneously implying that knowledge is set, even predetermined as a landscape exists and is the same for each person exploring it. Thus, it is not the result of creative activities of the scientist, which could take any shape, as in Flannery's (2001) quilt metaphor.

The above metaphors also relate to each other. For instance, if science knowledge is a landscape, then learning would be travelling over or exploring that landscape of knowledge. The buildings are a kind of landscape as well—an urban landscape. Land development often refers to the construction of buildings on a parcel of land. This connection exists in the data, as evidenced by these statements:

29. It helped other scientists to *dig into that field* and realize the importance of it.
30. In this manner, *developing* scientific knowledge *on whatever field* can be more useful if *founded upon* existing knowledge on *that specific area.*
31. I was reading, I was analyzing new information and *building it onto my prior knowledge* in the same *existing field.*
32. We must also understand the historical context *surrounding the development* of the knowledge.

The responses (excerpts 29–32) refer to digging into a specific area or field—breaking ground—and developing it, building on top of past knowledge. This metaphor therefore seems to acknowledge the constructed nature of scientific knowledge, in that it is built, constructed, or developed on top of old knowledge or the landscape, or maybe even from the old knowledge of the landscape. This leads into the next most applied metaphor code, CURRENT KNOWLEDGE IS BUILT ON PAST KNOLWEDGE (61 applications). This metaphor incorporates the THEORY IS A CONSTRUCTION (Lakoff, 1994c) metaphor while also acknowledging the role that previous knowledge plays in the generation of newer knowledge. Examples of this metaphor occur in the following responses:

33. It is important to remember that science is always evolving, and new hypotheses and theories are always being implemented or even *added on to old ones.*
34. Further scientific knowledge is also open to interpretation and *is based on previous understandings* and being open to new thoughts and beliefs.
35. New findings made by people are *building onto the previous arguments,* making the attempted question even more complicated as time goes on.
36. This research in turn helped me *to build off my previous knowledge* and *develop* a more comprehensive understanding of geological topics.
37. There should be no bias in science or ideology, but [rather,] *reliance on past theory or ideas to support new theories.*
38. Scientists are able to *develop* and discover *new theories based on what was known prior* and are able to *build on those previous concepts.*

The above excerpts (33–38) highlight that older ideas perform a function of support for newer ideas. Newer ideas are added onto older ones, as in (a) adding a new room to an already existing building, where that existing edifice supports the new construction, or (b) the land plays the role of previous knowledge or is the location or base on which the new knowledge is developed or built. This idea parallels the earlier notions (Chap. 6, Sects. 6.3.2.2 and 6.4.3) of science being both discovered truth (permanent, like a building) while at the same time being always changing as pieces of new knowledge are added on to the original knowledge (see this Chapter, Sect. 3.4), like bricks added to a building can change it over time.

7.3.4 Ideas Are Objects

Another commonly referenced metaphor is IDEAS ARE OBJECTS (61 applications), a framing that objectifies mental conceptions (Lakoff, 1994b). The following excerpts provide examples of this metaphor in use:

39. Scientific knowledge is always open for debate and inquiry to further extend *our grasp* on a certain subject matter.
40. *Finding* the best answer that is simple enough to understand is what science is about.
41. I *used* the concepts and ideas discussed in class in order to produce a final study guide, whilst *uncovering* the best possible explanations for each *topic at hand.*
42. Science is not static and unchanging; *new discoveries are made* and we *find out things* we never knew or *find that something* we believed to be fact is actually false.
43. It is most important to remember that *acquiring it* [science knowledge] stems from the curiosity of man [*sic*].
44. You can *gain scientific knowledge* in so many different ways.
45. Theories change all the time as the result of *uncovering new information.*

46. Everyone has a different point of view and there will always be people trying to *put down another's hypothesis*.

As expressed in the above excerpts (39–46), this metaphor treats ideas, which are mental constructions, as concrete objects. They can be found, discovered, or uncovered, and they can be given, acquired, or used to accomplish a task. Finally, they can put down or set aside when a better idea comes along. The metaphor also has the following entailments:

- Becoming known is emerging
- Understanding is holding/grasping
- The mind is a container
- Learning is discovering/finding/gaining
- Applying knowledge is using the object
- Invalidating knowledge is putting the object down

This metaphor has its foundation in the concrete experiences we have with things. From very early on in life, the way we come to know our environment is by grasping a hold of things and examining them. We have come to know different things by finding them, uncovering them, and discovering them. It highlights how we use things as tools to accomplish some work—like ideas to solve a problem—and when that tool is no longer useful to us, we put it down and stop using it. The metaphor, however, also hides the fact that ideas are constructed in the mind, and therefore different from person to person; they are not lying around, already formed, waiting to be discovered. It also hides the creativity necessary to construct the idea. A more complex example of the metaphor IDEAS ARE OBJECTS shows up in the Sect. 3.6, below.

7.3.5 Science Knowledge Is a Liquid/Ocean

Another metaphor coded in the responses is SCIENCE KNOWLEDGE IS A LIQUID/OCEAN (38 applications). This metaphor serves to highlight the changing nature of scientific knowledge by describing its fluidity. The following excerpts illustrate students' emphasis on this aspect of knowledge:

47. It is most important to remember [when studying/analyzing data] that information is constantly *adapting to fit within* the new parameters of what is *becoming mainstream*.
48. It is important to remember that science can *be fluid* and there is often not one singular *permanent solution*.
49. In order to contribute something new to *our current* state of knowledge you cannot be afraid to think for yourself.
50. It is important to understand that science is constantly changing. New data, studies and information is constantly *flowing* and evolving.

51. It is important to understand that science is an *everlasting source* of new information and that the occurrence of new discoveries is very common.
52. Science requires a continual curiosity, *thirst for knowledge*, and desire for improvement in order to further understanding and improve the quality and accuracy of our findings.
53. Science is always expanding, and we will *always reach deeper into the unknown*.
54. On the other hand, it is also vital to remain skeptical of the various ideas that are *brought up* and to scrutinize them as closely as possible to encourage the theories to be presented *in-depth* and based on data.
55. New technology is being discovered which will allow us *to dive deeper* into scientific facts.
56. Some is changed during experimentation and observation *as new information emerges*.

The metaphor highlights the behaviour of a liquid, like water, which is free-flowing and will fit into a new container via its fluid properties (excerpt 47). Narrow regions of strong flow come from a source (excerpt 51) and create a current, a somewhat stable directional flow or becoming mainstream (excerpts 47, 49, and 50). The metaphor also has such entailments as these:

- Change in science knowledge is fluidity
- Present stability of scientific knowledge is a current
- Scientific process is the source
- Researching is diving deeper
- Learning something coming up with or the emergence of an object

Excerpts 47–52 display a more general metaphor of scientific knowledge as a liquid. Liquids, by definition, take the shape of their container. They can be a solution and can also flow as a stream or current. In this way, the metaphor highlights the provisional aspect of science. The source of this fluid would be the process by which science knowledge gets generated, quenching the thirst to answer a question.

Excerpts 53–56 illustrate a more complex metaphor, SCIENCE KNOWLEDGE IS AN OCEAN, in which science knowledge is a larger body of water (Lakoff, 1994a)—identified here as an ocean. In this metaphor, the depth of the water is framed as the unknown (excerpts 53–55), as light can only penetrate so deep. Therefore, researching a problem is diving deeper, finding a solution is to find an object (see IDEAS ARE OBJECTS, above), and coming to know something is coming up with or emerging with the object (excerpt 56). In general, the metaphor highlights aspects of science very similar to SCIENCE KNOWLEDGE IS A LANDSCAPE, in that immersing oneself in a topic involves work, just as it takes energy to dive deeper in water. Also, unknowns abound, as the oceans cover 70% of the earth's surface. What is hidden with this metaphor, as with the others, is the constructed nature of science knowledge. It suggests that ideas, already formed, can be found and retrieved, as is, from the depths. The work of the scientist, in this case, is not the construction of the knowledge, but just retrieving it from the dark depths and bringing it to light at the surface.

7.3.6 Scientific Theory Is a Puzzle

A relationship exists between science knowledge and ideas, since ideas are what we assemble to construct knowledge or theory. Conversely, we can find, discover, or uncover pieces of the landscape of knowledge. The notion of ideas as part of knowledge represents a complex primary metaphor, CONSTITUENTS ARE CONTENTS (Grady, 1999). According to Grady (1999),

> When we look at a physical object that has distinct parts—differentiated by shape, color, function, etc.—these parts all fall within a visual boundary which defines the limits of the object. In other words, the parts appear to be contained within the space occupied by the object as a whole. (p. 8)

Mapping this metaphor illustrates that scientific ideas are part of broader scientific knowledge, similar to how stepping-stones make up paths, uncovered/discovered/found/dug-up objects are part of a landscape, or how bricks and other building blocks set in concrete are part of an edifice.

CONSTITUENTS ARE CONTENTS is the basis for another complex metaphor that shows up in the data: SCIENTIFIC THEORY IS A PUZZLE (10 applications). This metaphor maps experiences of putting together a puzzle onto constructing a theory, using various pieces of information, as illustrated in the following excerpts.

57. It would have been difficult to *piece together a story* about this plasmid DNA (which is what a geologist would do, I think?) so instead I used the scientific method.
58. I was doing research for projects in this class among different scientists throughout time and identifying how all of their observations and ideas impacted one another, and no one individual was completely right and *taking little pieces from a bunch of information to create a better, well rounded correct idea* is when I was acting like a scientist.
59. Hypotheses and stories regarding a question may be developed but it is only based off of the little data we can manage to gather …[which] scientists may assume in order to *fill in the missing pieces.*

This metaphor is grounded in our experience of completing jigsaw puzzles. It is a complex version of the primary metaphors IDEAS ARE OBJECTS and CONSTITUENTS ARE CONTENTS, where the former represents the individual puzzle piece, and the latter represents the pieces when they are integrated into the completed puzzle. The entailments for the metaphor include the following:

- Making connection will fit the puzzle pieces together
- Learning is a way to complete the puzzle
- Solving a problem is filling in the missing pieces

Some scientists also consider this a useful metaphor for the scientific process in that it highlights fitting pieces together or making connections based on the occurrence of patterns or themes (Harwood et al., 2005). In fact, it is quite a popular metaphor and famously used in the geosciences by Alfred Wegener when referring to how the continents could fit together like a jigsaw puzzle to form the supercontinent, Pangaea. Wegener relied on connecting evidence from fossils (biological disjuncts), structural and lithological congruities (geological disjuncts), paleo-climatic evidence, and continent shape. As with IDEAS ARE OBJECTS, this metaphor carries the limitation of objectification of science knowledge—that it is already preformed, predetermined, and needs no creativity or work by the scientist to construct the knowledge. The picture is already on the box. All puzzle pieces have predetermined places to fit, and the picture is set *a priori*.

7.3.7 Science Knowledge Is a Plant

The final metaphor I will describe here is SCIENCE KNOWLEDGE IS A PLANT (31 applications), which notably has less overlap with the metaphors discussed so far. Instead, this metaphor uses notions of growth and plant parts to describe science knowledge and processes, as illustrated by the following excerpts:

60. It's important to remember that there are so many *various branches* of science, meaning that we are always learning every day.
61. It is most important to remember that acquiring it [science knowledge] *stems from* the curiosity of man [*sic*].
62. Science is an ever *changing and evolving* thing, [and] it is important for us to *stem from the* history of science and previous theories, to *further grow our knowledge* in this area.
63. It is important to remember that the world is always changing, so our knowledge of science must *grow and adapt* with it.

Mapping this metaphor, science knowledge is an entire plant and disciplinary knowledge is represented by different branches off the main trunk (excerpt 60). The beginning of an investigation is a new stem forming from an established branch (excerpt 61), perhaps becoming a branch itself. New ideas are like seeds that can be sown, and if the environment is conducive and the soil fertile, the idea can germinate, take root, grow, and blossom (excerpts 62 and 63). I interpreted "growing" as part of the plant metaphor because even though many things, organic and inorganic do grow—puddles grow with rain, piles grow from dumping more stuff on top of them—the associations with other phrases seem to indicate growing as a living process (such as "grow and adapt" in excerpt 63). Moreover, its proximity to more explicit reference to the plant metaphor ("stem from the history" and "to further grow" in excerpt 62). Another view is that this might be a more specific version of the landscape metaphor, since plants are a natural extension of the landscape, are part of the landscape, and do derive at least some of their mass from the ground. However, "landscape" does denote more of a spatial orientation that is not so obvious with the plant metaphor, aside from the space taken up around the branches of science.

This metaphor does highlight the change that happens in science and to scientific knowledge. There can be interdisciplinary and collaborative activities, or cross pollinations. It also highlights the growth and eventual death of branches, or disciplines in science, like alchemy, phrenology, and astrology. However, the implied physical separateness between the plant—science knowledge—and the cultivator—scientist—hides the connection of that knowledge to the mind of the scientist. The scientist can influence the growth of the science knowledge by nurturing, pruning, weeding, and watering but the plant is separate. It has its own genetics, which predetermine what that plant will be.

7.4 Discussion: Metaphors of Science Processes and Products

When considered collectively, all the metaphors described are related. As is appropriate with phenomenographical analysis, I organized the codes and metaphors hierarchically. In Fig. 7.3, the hierarchy begins on the left with primary metaphors and moves towards the right with complex or specific metaphors (Grady, 1999), and then on to entailments (Lakoff & Johnson, 1980). In some cases, yellow ovals (Primary Metaphors) are not on the left side of the figure. This was an attempt to simplify the diagram by not having so many arrows crossing. The arrows designate the hierarchy. Interwoven throughout the themes are (a) objects that stand for the products of science—discoveries, findings, stepping-stones, building blocks, and so forth—and (b) actions that represent science processes—exploring, diving, digging, breaking ground, and the like. This makes sense as I divided the open coding in Chap. 6 into science products and science processes. However, I have chosen not to break up the present analysis into these two bundles because it will obscure the broader relationships among the metaphors. Instead, I look first at how students used metaphor as a framing mechanism for science and how metaphors operate in teaching and learning, and then turn to how my interpretations here relate to the interpretations in Chap. 6.

7.4.1 Framing Science Knowledge

Many participants described science knowledge as a broad, quasi-two-dimensional surface. I say "quasi" since the surface is far broader than it is thick, though it does have some depth. Participants spoke of this surface in terms of a landscape or a seascape. Importantly, framing the abstract idea of science knowledge in this way objectifies it, in that it treats it as something concrete (Jelec, 2014), so that the idea has an experiential basis in our extensive interactions with the surface of the earth—the seascape and the landscape.

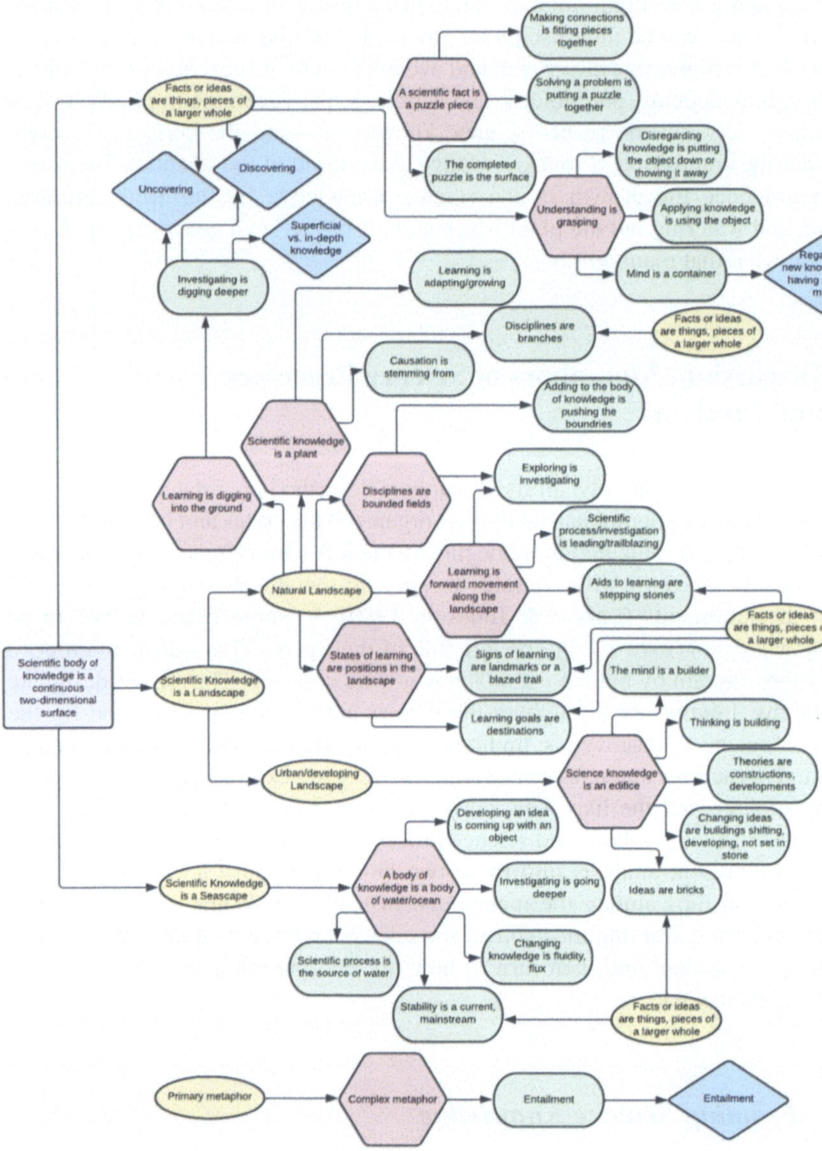

Fig. 7.3 Conceptual metaphor outcome space. (Note. A visualization of the outcome space derived from the pre- and post-instruction responses to the open-ended questions. The hierarchy of concepts moves in general from left to right, or primary and complex metaphors to entailments)

Familiar landscapes are also classified as either natural (undeveloped) or urban (developed). Upon this knowledge surface, we may set about exploring, learning more about it. Further, the landscape may be divided up into bounded fields of study. We may blaze a trail or use stepping-stones laid down by those who went before us. Landmarks or other destinations may indicate a desired state of learning. In terms of the seascape, we might chart a course for a destination or just drift, with no direction at all. These last entailments were not explicitly part of the participant responses, but they are nevertheless common entailments for this metaphor.

In addition to the learning that takes place by exploring the surficial landscape, students search below the surface. In this way, they move from focusing on a process ("exploring") to a product ("landscape"), and back to another process ("searching"). They can dig or dive deeper, uncover or find things (processes), and then come up with or emerge with new knowledge (product). Digging or diving reflects the energy needed to make the discovery or finding. An increased depth marks an increase in difficulty. This understanding also incorporates the metaphor FACTS OR IDEAS ARE PART OF THE LARGER KNOWLEDGE, a complex metaphor combining IDEAS ARE THINGS and CONSTITUENTS ARE CONTENTS (Grady, 1999). Since ideas are things, we can grasp those things—meaning we understand them. Because KNOWING IS POSSESSING, and we commonly consider that thinking and knowing happen in the mind, the mind becomes a bounded container for ideas; correspondingly, having an open mind is an important requirement for learning.

This relationship between knowledge and knower easily relates to science knowledge and scientists. If the knowledge is already in existence, and only needs to be found, dug up, uncovered, discovered, or emerge, then scientists need not be creators of that knowledge. Knowledge construction (Carey, 2009; Giere, 1988; Nersessian, 2008; Thagard, 2012) does not make any sense within this framing. Scientists only have to be present, with an open mind, and accept the facts as the universe presents them, already predetermined and fixed in nature. They speak for themselves. The only change in knowledge is the addition of the bits and pieces to the surficial whole.

Readers may recall my description earlier, in Chap. 6, Sect. 6.3.2.1, where student descriptions implied scientists are technicians. This perspective also makes sense here if all scientists have to do is uncover, dig deeper, come up with, push some boundaries, and prove or disprove hypotheses—essentially, have no involvement with the actual creation of scientific knowledge.

7.4.2 Metaphors for Teaching and Learning Science

The relationship between knowledge and knower also pervades our discourse about teaching and learning. As instructors, we often talk about *delivering* a lecture, *spoon-feeding* students, highlighting the *concrete points*, *filling gaps* in student knowledge, and so on. We are the keepers of the knowledge and are responsible for giving it out to the students. As for the students, we talk about them in terms of being *receptive* to what we teach, *taking* notes, and *soaking up knowledge like a sponge*. We hope the ideas *sink in* and that ultimately students just *get it*. They are the receivers of the knowledge we teachers dole out, a framing that emphasizes a transactional relationship in which students are passive recipients.

The research literature is quite clear about how our use of metaphor influences how we think about a concept, shaping our understanding of what it highlights and what it hides. This can both help and hinder understanding of the target concept. Talking about teaching as content delivery makes traditional, lecture-style teaching seem reasonable; at the same time, this traditional teaching behaviour reinforces how we understand teaching (Dolphin, 2016), setting up a reciprocal dynamic in which the process reinforces its own worth so that other avenues are less approachable or meaningful.

Sfard (1998) and Wegner and Nückles (2013) described two different metaphors of learning: acquisition and participation. The acquisition metaphor parallels what we have been describing here, wherein students acquire information from an expert source. In contrast, the participation metaphor aligns more with the social construction of knowledge perspective; in other words, within a situated cognition context (Anderson et al., 1996; Hutchins & Renner, 2012; Lave & Wenger, 1991), a group of people and the immediate context mediate the construction of knowledge.

Students have been enculturated into the acquisition metaphor, since it is the dominant way people, policy-makers, and even educational psychologists speak of learning (Donaldson & Allen-Handy, 2019). Students typically expect to be passive recipients of the knowledge in science classes and behave accordingly. They will be happy to *take* notes, *cram* for the exam, and then *regurgitate* the information on command. They can be so expectant of this transactional dynamic that any variance from it could end with students feeling the instructor is not doing their job. This is why it became necessary for me to pre-empt such attitudes by explaining to students my theory of learning and the implications it has for my teaching practice.

Two metaphors from the data suggest learning as the construction of knowledge. The first is SCIENTIFIC FACTS ARE PUZZLE PIECES, and the second, SCIENCE KNOWEDGE IS AN EDIFICE. In the first, construction of a theory is the completed puzzle, remembering too that a completed physical jigsaw puzzle is also traditionally a quasi-two-dimensional surface, like a landscape or seascape. This metaphor may seem to represent a more constructivist view, with the scientist putting the pieces together to construct the greater whole.

However, where do these pieces come from? In a real jigsaw puzzle, the pieces are ready-made and have a predetermined place within the entire puzzle. For instance, the upper right-hand corner piece cannot fit in any other position. What the theory (puzzle) will look like, then, is already predetermined. It is not a construction of the scientist, but a reproduction. The scientist is like a technician merely fitting the pieces together, making the appropriate connections to end up with a final picture.

The second metaphor of construction, SCIENCE KNOWEDGE IS AN EDIFICE, seems even closer to the idea of the true construction of knowledge. A newly constructed theory starts with breaking new ground, then developing it, usually founded on or derived from past knowledge. Bricks—that is, facts or ideas—structure the edifice, constituents of the greater whole. Doors and windows provide access to new views or ways of thinking. However, such buildings usually begin as plans or blueprints that determine what the edifice will look like eventually. Anyone who has tried to build a house knows that generally not much room remains for negotiation from the plan to reality once the project has started.

Similarly, the knowledge landscape or developed knowledge is predetermined, with not much chance for scientists' creativity, in which they can make something up from scratch. An opportunity may exist for the scientist, as builder, to make an addition here and there, but it is always vis-à-vis the structure of the previous knowledge. It is always an accumulation of knowledge, with little opportunity to turn the original structure into something completely different—what we might call a paradigm shift (Kuhn, 1962).

7.4.3 Summary: The "What" and "How" of Science

The metaphors representing the "what" of science and the "how" of science—science product and science process, respectively, align with the emergent themes in the open coding and data analysis outlined in Chap. 6, where students commented on the nature of scientific knowledge as well as its process. Though I have used two different lenses to look at and analyze the data, the findings from both support each other.

In general, students think that scientific knowledge is additive, with little bits and pieces constantly being added to the larger body—puzzle pieces to the jigsaw puzzle, bricks to the building, stepping stones to the blazed trail, and so forth. These ongoing additions are responsible for the "constant change" they emphasize, but they are also predetermined bits of knowledge (truths), needing only to be discovered as is and added to the existent body of knowledge. The purpose of experimentation, and indeed the scientific process, is to test the veracity of these bits of knowledge, to prove or falsify them. In this context, *reality determines the winner* of controversies, as opposed to the winners of controversies *determining* reality (Latour, 1987). Participants did not significantly emphasize the constructed nature of science knowledge, based on their use of the metaphors SCIENCE KNOWLEDGE IS AN EDIFICE, with 160 applications, and CURRENT KNOWLEDGE IS BUILT ON PAST KNOWLEDGE, with 61 applications. These application numbers were relatively stable from pre-instruction to post-instruction.

In opposition to these metaphors, Flannery (2001)'s quilting metaphor, discussed above in Sect. 2.2, offers an example of a more constructivist metaphor for scientific processes and products. She asserted that quilting highlights aspects of science processes and products, as follows:

- Bringing order to many individual pieces
- Learning by doing
- Creative assembly of constituent pieces within a limited set of parameters
- The role of tinkering; doing the best with what you have
- Incorporating work that is tedious, frustrating, and drudgery
- Being a communal activity
- Using the backing to hide the rough edges, as no uniform process is available
- Ideas are often reworked, repaired, and remade

Could using such a metaphor more often when talking about science processes and products influence how we perceive the larger scientific process? Certainly, this idea for teaching and learning science deserves further study, especially as it is likely rare to find students who have had the embodied experience of making a quilt. Giving plenty of examples of the creative process in science and discussing how scientists took initial observations and constructed explanations within limited parameters could help this metaphor take hold. The three narratives in the course (see Chap. 3) do just this. Initial observations help form an explanation, further information serves to enhance the explanation, and so on. Discussing how communicating ideas in a social group—among conference attendees or peer reviewers—helps to refine and validate science knowledge, even when these actors are not repeating the procedures or conducting controlled experiments.

Science knowledge is not predetermined. It is at once personal *and* socially negotiated. In the controversies of ice ages, the age of the earth, and the cause of continents and oceans, argumentation between and among competing camps and using the different research strategies finally led to a resolution (Frankel, 2012). Examples of such strategies, as I have discussed earlier, include solving difficulties in one's own explanation, pointing out difficulties in a competing explanation, and demonstrating that one's own explanation is better than the competition. That resolution often contained aspects of the earlier competing explanations, like Airy's isostasy and Pratt's isostasy, uniformitarianism and catastrophism, and the resolution of the great Devonian controversy (Rudwick, 1985).

Importantly, explanations must be constrained by the empirical world. After that, anything goes. Perhaps if students perceive the creativity involved in scientific knowledge, including that it is not just a list of stuff to memorize, more of them may like this domain enough to consider career paths in science (Harper et al., 2019). My hope was that students would start to better understand the process if I gave them experience developing their own answers in response to presented data, which they achieved in collaboration with a small group of fellow (student) scientists; through this process, they could then see in the narratives how the knowledge incrementally developed over time.

Constructive realism (Giere, 1988; Madzia, 2013) ostensibly aligns nicely with the some of the constructivist aspects of the metaphors SCIENCE KNOWLEDGE IS A PUZZLE and SCIENCE KNOWLEDGE IS AN EDIFICE, although in the end this falls apart as I discussed in Sect. 4.2 above. Likewise, the metaphor SCIENCE KNOWLEDGE IS A PLANT fits into this more constructivist stance, highlighting the impact that scientists—as cultivators—can have on the growth of knowledge. The genotype of the plant may be ready-made, but the phenotype (the expression of the genotype in response to the environment) does allow some difference from garden to garden, or lab to lab.

Most of the metaphors expressed by the students, however, belie a much more objectivist stance, supporting the idea that they have reified science ideas into objects (Reynolds, 2021), pieces of a larger surface. This objectification—again, the purpose of metaphor—still emphasizes a sense of predetermined science knowledge, making it something that is just waiting to be uncovered and brought into the light, rather than being the product of some creative act or the outcome of a negotiation within a social group of experts. The creativity of the scientist is therefore discounted. Ultimately, the students' metaphors demonstrated little in the way of an epistemological shift, like that described in the previous chapter, from pre-instruction to post-instruction.

As for the process of science, students described it as searching, exploring, digging, or diving, all of which imply work on behalf of the scientist, but that work is technical rather than creative. It is routine and redundant, and where some science incorporates the routine and redundant—culture growing, sample analysis, and the like—it is really a small part of the process itself. It is, however, the aspect of science that students get most practical experience with. This standard procedural teaching approach to the processes of science in combination with the *delivery* of science facts that are immutable, as if they had always been, is the common experience for students of science. I have interpreted this from the student responses in the previous chapter as well as in this chapter, strengthening the reliability of both interpretations.

How might we explain the apparent shift in NOS understandings documented in the previous chapter and the lack of a shift here? The answer could be that such a shift is very complicated and more time consuming than traditionally thought. For instance, Korthagen (2004) used an onion model (Fig. 7.4) for levels of change in science teacher development. In Korthagen's model there is an external region identified as the "environment" surrounding a series of concentric rings decreasing in diameter. In contact with the environment is the outermost ring of the onion, "behavior," which responds to the environmental conditions, but only within the scope of the next internal ring, "competencies," the skills and procedures the person knows. Competencies are behavior potentials. Within competencies is "beliefs." Within beliefs is "professional identities," and finally "mission." I will not be referring to identity or mission, so I will not be describing them here.

With respect to the model, influence can occur in both directions. From the outside in, the environment influences behavior and successful behaviors can develop effective competencies, influencing beliefs in a particular direction. From the inside out, beliefs guide the learning of competencies, that in turn will limit the behaviors that are possible, with that behavior changing the environment.

That said, layers or aspects of the person that are deeper within the onion—beliefs, identity, and mission—are much more stable, or resistant to change. Where students in science class have mainly experienced objectified knowledge,

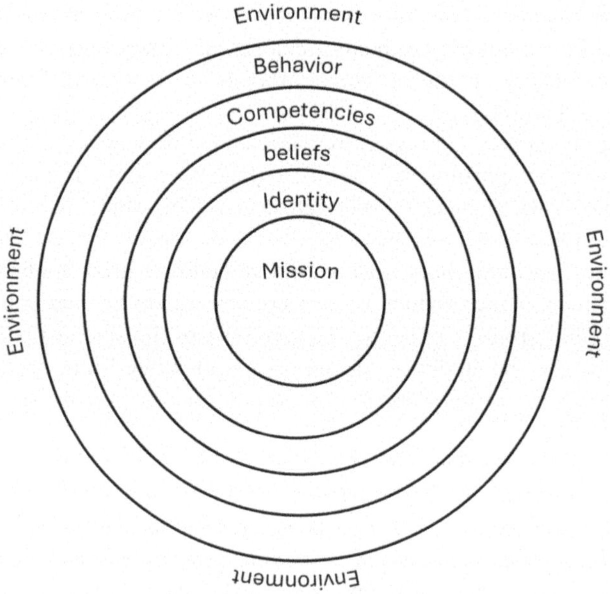

Fig. 7.4 The "Onion Model" of levels of change. (*Note*. The model illustrates levels of identity of a person surrounded by the environment. Influence of one layer on another can proceed in both directions. Change within a layer gets more difficult when influencing deeper and deeper layers (After Korthagen, 2004))

confirmatory labs, traditional lecture-style teaching, and the apparent transfer of knowledge from teachers or from books, their beliefs and competencies would most likely be coherent with that worldview. This is borne out in the pre-course responses described in the previous chapter. When the students enter a different environment—like GLGY 209—that emphasizes facets of a worldview radically different from what has previously structured their beliefs, it will take time to first teach some new behaviors that eventually build into a repertoire of competencies that align with this new worldview.

Changing beliefs, the layer guiding such implicit metaphor use as described above, would take a long time of intensive experiences. The data seem to support the idea that from pre-course to post-course student metaphor use did not seem to signify a change in terms of how they considered NOS. This is contrary to what the data in Chap. 6 indicate. If you remember from Chap. 6, based on the applied codes, students did demonstrate a shift in understanding of how science happens, the nature of the products, and who does science. The course supported students in developing new competencies and behaviors, and this is what is reflected in the data in Chap. 6. On the contrary, the analysis of metaphor use in this chapter supports the notion that students' beliefs have not had enough time and/or pressure to change in line with the epistemological stance favoured in the course.

The take-home message (metaphorically speaking, of course) here is that if we want students to have a more nuanced understanding of (or beliefs about) science, they need to have many experiences that emphasize that intended understanding. They need to reflect on those experiences, and they need to encounter this kind of pressure from the environment for long periods of time. We cannot expect students to change after a single lesson, or series of lessons, or, apparently, a whole course devoted to HPS. In addition, changing our language—especially the metaphors—that pervade teaching, learning, science processes, products, and scientists could also help students think differently about these concepts, and even help us to *act* in ways that further reinforce these intended learning goals. There were about 300 more student responses in the post-course data, yet the number of metaphor code applications was not greater in most cases compared to pre-course usage. This could also reflect students reporting based on their competencies and behaviors they learned during the course. By having the concrete experiences to report on, they did not need to use metaphors to represent an otherwise abstract behavior or competency. These points will be part of the discussion in the final chapter.

The final chapter will wrap up the main themes of the book—my rationale for restructuring GLGY 209, the importance of transformative teaching, the role of the history of geology in a geology class, and the impact of such a class on student learning. It will also set sights for future work.

References

Allchin, D. (2013). *Teaching the nature of science: Perspectives & resources*. SHiPS Education Press.

Anderson, J. R., Reder, L. M., & Simon, H. A. (1996). Situated learning and education. *Educational Researcher, 25*(4), 5–11.

Barsalou, L. W. (1999). Perceptions of perceptual symbols. *Behavioral and Brain Sciences, 22*(4), 637–660. https://doi.org/10.1017/S0140525X99532147

Carey, S. (2009). *The origin of concepts*. Oxford University Press.

Clark, A. (2011). *Supersizing the mind:Embodiment, action, and cognitive extension*. Oxford University Press.

Demirbilek, N., & Korkmaz, C. (2021). Study concept as a metaphor from the lenses of university students. *African Educational Research Journal, 9*(1), 227–236. https://doi.org/10.30918/AERJ.91.21.028

Dolphin, G. (2016). Toward an understanding of "teaching in the making": Explaining instructional decision making by analyzing a geology instructor's use of metaphors. *Geosphere, 12*(6), 1725–1743.

Dolphin, G., & Benoit, W. (2016). Students' mental model development during historically contextualized inquiry: How the 'tectonic plate' metaphor impeded the process. *International Journal of Science Education, 38*(2), 276–297.

Dolphin, G., & Tillotson, J. (2015). "Uncentering" teacher beliefs: The expressed epistemologies of secondary science teachers and how they relate to teacher practice. *International Journal of Environmental and Science Education, 10*(1), 21–38.

Donaldson, J. P., & Allen-Handy, A. (2019). The nature and power of conceptualizations of learning. *Educational Psychology Review, 34*, 545–570.

Fabrikant, S. I., Montello, D. R., & Mark, D. M. (2010). The natural landscape metaphor in information visualization: The role of commonsense geomorphology. *Journal of the American Society for Information Science and Technology, 61*(2), 253-270. https://doi.org/10.1002/asi.21227

Flannery, M. (2001). Quilting: A feminist metaphor for sientific inquiryc. *Qualitative Inquiry, 7*(5), 628–645.

Frankel, H. L. (2012). *Wegener and the early debate* (Vol. I). Cambridge University Press.

Giere, R. N. (1988). *Explaining science: A cognitive approach*. University of Chicago Press.

Grady, J. (1999). The "conduit metaphor" revisited: A reassessment of metaphors for communication. In J. P. Koenig (Ed.), *Discourse and cognition: Bridging the gap* (pp. 1–16). CSLI/ Cambridge. Retrieved from http://terpconnect.umd.edu/Israel/Grady_Conduits.pdf. Accesses 15 Nov 2012.

Haack, S. (2019). The art of scientific metaphors. *Revista Portuguesa de Filosofia, 75*(4), 2049–2066.

Harper, R. P., Weston, T. J., & Seymour, E. (2019). Student responses to problematic STEM teaching methods. In E. Seymour & A.-B. Hunter (Eds.), *Talking about leaving revisited: Persistence, relocation, and loss in undergraduate STEM education* (pp. 149–196). Springer.

Harwood, W. S., Reiff, R. R., & Pillipson, T. (2005). Putting the puzzle together: Scientists' metaphors for scientific inquiry. *Science Educator, 14*(1), 25–30.

Hauser, D. J., & Schwarz, N. (2019). The war on prevention II: Battle metaphors undermine cancer treatment and do not increase vigilance. *Health Communication, 35*(15), 1698–1704. https://doi.org/10.1080/10410236.2019.1663465

Hutchins, E., & Renner, N. (2012). Situated and embodied learning in the field. In K. A. Kastens & C. A. Manduca (Eds.), *Earth and mind II: A synthesis on research on thinking and learning in the geosciences* (Geological Society of America special paper 486) (Vol. 2, pp. 181–182). Geological Society of America.

Jelec, A. (2014). Are abstract concepts like dinosaur feathers? Conceptual Metaphor Theory and the conceptualisation strategies in gesture of blind and visually impaired children.

Kahneman, D. (2011). *Thinking, fast and slow* (1st ed.). Farrar, Straus and Giroux.

Korthagen, F. A. J. (2004). In search of the essence of a good teacher: Towards a more holistic approach in teacher education. *Teaching and Teacher Education, 20*(1), 77–97.

Kuhn, T. (1962). *The structure of scientific revolutions*. University of Chicago Press.

Kuhn, T. S. (1993). Metaphor in science. In A. Ortony (Ed.), *Metaphor and thought* (2nd ed., pp. 533–543). Cambridge University Press.

Lakoff, G. (1994a). *A problem is a body of water*. University of Berkeley. Retrieved 15 January 2022 from http://www.lang.osaka-u.ac.jp/~sugimoto/MasterMetaphorList/metaphors/A_Problem_Is_A_Body_Of_Water.html

Lakoff, G. (1994b). *Ideas are objects*. Berkeley University. Retrieved 15 January 2022 from http://www.lang.osaka-u.ac.jp/~sugimoto/MasterMetaphorList/metaphors/Ideas_Are_Objects.html

Lakoff, G. (1994c). *Theories are constructed objects*. Berkeley University. Retrieved 15 January 2022 from http://www.lang.osaka-u.ac.jp/~sugimoto/MasterMetaphorList/metaphors/Theories_Are_Constructed_Objects.html

Lakoff, G., & Johnson, M. (1980). *Metaphors we live by*. University of Chicago Press.

Lakoff, G., & Johnson, M. (1999). *Philosophy in the flesh: The embodied mind and its challenge to western thought*. Basic Books.

Latour, B. (1987). *Science in action: How to follow scientists and engineers through society*. Harvard University Press.

Lave, J., & Wenger, E. (1991). *Situated learning: Legitimate peripheral participation*. Cambridge University Press.

Madzia, R. (2013). Constructive realism: In defense of the objective reality of perspectives. *Human Affairs, 23*(4), 645–657. https://doi.org/10.2478/s13374-013-0155-z

Meriam-Webster Dictionary. (n.d.). Field. *In Meriam-Webster.com dictionary.* Retrieved May 19, 2022, from https://www.merriam-webster.com/dictionary/field

Müller-Wille, S., & Rheinberger, H.-J. (2012). *Cultural history of heredity*. Chicago University Press.

Nersessian, N. J. (1985). Faraday's field concept. In D. C. Gooding & F. A. J. L. James (Eds.), *Faraday rediscovered: Essays on the life and work O Michael faraday* (pp. 377–406). Macmillan.

Nersessian, N. J. (2002). Maxwell and the "method of physical analogy":Model-based reasoning, generic abstraction, and conceptual change. In D. Malement (Ed.), *Reading natural philosophy: Essays in the history and philosophy of science and mathematics* (pp. 129–165). Open Court.

Nersessian, N. J. (2008). *Creating scientific concepts*. MIT Press.

Palma, H. A. (2018). Metaphors in science: A change in perspective. In J. Horne (Ed.), *Philosophical perceptions on logic and order* (pp. 242–254). IGI Global

Reddy, M. (1979). The conduit metaphor: A case of frame conflict in our language about language. In A. Ortony (Ed.), *Metaphor and thought* (pp. 284–324). Cambridge University Press.

Reynolds, A. S. (2021). Science is based in metaphor. *IAI News, 96*. https://iai.tv/articles/all-science-is-based-in-metaphor-auid-1809

Rivard, S. (2020). Theory building is neither an art nor a science. It is a craft. *Journal of Information Technology, 36*(3), 316–328. https://doi.org/10.1177/0268396220911938

Rudwick, M. J. S. (1985). *The great Devonian controversy: The shaping of scientific knowledge among gentlemanly specialists*. University of Chicago Press.

Sfard, A. (1998). On two metaphors for learning and the dangers of choosing just one. *Educational Researcher, 27*(2), 4.

Shapiro, L. (2011). *Embodied cognition*. Routledge.

Steen, G. J., Dorst, A. G., Herrmann, J. B., Kaal, A. A., Krennmayr, T., & Pasma, T. (2010). *A method for linguistic metaphor identification* (Vol. 14). John Benjamins Publishing Company.

Thagard, P. (2012). *The cognitive science of science: Explanation, discovery, and conceptual change*. MIT Press.

Thibodeau, P. H., & Boroditsky, L. (2013). Natural language metaphors covertly influence reasoning. *PLoS One, 8*(1), e52961.

Tobin, K., & LaMaster, S. U. (1995). Relationships between metaphors, beliefs, and actions in a context of science curriculum change. *Journal of Research in Science Teaching, 32*(3), 225–242.

Wegner, E., & Nückles, M. (2013). Knowledge acquisition or participation in communities of practice? Academic's metaphors of teaching and learning at the university. *Studies in Higher Education, 40*(4), 624–643.

Zarkadakis, G. (2015). *In our own image: Savior or destroyer? The history and future of artificial intelligence*. Pegasus Books.

Chapter 8
Concluding Remarks

Logical, scientific thinking is essential to helping me arrive at my conclusions, but putting on a creativity hat is equally important. I use both good and failed examples as clues to observe new patterns. It's all about "coded intelligence."

Ben Chu

Finally, in conclusion, let me say just this.

Peter Sellers

8.1 Introduction

A key goal in Part I of this book was to outline the design process for creating a course that historically contextualized the geoscience content. I was especially interested in having students develop a more nuanced and robust understanding of NOS while also simultaneously learning core geoscience concepts. As noted in Chaps. 5 and 6, this historical approach effectively teaches both. Traditionally, these two areas are taught separately, or NOS content is not taught accurately or effectively since it is thought to be something extra to cover after the disciplinary content. However, as addressed in this book, teaching NOS is not something added on; rather, it is intrinsic to teaching the disciplinary material.

However, research literature makes clear that most instructors are hesitant to sacrifice science content for NOS. This book illustrates how teaching the foundational geology content and NOS at the same time are not at cross-purposes, and that students show learning gains in both domains. The research shared in Part II of this book provides evidence of the efficacy of seamlessly incorporating HNOS with geology content.

The rest of this chapter will be a reminder of the need to approach geoscience education in a new way. It will recount the design process for transforming traditional teaching into an engaging learning environment for students. I also provide a general summary of the learning that took place during the course and close with a look to the future.

G. Dolphin, *Teaching Geology Using the History and Philosophy of Science*, Science: Philosophy, History and Education,
https://doi.org/10.1007/978-3-031-68667-2_8

8.2 The Importance of Geoscience Education

Humanity (and other earth-bound organisms) currently face many earth-related challenges. As we continue to occupy more and more of the surface of the earth, we are increasingly affected by surficial circumstances and processes—as these examples of persistent or continual development in problematic regions demonstrate:

- Development on fluvial floodplains and coastal lowlands, putting those who dwell there at greater risk of flooding
- Development in places that are seismically active
- Increasing land purchases in fire-prone areas, while wildfires are becoming larger and more common

These hazards impact more vulnerable, poorer populations as well as those with more resources. For instance, close to 30% of the global population lives in the near-coastal zone within 100 km of the coast and up to 100 m elevation.[1] Any significant rise in sea level could cause the mass migration of hundreds of millions of people, causing major socio-political instabilities globally. Other areas of the world are subject to increasing heat and wildfires and drought. These conditions are most devastating to those without the resources to mitigate such threats.

Further, as the global population continues to grow at an ever-increasing pace, and as the standard of living rises globally, our need to find and extract more natural resources increases. Anything that we cannot grow, we must take out of the ground. This includes the mineral resources that form the foundation of our local, national, and global economies, as well as the energy resources that power those economies. Though we see some movement (ever so slightly) away from dependence on fossil fuels for greener energy resources, alternatives—like solar panels, wind turbines, hydro-electric dams, geothermal pumping stations, and nuclear power plants—still come with a heavy natural resource price tag, and a significant environmental and ecological cost.

Fresh water represents just tenths of 1% of the total volume of water globally and is another resource in jeopardy. Relatively speaking, very little water exists at the earth's surface for human consumption; moreover, increasing populations and increasing pollution continue to decrease that already limited resource. Though water is technically considered a renewable resource, it frequently takes considerably more time to renew this resource than it does to pollute it or consume it. At present, agriculture and industrial processes are on the rise and consume the majority of the earth's fresh water. Large scale, regional droughts are also taking their toll. Though each human only requires about 7 L of water per day to survive, most of us (especially those in industrialized nations) live well above that minimum. The loss of a sustainable source of water certainly poses a threat to global, social stability.

---

[1] Reimann, L., Vafeidis, A. T., & Honsel, L. E. (2023). Population development as a driver of coastal risk: Current trends and future pathways. *Cambridge Prisms: Coastal Futures*, *1*, e14. doi:10.1017/cft.2023.3

Finally, anthropogenic global climate change exacerbates many of the above concerns. Our ever-increasing use of fossil fuels has changed the chemistry of the atmosphere, causing a rise in temperature, globally. The rise in temperature, however, is only a part of the problem: its downstream implications are impacts on global circulation and moisture patterns. Large, wave-cyclone related storms and cyclonic tropical storms are becoming more common and more severe, causing more damage. This includes higher storm surges and heavier rains, which increase flooding, as well as wider swings in the jet-stream, causing more variable weather and more tornadic activity. Droughts are more common, longer, and harsher, leading to more fires and making farming more difficult. We live on a world that is dynamic, but we also influence those dynamics in ways that are detrimental to our survival on the earth.

Mitigating these negative influences requires a good working knowledge of the systems of nature—how they operate and interact to generate the resources we need. Until only recently, we have maintained a way of thinking about the earth as a machine (Marques & Brito, 2014; Norwick, 2002).[2] Indeed, this has been an important metaphor for the past few of centuries, however it diminishes the role of the complex and chaotic actions and interactions of the earth. This simplification leads to thinking if we can just fix this one component of the machine, then the whole thing will run as it should. This way of thinking also places humans outside the system, as the ones maintaining the system, the mechanic, or technician. Maintaining that humans are outside natural systems is naïve and hinders efforts to address SSIs that impact personal and societal well-being.

Perhaps new ways of thinking about Earth systems are necessary and a tall order. Most people do not get the kind of education that encompasses such detailed information about the earth and how it operates. However, a good science education can teach students several important skills: how to identify a problem; how to understand what kind of information is needed to solve the problem; knowing where to look for information and accurately determine its credibility; and knowing what to do with the information once they have it. A good earth science education should teach about geoscience including how we know what we know (i.e., the process of geoscience). Specifically, how did scientists and the scientific community come to understand today's well-established fundamental Earth Science ideas.

A robust geoscience education that addresses how credible geology research is done must play a more serious role in school science instruction so that students better understand the earth as well as many fundamental biology concepts that are tied to an understanding of Earth. This understanding among the general public would place them in a better position to separate authentic experts and vetted knowledge from the extensive and growing science mis/disinformation efforts. Accurately understanding how geologists go about their work and how knowledge of Earth is

---

[2] Marques, V., & Brito, C. (2014). The rise and fall of the machine metaphor: Organizational similarities and differences between machines and living beings. *Verifiche*, *XLIII*(1–3), 77–111.

Norwick, S. A. (2002). Metaphors of nature in James Hutton's "Theory of the Earth with Proofs and Illustrations". *Earth Sciences History*, *21*(1), 26–45.

generated would likely bolster interest in geology careers; thus, assist in mitigating the extensive retirements of earth science professionals in the next 10 years. Unfortunately, the state of geoscience education around the world is currently quite inadequate to meet these challenges. Many calls have been made over the past few decades to enhance earth science education. Though some progress has been made to improve earth science education, including the development of multiple standards documents devoted specifically to the earth sciences, the need for enhanced earth science education persists today.

However, geoscience literacy is only part of this complex terrain. As previously noted, but worth emphasizing again in this closing chapter—beyond understanding the products and facts of geoscience is understanding the *processes* of geoscience, which are how those facts come to be. Taking an in-depth look at these processes is central to developing the capacity to discern who is speaking for geoscience and determine the reliability of the facts they generate. This book has been an argument for just this proposition. If we can engage students in the stories of geoscience—what we know and how we figured it out—they will learn both the geoscience and the process of geoscience.

Rather than merely covering large swaths of factual information (that research makes clear is often not understood or remembered), a robust understanding of the practices that geoscientists incorporate into their research promotes deeper examination of fundamental earth science content—such as their being able to substitute space for time, and comprehending and visualizing the immense scale of time and space over which geologic phenomena take place. Awareness has been growing about how using the history and philosophy of science as a tool for teaching science is effective to help students develop a more robust and nuanced understanding of the process of science, and in this case, geoscience. Indeed, acquainting students with "conceptual biographies" is one way to explore how present knowledge is generated so that they become more adept at judging the reliability of geoscientific claims, develop a geoscientific way of thinking, and approach earth-related problems with the tools to solve them.

The history is not enough, however. Effectively engaging students in the learning process will take other, transformative teaching strategies, along with a pedagogical design that creates an environment where students participate in activities that promote useful learning. This, of course, has been the second key focus of this book.

8.3 Design Engineering and Reflective Practice

To set about creating such an environment in GLGY 209, I partnered with a colleague who helped lead me through the design engineering process: discovery, interpretation, ideation, experimentation, and evolution, described below. I described this in detail in Chap. 2, and only reiterate the most salient aspects here with the addition of some observations and possible paths for future research.

In designing the course, I distinguished four design constraints that I would need to find solutions for:

1. I had a large-lecture course with about 300 students.
2. Many students were unfamiliar with geology.
3. Many students were only taking the course because they needed to take a science course and they did not want to take chemistry or physics.
4. Students came from across the campus, and thus many of them likely felt they could not be a scientist or do science.

Those solutions came in the for of the following social learning goals:

- ***Level the classroom:*** I needed to get all students, regardless of background, to feel they could contribute to classroom knowledge development.
- ***Develop a small-class feel:*** In large-lecture environments, students often feel anonymous, because they are one face in a sea of faces. The physical environment also emphasizes a transactional format of teaching and learning, where the student is a passive receiver of knowledge, which is delivered by the instructor.
- ***Decrease anonymity:*** To get students to engage more fully in the course, they need to develop a small group of colleagues, people they can count on for help in the course. Concomitantly, they have to believe those colleagues are also counting on them for help.
- ***Have students discuss geology without realizing it:*** Learning is much more likely if students are engaged in multiple, short, but fertile, discussion of the content material during the class time. I needed to facilitate these discussions.

I researched, designed, and implemented multiple instructional strategies that would help achieve the above listed goals. Most importantly these included team-based/problem-based learning and the use of historical case studies for teaching the science content and NOS. I then set about distilling the concepts most useful for this population of students to know, students who were unlikely to become geoscience majors and whose attendance in this course is likely to be their last formal exposure to science? From this, I developed these four learning goals:

- ***Understand the historical nature of the earth:*** The fact that the earth's change is unidirectional and contingent upon all that came before the present moment.
- ***Comprehend the immense age of the earth:*** The fact that the earth is tremendously old—4.6 billion years—and that most of the history of life occurs in the last 20% of that time.
- ***Understand the dynamic nature of the earth:*** The fact that we can (a) visualize the earth as a complex system, (b) use signs in the rocks to help us to understand the processes that have taken place in the history of the earth, especially within the context of continental displacement and (c) understand the theory of the earth, which is plate tectonics.
- ***Appreciate the process of science and judge the reliability of claims (NOS):*** The fact that (a) knowledge is constructed based on observations, (b) but it also develops in response to social and temporal context, and (c) the investigator's subjectivity.

As part of my reflective practice, I have used formative and summative assessments to understand student learning and then modify the curriculum as needed based on those results—the ones I reported Chaps. 5, 6, and 7. These results have helped to guide this design experiment making the course more effective. What I have found here, which certainly merits future investigation is the fact that though I normally have 20–30 students per course offering (approximately10% of students) request academic accommodations from our student accessibility services—accommodations like extra time for taking exams or having a special location for doing exam work—I have never had a student of this course ask to take advantage of such accommodations. Knowing they can rely on their team mates, students seem to feel prepared enough and are confident enough in their work that they do not need to use the extra time or the special room allotted. In this way, it seems that the small group work appears to be an effective strategy for student inclusion in all aspects of course activities. Currently this is only anecdotal, though many course feedback responses do include how working in groups, especially groups maintained throughout the term, fostered more discussion during study sessions and especially during the exams. It would be worth researching to understand how students with accommodations perceive not having to take use those accommodations.

8.4 The Three Narratives

One of the important strategies—and largely the focus of Part I of this book—is the use of historical case studies in presenting the development of geological ideas. While separating the history into these three narratives is a simplification, in that oftentimes the same actors in one narrative are also in another narrative. Notably, they were not working on different problems at different times; rather, they were working on *their* problem, but their work, for the purposes of my course, could be broken into two different lines of inquiry. I divided these narratives to suit my purposes: I parsed the history based on what I was using it for, which was teaching towards the three content learning goals.

The first concept was the notion that the earth is a historical body. In the early days, in the Western European tradition, most people considered the earth to be static, a reflection of its form "in the beginning." This idea changed to one where signs in the rocks—cross-cutting relationships, fossils, and so forth—could be used as clues to understand a history of change through time. They were clues to "previous worlds." This idea then branched into two possibilities: (1) The earth is in a state of dynamic equilibrium where very small changes occur and only mount up into big changes over long periods of time—uniformitarianism; and (2) The earth accommodates revolutions, or large catastrophic events occurring over very short periods of time—catastrophism. In the end, neither perspective was completely correct or incorrect. Our current consensus view is one of small changes over extreme lengths of time interrupted by massive events every now and again—millennia of complete boredom punctuated by moments of absolute terror.

For the age of the earth, the second concept, the quest for figuring this out started with texts, as they were thought to contain the sum of all knowledge. Many of those texts happened to be religious in nature. Importantly, calculating the earth's age was not a science versus religion endeavour, but was a scholarly investigation using the best knowledge at the time. As people started to rely more and more on empirical observations of the natural world, it became increasingly difficult to reconcile these facts with the texts. In fact, the default strategy was to change the interpretation of the text to fit the empirical observations.

Investigators of the earth's age used many different approaches to solve it. Some were *bottom-up approaches* (abductive reasoning), such as using the accumulation rates of salts in enclosed basins, and sedimentation rates and the thickness of sedimentary rocks to calculate the age of the earth. Others started with universal propositions, a *top-down approach,* like the laws of thermodynamics to calculate the age of the earth. Both systems of thought included their own assumptions about the natural world and the way it works; both resulted in very different ages for the earth. The controversy ended with the discovery of radioactivity and the ability to derive absolute ages of rocks based on their composition, given their special radioactive atoms.

Finally, the dynamics of the earth came into play when trying to understand why mountains exist on earth. Numerous explanations have emerged. Important to the story here was the notion that the earth began as a molten ball and was hence cooling and contracting, wrinkling like a desiccating apple. This was the consensus view at the start of the twentieth century. Serendipity had a meteorologist seeing a map of the South Atlantic Ocean and positing that the two coasts on either side of that ocean were once together and had since broken and moved horizontally to their (then) present positions. Publication of this idea and data supporting it caused a bit of a stir in the geoscience community.

Issues such as regionalism, disciplinary inertia, and social power became obvious in the scientific process, evident in who decides who can publish and what they can publish. Such aspects highlight the importance of a champion for an idea. According to many science textbooks, scientists are their own worst critics; in contrast, scientists are usually cheerleaders for their ideas over others' ideas. This why the wider scientific community is essential for vetting ideas, a kind of collective objectivity. In the case of plate tectonics, the consensus view resulted from numerous and diverse research groups working on several unique problems, that appeared separate from the drifter/fixist controversy, and then putting their ideas together. The importance of subdisciplines of geophysics—seismology, geochronology, paleomagnetism, and the like—became visible, even though many geologists at the time were not interested in these burgeoning fields. As with various other controversies, the current answer took bits and pieces from each of the different competing conceptions to create a model that worked best at the time.

For each of the above conceptual biographies, I pulled forward some major examples of NOS, which I highlighted within instruction and again at the end of the course during the explicit and reflective portion of the course material. Some of the more salient examples include:

- geology is a historical and hermeneutic science
- the role of assumptions
- resolution of controversy
- cognitive resources
- serendipity
- social, political and economic influences on science
- prematurity in scientific discovery
- models and modeling

Significantly, I did not pick these specific facets of NOS because I find them particularly important over other possible facets. I highlighted them because they stood out as important *within the context of the narrative*. Thus, I took a more bottom-up approach to NOS as opposed to a top-down approach.

8.5 Transformative Teaching Strategies

In addition to the historical narratives, the most consequential transformative teaching strategy I implemented into classroom instruction was *team-based learning*. Putting the students into teams that remained stable throughout the term was crucial, as it became the basis of engagement in the course. Each member of the group could count on the other members to help them with understanding content, completing assignments, arguing ideas, and answering exam questions. As well, they came to know that their teammates also counted on them. This placed the responsibilities of the course square on each student's shoulders. Thus, for these students, the class transmuted to a smaller community of interdependent actors who collectively contributed to the betterment of the group. This strategy of team-based learning therefore addressed all the learning aims.

Problem-based learning is a key adjunct here. Basically, all of the course content was problematized. Historical interludes introduced observations, approaches, and social context, and students had to make meaning of those inputs. After time spent trying to solve the problem, a new interlude would continue the narrative for students, adding new data and more context to the particular problem. Because the "right" answer was not immediately forthcoming, students became invested in the problem-solving aspect. It became a challenge for them as a group to try and put all the pieces together, even pieces they might not have yet.

Inquiry-based learning, which is related to problem-based learning, was also a component of the course. This approach entails students asking their own questions and searching for an answer. This open-ended approach allowed students the freedom to explore based on their interests, thereby encouraging higher engagement levels. While couching the inquiry in the history, students were nevertheless free to approach the inquiry as they saw it, attributing presented data in a way that was personal to their group, while always remaining faithful to the history. In other words, students approached the topic in the way that seemed best to them; then, any redirection of thoughts occurred in response to what actually happened historically. Students consistently maintained the experience of open inquiry, but the history simultaneously restricted where that inquiry ultimately went—which is in the direction of the learning goals. This process highlights learning in both science and education, which unfolds through the iterative development of models. The difference is this: science knowledge development *in science class* is constrained by data and hindsight, but science knowledge development *in science* is only constrained by the data.

Model-based learning was another strategy important to the structure of the course. By breaking down the integrated target concepts, what I have called the learning goals, into a hierarchy of elements classified as primary, secondary, tertiary, and so on, I mapped a direction for presenting information (see Fig. 4.1). This trajectory moved from most concrete to more abstract, affording a kind of scaffolding for students to develop their mental models by satisfactorily explaining the data they were using as the basis of their models. In this way, the culminating models were founded on more concrete conceptual understandings, built in a way that made most sense to them; they were not presented as finished products, which they would have to accept as is, on faith.

Finally, *metacognition* played a part in the course. By including an explicit and reflective portion concerned with the aspects of NOS highlighted in the narratives, students made connections on a metacognitive level to these aspects. The students became sensitive to these aspects, identifying them in the different scenarios. Of course, when handed a hammer, everything around looks like a nail. Thus, the possibility exists that by giving students this limited number of hammers—that is, NOS aspects—they will be the only things the students recognize as examples of scientific development.

However, I maintain that this is not necessarily a bad thing, certainly not any worse than learning a list of tenets. I gave students experience with this analytical tool along with a *process* to approach such analysis from a higher, critical level. This practice emerged from the way I revealed NOS aspects to them in class: I presented the stories first and analyzed them from the bottom up, drawing them out organically, showing students how to discern the different aspects, and to identify a cause or causes for a concept's particular developmental trajectory. This approach worked in opposition to shoehorning a single, particular tenet into the narrative, or amending the story to "fit" the NOS facet.

8.6 Measuring Learning

Using click-on diagrams (COD) in class was a great way to get students engaged, while also enabling me giving data I could use as a diagnostic tool to judge the efficacy of the instruction. From the results of the COD responses, students made significant gains in their answers to questions related to each of the three main content learning goals. In particular, they were better able to infer relative ages of layers or events based on logical reasoning. They also had a better sense of the immensity of geologic time and our relative position within that time. As well, they were better able to project the implications of lithospheric dynamics.

In places where the COD data revealed little significant change in understanding about particular aspects, I could subsequently address why this might be. The design of the question is a more important factor affecting student responses than we often acknowledge. A few examples demonstrate this process:

- ***When humans evolved within the span of the age of the earth:*** Learning gaps here could be due to students' inability to reconcile the sheer immensity of the age of the earth with perceptions of our own existence. Or, the reason could be merely the way we present the relationship, such that students were unable to assimilate this information in the intended manner.
- ***Subduction-related earthquake question:*** Little evidence of improvement here may be due to so many students initially choosing the correct answer, thus foreclosing the possibility of significantly *improved* understanding. Also, the question itself allowed for a lot of interpretation; the intended correct answer along the contact between the two lithospheric plates was a far narrower region than the area of subduction-related earthquakes in reality.

The use of the Geoscience Concept Inventory (GCI) questions afforded a way to further reinforce the results from the COD questions. The questions used a different approaches to evaluate student understandings within the context of the same learning goals. The results from the multiple questions and different kinds of instruments support the notion that students learned content within the domain of the content learning goals. The GCI responses also did a better job of indicating students' understanding of the timing of humans within the context of the age of the earth; specifically, I saw a significant shift towards the understanding that humans evolved very close to modern day—the scientific consensus—compared to the pre-instruction location, which was much closer to the beginning of the earth. Thus, the students demonstrated geoscientific conceptual development.

They also showed development within the domain of NOS. For instance, the Views on Science–Technology–Society (VOSTS) survey helped me to generalize a profile of students' understanding of NOS prior to instruction and after the completion of the course. Students' understandings of the processes of science shifted from thinking that only one single method was used to do science to thinking there *was no one method*. This fresh understanding contrasted with the singular "scientific method" that they had largely been exclusively exposed to throughout their prior schooling in science. The shift recorded by the VOSTS also reflected that found I the qualitative analysis of responses to the open-ended questions.

Examples of such a shift include pre-instruction descriptions of "doing science" as usually involving something situational, like being in a science class, following directions, mixing chemicals, or doing dissections—but without any mention of a purpose or the basis of inquiry for those activities. However, post-instruction, students were more apt to describe the geology-related activities they did in class. To be clear, they were not just affirming their location in geology class, but used words such as *researching, solving a problem, formulating a question, correlating layers*, and similar; these were all actions they participated in during class activities.

This pre-instruction to post-instruction shift from rigid procedures to a more open and cognitive exercise in inquiry, resonated with the historical contextualization. It also persisted in students' descriptions of scientists and was linked to their broadening notion of scientific knowledge. They initially predominantly held the view that each bit of *uncovered or found knowledge* gets added to the larger body of knowledge, which in turn continually changes and evolves. From this perspective, scientific knowledge is understood as predetermined or even fixed, a view that relieves scientists of the creative act of knowledge construction. With exposure to the learning provided in the course, however, the lines around scientific knowledge have been blurred. The students shifted to embrace the understanding that knowledge is, rather, *created.* These findings were consistent across the VOSTS and the open-ended responses, reinforcing the conclusions I drew from the phenomenographic interpretations of the answers to the open-ended questions.

In particular, within the collective, pre-instruction minds of the students, scientists were more like technicians: wearing particular clothing, using particular equipment, finding knowledge, following directions, mixing chemicals, and so on. However, the post-instruction responses unquestionably revealed that students' thinking about scientists had shifted. While they acknowledged the additive version of scientific change described above, they now considered that scientists did not simply *find* data, they *interpreted it*, and that those interpretations were necessarily personal to some extent—founded on the prevailing scientific knowledge, the scientists' past experiences, assumptions they made during the data collection and interpretation periods, and their unique social and temporal circumstances.

The students thus now acknowledged that scientists' subjectivity, and even bias, are part of the scientific process. Continuing this train of thought, they no longer held an idealized stereotype of scientists as being without bias, open-minded, their own worst critic, and so on; this was revised to a representation that allows for the humanness of scientists—individuals with creative capacity, agency, and bias. Scientists not only exist in the world of science but can be crucial to its processes. It is only through things like peer-review and other communications between scientists (such as at conferences) that we can mitigate some of the limitations of such biases and subjectivity. It reflects a shift to a far more nuance view of science and the scientific process.

The overarching outcome of these shifts in understanding is that the students transitioned from an epistemological stance of naïve or scientific realism to one of constructive realism. Blind acceptance of scientific propositional statements shifted to a more critical stance, with the aim being to understand and judge the reliability of the expert and their message.

My analysis of metaphor use by the students in their open-ended responses further supports this fresh stance. In general, students objectified science ideas: they were described as things that one could dig deep for, or which were found, discovered, brought to the surface, and brought into the light. These smaller bits, like facts, belong to larger complete wholes, conceptualized as surfaces (landscape, cityscape, or seascape). This obviously reflects our vast concrete experience with things, surfaces, and constituents as contents. Despite the overriding reference to ideas as things, students also acknowledged some constructive aspects to the process, building upon a foundation of old knowledge or putting pieces together to complete a larger puzzle.

One other result that merits future investigation is the observation of extraction and colonial metaphors that the students used and how this might influence how they think about the process and products of science. There were instances of colonial-relating metaphors such as *pioneering work*, *trailblazing*, *exploring uncharted territory*, and *conquering/overcoming/taming the unknown*. Also, there were instances of extractive language–harkening to the *IDEAS ARE OBJECTS* metaphor– such as *ground-breaking*, *drilling (or digging) down into a problem*, *data mining*, *discovering/uncovering/digging up* science knowledge. Seeing how the use of metaphor profoundly impacts one's perceptions, it would be worth looking at how people perceive the role of science in society (both its processes and its products) and how that perception is related toward their behaviors around science and scientific knowledge.

In the end, about a third more students responded at the post-instruction stage compared to the pre-instruction stage. It is noteworthy that the numbers of applications of metaphor codes were very similar from pre- to post-instruction, which reveals that, proportionally, the post-instruction cohort relied less on metaphor than the pre-instruction cohort did. Metaphor use is a strategy for making the abstract concrete; thus, the lower reliance on metaphor at the post-instruction stage could mean that students were, by then, more able to concretely describe science knowledge and processes using the concrete, classroom experiences, as compared to their ability at pre-instruction. In other words, they did science and then described it in concrete, personal terms.

8.7 Final Words on the Results

Despite persistent calls to overtly address accurate NOS content in science classes, even within the national science education standards documents (also cited in Chap. 1), a large body of literature exists expounding on why teachers do not do so. This is the case even with those teachers who are interested in teaching NOS material to their students. Some of their major reasons, as noted, are the following:

- A concern with it taking extra time
- Little experience with NOS content or with teaching it
- A belief that it is not science
- A belief that it would confuse students
- The view that no NOS consensus exists

By historically contextualizing the content material, this book supports the contention that students can learn content and make meaningful shifts in their NOS understandings. Because the content and the historical framing inform and support the other, the HNOS approach advocated in this book does not require more time.

Certainly, an instructor who has never taught in this manner or taught this kind of content cannot be expected to pivot to this in just a few days. Instructors will want and need some experience, through high-quality professional development, to feel comfortable with this broader approach. And they would need to acquaint themselves with the history and the philosophy of science, at least superficially, to be able to conduct discussions on the topics, as I have argued earlier. Once they become familiar with such tactics and strategies, it will be obvious that this humanistic part of science—the part we try so hard to dissociate from what we imagine as ideal science—is inseparable from that science. The narratives give example after example of human involvement in making science happen, demonstrating true science-in-the-making.

From the students' perspective, following and participating in the development of scientific concepts is certainly the way to learn this material; substantial literature promotes the use of inquiry as an effective tool in science learning, as I reviewed in Chap. 4, Sect. 4.1, in my discussion of braiding history and inquiry. In essence, students became invested in the problem-solving, and engaged in doing the necessary rationalization. They participated in the dialogical negotiation, which entailed the very activities that support their learning on a level deeper than that offered by the traditional, transactional model of teaching and learning. It was the story and the historical aspects that helped students, inviting them into the learning experience, guiding them through an otherwise open inquiry to the intended learning outcomes. As NOS content was *part* of science in this model of teaching, there was nothing "extra" for instructors to teach, nothing extra for students to learn. It was all part of the same endeavour and showed up as part of the eventual, expected learning outcomes.

As I have outlined earlier in Chap. 1, Sect. 1.3, various perspectives exist about what constitutes NOS and its important aspects, and what we should teach students. I have not structured this course, and the narratives therein, according to some initial specific NOS aspects that I had in mind. Instead, I have drawn on the history, focusing on the points that seemed most salient and obvious to students; they can then experience and reflect on them. My aim was never to enable these non-science majors to list off this or that aspect or nature of science. Rather, my goal was that they become critical of the knowledge being presented to them and seek to determine whether both the knowledge and its presenter are reliable. They developed the skills to search for clues about the back story—to inform themselves about possible conflicts of interest, potential bias, the reasonableness of embedded assumptions, and the like. Going forward, accepting claims at face value becomes insufficient.

My argument here is that it is possible to achieve these lofty goals. It does take work, as anything worth doing does, but it is possible to engage students in science—even non-science majors—and enhance their learning of the content as well as the process of how that content knowledge came to be. The data I collected and analyzed from over five iterations of the course incontrovertibly support such a proposition.

In addition to the measured shifts in understanding, I have also received numerous emails and other course feedback from students expressing their delight in how I present the course. I will insert a few of those correspondences here as examples of the impact on student affect for the course and science in general.

> I also wanted to add that I do not traditionally have much of a "science" brain, so the broader subject has always been hard for me, but you made the class so approachable and impactful for me, that I really enjoyed it. I am also neurodivergent, so it really makes me happy to see instructors who are "breaking the mold" of the traditional format to make their material accessible, so it was a huge help for me [unsolicited email from a student].

> Prof, your class is really making me dream to learn more about science which I have no clue of previously, I am beginning to appreciate the planet am living in and why, I guess. I hope I will continue this interest/curiosity even after your class, who knows! but do you think that a mature student like me, can actually break into knowing more-in-depth about science—Geology on a higher level, since I don't have any science background? Thanks" [unsolicited email from a student].

> I just thought of you because I am currently abroad with the university in Hokkaido, Japan and we just experienced a 6 magnitude earthquake while eating dinner. I was shocked by how quick the warning was before it happened! I wanted to let you know that your instructions and lectures on earthquakes came into mind immediately, I could hear you in my head during it haha.. everyone is OK so overall it was kinda fun" [unsolicited email from a student].

> Just wanted to say thank you for teaching me this semester, it was a pleasure being in your class, I really do appreciate the enthusiasm that you brought to your lectures, I would not have imagined that study of rocks would be my favourite class this semester. Also, your "pulling out the rug" on the scientific method over the past few months has truly changed my perspective on what we see as concrete evidence, fact, and what value speculation still has. Thank you! ["pulling out the rug" was the metaphor I used when dispelling the myths we have about science]

> I just wanted to send over a note of appreciation for the tone set in class today about learning over the "construct of grades". I have a learning disability and have always been terrified to complete my science requirements because of the harsh rhetoric around the classes themselves and my ability to complete them- I just wanted to thank you for having such a wonderful set up that I'm so excited and confident to start this class. I hope that over time, more professors will structure their classes like yours [unsolicited email from a student].

> I wanted to send you a quick note to thank you for all your time, effort and enthusiasm during glgy 209 this semester. I am studying to be a teacher and feel so enthused after taking your class. Your passion for teaching and creating inspired, engaged learners is truly motivating. I have left the course with a better understanding of geology and some life skills too ☺ Thank you for your vivacious spirit and ambition for your students' learning [unsolicited email from a student].

> I wanted to write this email earlier but never had the chance to process the words. I wanted to discuss my feelings/thoughts on science. I've never had an interest towards science. I've always been turned off by the idea of processes, research and the formalities of science. These feelings changed this semester. I wanted to give thanks for a new insight into the way of understanding what is true scientific knowledge. I'm now able to look at an assumption, stat or process and challenge it. As a commerce student, this class has both resonated with me and encouraged me to step outside the boundaries of business and look for science related interest [unsolicited email from a student].

Hopefully, these communications emphasize how even students with no prior interest in science could find science interesting and part of their world—because, after all, it is. The humanization of the process made it inviting and worth the risk of engagement. It was welcoming to those who did not think they had the mind for science. They could see how they could leverage the information I presented in class into various aspects of their own lives. I find this to be far more important than having students memorize a bunch of jargon to regurgitate on and exam and then promptly forget. I have opened a door for them and although it is up to them whether to enter or not, I made going through that doorway as attractive as possible.

8.8 Final Words on the Process

The journey to writing this book started decades ago. As a ninth-grade earth science teacher, I became interested in why I would say something to the class and the students would somehow hear something different—sometimes disturbingly different! Learning about learning became my focus. Eventually, I realized that my motivation was not simply to teach better, but to come to know the content better myself. I derived great satisfaction as I came to a deeper understanding of both geoscience and NOS. So, it was a selfish motivation, but with that richer grasp of this material, I could present content more effectively to my students. Regardless of that motivation, my interest in the learning process has certainly had a positive impact on my students' learning. That said, I wish to make two points here.

First, transformative teaching is a tremendous amount of work: I'm not going to pull any punches about that. I have developed an abiding love of history because I have read so much of it now, with numerous and varied interpretations of the same developments. However, it takes a good deal of time and energy to read and then assimilate the history into something that is presentable in class. Reading the science education literature on topics like team-based, problem-based, and model-based learning is similarly demanding and time-consuming, to absorb the material and integrate it into the instructional structure of the course. As well, it was a lengthy and taxing process to read the geoscience education literature. I had to consider both the best teaching strategies in geoscience courses, and the key skills and knowledge that experts declare most important for students who are science majors, and for those who are not.

Taking all these independent pieces—the history, learning science, and geoscience education research—and purposely shuffling them together to make a cohesive whole was no small task. As I stated earlier in the book, I began my historical contextualization of content in the late 1990s—an eternity ago, it now seems. In 2015, I was preparing to make an entire course with this structure, but many colleagues told me I should just make a slow transition, implement small changes gradually. However, I saw things differently: the change had to happen all at once. Then afterwards, the design engineering process took care of the fine-tuning.

Teaching the course each time required that I launch a bunch of balls in the air simultaneously, which was a tricky process at times. Getting groups assigned as soon as possible was particularly important; then staying on top of the inevitable changes that take place as students dropped and added the course over the ensuing few weeks was its own special challenge. It was difficult work, but at the same time, it was so satisfying to see my plans in action and to observe the students responding and engaging.

The second point I want to stress is that I am under no illusion that the shift in NOS understandings is a permanent one, though it is an important sign of the efficacy of this instructional approach. Most research that records the change in students' NOS understandings involves some intervention, sometimes as short as a single or a few class periods. The researchers do a pre- and a post-test, find a significant change in understanding and then pat themselves on the back for a job well done.

My example entailed 26, 75-minute-long class periods over a full 13-week term, improving over five iterations. I have documented a shift in students' NOS understandings, but it is valid to wonder how long-lasting such a change will be. Scientism and scientific realism pervade science teaching and popular media. The ideas the students came into class with reflect the mixture of these two influences, which will continue to be influential once the student leaves the class. It is very likely that much of the gains documented in this book will revert back in a matter of time without periodic reinforcement.

By documenting my process and my rationales, I hope to encourage others to develop their own transformative teaching, so they structure their own courses with similar strategies and historical contextualization. It is also important that publishers incorporate more of these humanistic aspects of science within their textbooks. As I've already discussed (see Chap. 4, Sect. 4.7.1), textbooks are written for the purpose of knowledge transfer, based on a folk theory of how learning happens. Textbook writers and publishers need to be more transparent about how the science knowledge, which exists between its covers, developed. The periodic sidebars of history, written in hindsight, are ineffective at illustrating the scientific process; in fact, they often portray the process of science incorrectly.

In discussions with publishers about this very matter, I have heard that what they publish reflects what instructors want. But we also know that instructors most often use the textbook as the default course curriculum. Thus, a feedback loop is in play: what gets published becomes the norm for what to teach and what is taught remains the goal for publishing. I am convinced that if the textbooks contained substantial elements of the context surrounding knowledge development, then that would likely make it into the instruction. The question is how to break the cycle. Either instructors need to start demanding that more accurate NOS become part of the text, and/or publishers must take a little risk and start creating books that contain more of this social context.

Overtime then, students might encounter multiple examples of such teaching throughout their schooling. They will still learn the content, but their NOS understandings will develop, becoming more complex and nuanced. At the very least, when they later hear or read scientific propositions, they can be more thoughtful and reflective about the messages they are receiving. However, my hope is that more teaching about the human aspects of science will invite a larger and more diverse group of people to become scientists, or, at least, to relate science to what they do in their lives and professions. When scientists are thoughtful and reflective, they are better able to find solutions to problems that threaten our existence on this planet.

Index

C
Case studies, 20, 23, 29–33, 45, 55, 66, 99, 105–108, 110, 119, 126, 219, 230, 279, 280
Click on diagram (COD), 33, 147, 149, 151, 153–164, 167, 170, 192, 283, 284
Climate literacy, 14, 15
Cognitive resources, 133–134, 248, 282
Conceptual metaphor theory (CMT), 34, 127, 180, 204, 240, 246–247
Conduit metaphor, 127, 128, 248
Contractionism, 181, 188
Controversies, 20, 33, 68, 73, 80, 82, 107, 114, 128, 131–133, 135, 160, 190, 267, 268, 281, 282

D
Deep time, 14, 15, 17, 77–83, 123
Design process, 33, 44, 45, 47, 49, 50, 275

E
Elastic rebound theory, 18, 106, 108, 133, 249
Embodied cognition, 127, 246
Embodied experiences, 18, 19, 108, 115, 246, 247, 268
Entailments, 246, 253–256, 259–261, 263–265
Eternalism, 70–71, 74
Expanding Earth, 94, 113

G
Geological Concept Inventory (GCI), 147, 149, 151, 164–175, 192, 284
Geomagnetism, 94–97, 116, 134
Geoscience education, 12–26, 44, 154, 206, 275–278, 289
Geoscience education research (GER), 16–19, 289
Geoscience learning standards, 15
Geoscience literacy, 13–15, 19, 26, 278
Geoscience teaching standards, 15

H
Horizontal displacement, 86, 87, 94, 95, 113, 114, 124, 125, 134, 135, 181, 188, 190
Horizontal displacement (continental drift), 86, 87, 94, 95, 113, 114, 124, 125, 134, 135

I
IDEO, 47, 58
Individual-Team Performance (ITP) Metrics, 105
Inquiry, 15, 45, 66, 108, 175, 205, 280
Inquiry-based learning, 108–114, 282
Integrated target concepts, 52, 54, 66–68, 115, 283

G. Dolphin, *Teaching Geology Using the History and Philosophy of Science*, Science: Philosophy, History and Education,
https://doi.org/10.1007/978-3-031-68667-2

M
Metaphor, 19, 73, 127, 180, 203, 245, 277
Metaphor identification, 203, 250
Model-based learning (MBL), 54, 55, 58, 66–68, 99, 114–119, 138, 283, 289
Models, 4, 47, 67, 105, 160, 206, 247, 281
Multiple working hypotheses, 27, 87, 129

N
Nature of science (NOS), 15, 19–26, 29–34, 45, 50, 51, 55–58, 107, 120–122, 125–129, 136–138, 149, 150, 173–193, 197, 249, 250, 260, 267, 269, 270, 275, 279, 281–284, 286, 287, 289, 290
NGSS standards for geoscience, 15

O
Ocean Literacy, 14
Outcome space, 237–240

P
Paleomagnetism, 94, 156, 281
Permanentism, 86, 95, 114
Phenomenography, 180, 200, 201, 237, 240
Prematurity in scientific discovery, 135, 282

R
Radioactivity, 81–83, 281
Research strategies (RS), 28, 268

S
Science literacy, 12–20, 46, 51
Scientific literacy, 12, 19–20, 46, 111
Scientific method, 20–22, 25, 26, 51, 109, 121, 129, 131, 182, 183, 189, 200, 212–214, 217, 220, 232, 233, 237, 261, 284, 288
Serendipity, 134, 135, 281, 282
Story telling, 56

T
Team-based learning (TBL), 54, 58, 104, 105, 137, 282
Transformative teaching, 44–59, 103–138, 150, 170, 191, 198, 199, 271, 278, 282, 289, 290

V
Views on science technology and society (VOSTS), 149, 151, 165, 170, 173–192, 284, 285

W
Watch dial painters, 82
Wikipedia, 54, 56–58, 119–121, 138, 230

SPRINGER NATURE

GPSR Compliance

The European Union's (EU) General Product Safety Regulation (GPSR) is a set of rules that requires consumer products to be safe and our obligations to ensure this.

If you have any concerns about our products, you can contact us on ProductSafety@springernature.com

In case Publisher is established outside the EU, the EU authorized representative is:

Springer Nature Customer Service Center GmbH
Europaplatz 3
69115 Heidelberg, Germany

The manufacturer's authorised representative in the EU is Springer Nature Customer Service Centre GmbH, Europaplatz 3, 69115 Heidelberg, Germany. If you have any concerns regarding our products, please contact ProductSafety@springernature.com

Printed and bound by CPI Group (UK) Ltd, Croydon, CR0 4YY
01/10/2026
02225906-0008